Craft Production in Complex Societies: Multicraft and Producer Perspectives

Foundations of Archaeological Inquiry

James M. Skibo, series editor

Living with Pottery: Ethnoarchaeology among the Gamo of Southwest Ethiopia
John W. Arthur

Complex Systems and Archaeology: Empirical and Theoretical Applications
R. Alexander Bentley and Herbert D. G. Maschner, editors

Material Meanings: Critical Approaches to the Interpretation of Material Culture
Elizabeth S. Chilton, editor

Pottery Ethnoarchaeology in the Central Maya Highlands
Michael Deal

Archaeological Perspectives on Political Economies
Gary Feinman and Linda Nicholas, editors

Archaeology beyond Dialogue
Ian Hodder

The Archaeology of Settlement Abandonment in Middle America
Takeshi Inomata and Ronald W. Webb, editors

Evolutionary Archaeology: Theory and Application
Michael J. O'Brien, editor

Style, Function, Transmission: Evolutionary Archaeological Perspectives
Michael J. O'Brien and R. Lee Lyman, editors

Race and the Archaeology of Identity
Charles E. Orser Jr., editor

Unit Issues in Archaeology: Measuring Time, Space, and Material
Ann F. Ramenofsky and Anastasia Steffen, editors

Behavioral Archaeology: First Principles
Michael Brian Schiffer

Social Theory in Archaeology
Michael Brian Schiffer, editor

Pottery and People: A Dynamic Interaction
James M. Skibo and Gary M. Feinman, editors

Expanding Archaeology
James M. Skibo, William H. Walker, and Axel E. Nielsen, editors

Essential Tensions in Archaeological Method and Theory
Todd L. and Christine S. VanPool, editors

Craft Production in Complex Societies

Multicraft and Producer Perspectives

Edited By Izumi Shimada

The University of Utah Press
Salt Lake City

Foundations of Archaeological Inquiry
James M. Skibo, editor

The Defiance House Man colophon is a registered trademark of the University of Utah Press. It is based on a four-foot-tall, Ancient Puebloan pictograph (late PIII) near Glen Canyon, Utah.

11 10 09 08 07   1 2 3 4 5

Library of Congress Cataloging-in-Publication Data

Craft production in complex societies : multicraft and producer perspectives / Izumi Shimada.
p. cm.
Includes index.
ISBN 978-0-87480-921-3 (cloth : alk. paper) — ISBN 978-0-87480-902-2 (pbk. : alk. paper) 1. Handicraft. 2. Industrial arts. 3. Artisans. I. Shimada, Izumi.
TT149.C732 2007
745.5—dc22   2007025626

Contents

List of Illustrations

List of Tables

Introduction

IZUMI SHIMADA

RESEARCH BACKGROUND AND ISSUES

The study of craft production is complex and challenging, addressing questions on the material, organizational, and ideological interests, values, and capacities of a given culture: How did people perceive, acquire, develop, manage, and transform raw materials into desired finished products? What roles did the technologies, potential and limitations of raw materials, and skills, desires, and beliefs of both the producer and the consumer play in how products were conceived, made, and used? How were these products distributed within and outside their own social boundaries for their mundane and symbolic needs and desires? What was the dynamic and creative interplay between the development, intensification, and control of craft specialization, on the one hand, and social inequality, political economy, and political and religious dogma, on the other?

To answer these questions, we must consider material, behavioral, technological, symbolic, social, and environmental factors and integrate a wide range of academic approaches and tools. Not surprisingly, a wide array of perspectives and approaches has been developed, including those of behavioral chain (Schiffer 1976, 1992); the biography of objects (Godsen and Marshall 1999; Jones 2002); ceramic ecology (e.g., Arnold 1985; Kolb 1988); *chaîne opératoire* (sequence of operations) and related technological choice (e.g., Lemonnier 1992, 1993; Sellet 1993); holistic or systemic understanding (Costin 2001; Shimada 1985; Shimada and Wagner 2007); political economy (e.g., Brumfiel and Earle 1987); social agency and dynamics (e.g., Dobres 2000; Dobres and Hoffman 1994, 1999); sociology, social anthropology, or the social construction of technology (e.g., Bijker et al. 1987; Costin and Wright 1998; Pfaffenberger 1992; Smith 1965, 1970); and technological style (Lechtman 1977; Lechtman and Steinberg 1979).

These approaches utilize various combinations of archaeometry (e.g., Bronitsky 1986; Sillar and Tite 2000; Wagner et al. 1998), ethnoarchaeology (e.g., Costin 2000; David and Kramer 2001; Stark 2003), and experimental archaeology (e.g., Schiffer and Skibo 1987; Shimada 2005; Skibo 1992; Tringham 1978) with a good deal of overlap in their aims and premises. In fact, all embrace the fundamental premise that production and associated technology as well as the use, value, and meaning of their products are socially embedded, that how and why certain raw materials and products are used, manufactured, and valued are socially conditioned. Production and related technology are thus envisioned as operating within specific historical and social contexts of use and meaning (e.g., Childs 1994; Hosler 1994; Kingery 1993; Lemonnier 1993; Sillar and Tite 2000).

Most of the above approaches also recognize that production and related technology develop and operate as a dynamic and creative system of interacting environmental, material, behavioral, and social, symbolic, and other cultural factors; in other words, such systems must be seen as both material *and* social products and processes. These dynamic and multifaceted systems demand that we not only elucidate the chaîne opératoire that transforms a given raw material into a manufactured product but also offer a balanced evaluation of the producer and the consumer. The first illuminates the material process of manufacture including the potential and limitations of raw materials and technology, as well as behaviors and choices made. The second focuses on the social dimensions of production, complementing the former in clarifying how the nature of the producer–consumer relationship and the skill, gender, and social identity and status of the producer affect the production, use, and value of the product.

Analysis of finished products and associated manufacturing debris outside of production context has predominated the archaeological study of craft production. The excavation of production loci (including workshops) within comprehensive regional studies, complemented by field and laboratory experiments, archaeometry, ethnoarchaeology, and the study of archives and existing artifact collections, offers an effective alternative (and complement). Workshops are defined here as clearly demarcated indoor and outdoor spaces where a major portion or the entirety of craft production occurs on a regular basis. They are dedicated to craft activities, although their size and organization may vary from a single built space to an interlinked set of spaces. Commonly, workshops are situated away from domestic areas and manned by differentially skilled workers, with varying degrees of independence from the central authorities. Craft production, however, is by no means restricted to the workshop as defined here. The importance of household or domestic production and kinship in recruiting and organizing such production has been well documented archaeologically and ethnographically, as seen in this volume (also Ames 1995; David and Kramer 2001; Feinman 1999; Peacock 1981; Rice 1987; Santley and Hirth 1993; van der Leeuw 1977, 1984). In this regard, production loci may be found in or near the household. In fact, where appropriate spaces are limited in extent or availability, craft activities may be carried out in spaces that are used for other purposes at different times of the day or year (Shimada 1994a:199). In general, workshop and other production loci must be seen in terms of the specific physical, social, economic, and political factors that prevail.

A few decades ago, the difficulty of finding craft production loci was a major preoccupation of archaeologists working in the Americas, particularly in nonurban settings, and their defining markers were debated (e.g., Connell 2002; Moholy-Nagy 1990; Stark 1985; Urban et al. 1997). This was a notable contrast to the seeming ease with which such loci were identified at urban sites (e.g., Shimada 1978; Spence et al. 1984; Topic 1982; Turner 1988). Craft production loci were not so much scarce as unidentified. The past two decades or so have seen a remarkable increase in excavations of these loci of varied crafts, age, size, and organizational complexity and form. In the Andes, for example, we can now readily name some two dozen well-documented workshops of different crafts and different eras (e.g., Janusek 1999; Shimada 1994b, 1998; Isbell, this volume).

More excavations of workshops and other production loci are needed as they afford us primary data on the sequence of operation, interrelationships among coterminous crafts, and the organization and identity of the producer as well as his or her relationships with the sponsor/consumer. However, such excavations should be "holistic" in scope and properly contextualized. In other words, we need to (1) define the organization and technology of the entire craft production process from the acquisition of necessary ingredients to the performance quality of finished products and (2) understand social conventions and institutions, the value system, distribution mechanisms, and both latent and manifest functions of products, all of which guide the design, distribution, use, and meaning of products (Shimada and Wagner 2007). This contextualization of the workshop in turn calls for the elucidation of its local and regional technological antecedents, environmental conditions, and resources, together with the social factors that affect craft production. In essence, to understand production as a synergistic social and material process, we must holistically approach the study of its context and processes (Costin 2001; Shimada 1985, 1994b; Shimada and Wagner 2001, 2007).

Adding to the mix, the political economy perspective on craft production (e.g., Brumfiel and Earle 1987; D'Altroy and Earle 1985) has tended to overemphasize the needs and preferences of elite sponsors or consumers without due attention to the role of artisans in production. This same perspective has created a misleading picture of independent and attached specialists as mutually exclusive in product quality and repertoire, as Costin (1991), Hayashida (1995, 1999), and others have pointed out.

Other perspectives tend to be similarly skewed. Mills (this volume), for example, justifiably criticizes that "many recent discussions of the biographies of objects...have, surprisingly, veered away from that part of object biographies that begins with production to an almost exclusive discussion of the many contexts of consumption in which objects were used." The elite or consumer-centric "top-down" vision of craft production, product distribution, and product use needs to be balanced with a producer or "bottom-up" perspective. The nature of both consumers and producers as well as their interrelations and other factors (e.g., intended usage, traditions and values) affecting both groups need to be considered if we are to understand why and how certain craft products were made in a certain manner (Costin 1991, 2001).

In spite of recent advances in various facets and issues in craft production, most of the available approaches and models noted above and the craft studies carried out focus on single crafts (e.g., ceramics, metallurgy, and textiles). Each craft—or even category within a craft (e.g., blackware ceramics)—is treated as analytically distinct. Moreover, many earlier studies assume that their etic categorization of crafts is valid and meaningful and do not allow for the possibility that ancient producers conceived of their crafts differently (Shimada et al. 1999; Shimada and Wagner 2001, 2007).

The organization of pottery production in middle-range and stratified societies has received a lion's share of archaeological attention (e.g., Bey and Pool 1992; Rice 1996; Shimada 1998). Though there are legitimate reasons for its disproportionate share (e.g., ubiquity, multifunctionality, durability, and its role as a sensitive indicator of cultural and natural changes), its predominance has unintentionally slanted our approaches and models so that craft production is viewed as a discrete, single-medium affair. We need to recognize that we may have unintentionally neglected or misrepresented the full organization and social and ideological significance of multicraft and other related production (see below for definitions) and their products.

The above vision of discrete craft production contrasts sharply with much of modern industrial production that integrates multiple components of different substances derived from different sources near and far. In fact, preindustrial material culture in both the Old and New Worlds "is replete with examples of what are in reality composites of products, knowledge and/or techniques derived from what are analytically treated as distinct crafts" (Shimada 1996:28). Further, many comprehensive studies of preindustrial urban centers throughout the world, as noted earlier, have documented the clustering of diverse craft workshops (e.g., Clark 2001; Janusek 1999; Shimada 2001). Kenoyer and Miller (this volume) remind us that the interrelationships among many crafts at the ancient cities of Mohenjo-daro and Harappa were considered in the excavation reports back in the 1920s. We need to examine such interrelationships in greater depth and cross-culturally.

Most recent craft studies have not adequately considered the relationship of the craft under study to other crafts practiced concurrently at local or regional levels, a point that is effectively made by Feinman and Nicholas in their work in the Ejutla Valley in Mexico (Feinman 1999; Feinman et al. 1993; also McGovern 1989) and by me in my study of Middle Sicán craft production on the north coast of Peru (Shimada 1996). Feinman's (1999) criticism that most models of craft production organization overemphasize the degree of complexity is well justified; the significance of multicrafting and other similar production forms is largely ignored. In fact, these production organization forms raise a series of important questions in regard to labor and material organization and management. Speaking of barrio or commoner residence workshops at the Chimú capital of Chan Chan, Topic concludes that "workshops were specialized spaces but did not necessarily specialize in one product" and "horizontal organization" allowed for more immediate cooperation among diverse artisans to produce compound products (1990:164–165). This organization then would have (1) offered productive flexibility in responding to variable demands for different products, both prestige and nonprestige, and (2) promoted artistic and technical transfer and creativity through collaboration or competition. To what extent did multiple craft production coevolve in the sense of the collaboration or competition that Topic infers? How was such collaboration or competition mediated, and by whom? How were craftspersons recruited? Are models developed strictly for one craft (i.e., ceramics) applicable to the multicraft cases? Of particular interest to us are the social identity of the craftsperson and the social and natural conditions and contexts under which multicrafting and multicraft production exist.

The above situation is understandable considering the complexity and quantity of data, our academic formation, and a scarcity of research in which multiple crafts are investigated in parallel or in sequence and in sufficient depth. Some archaeologists and materials scientists have explored the processes and consequences of the transfer of knowledge and cross-fertilization among various crafts concurrently practiced in close proximity. F. Matson (1989) sees ceramics as the "hub of ancient craft interplay," whereas others (e.g., Lechtman and Steinberg 1979; Smith 1965, 1970; Wertime 1964) point to the wide range of knowledge, materials, and processes shared among pyrotechnologies, such as glass, metals, and ceramics. Still others speculate that the rectilinear character of representations in stone or ceramics may have been influenced by the structures found in textiles (e.g., Lyon 1991). The foci of these earlier studies, however, are relatively narrowly conceived material and technological issues, not the broad systemic conception embraced in this volume of production and related technology as a material and social product and process. These studies, though they raise stimulating and worthy questions, did not spawn field research (i.e., excavations of workshops or other production loci) that examined in depth the material and social components of multicraft and other related production organization. Multicrafting, multicraft production, and multicraft interaction or, more generally, linkages among a wide array of productive activities (e.g., crafting, farming, and house building) that may be underlain by shared organizational principles and beliefs (Shimada 2001) have been largely ignored in the recent anthropological study of production and associated technology.

AIMS OF THE BOOK

The above considerations led me to organize a symposium entitled "Rethinking Craft Production: The Nature of Producers and Articulation among Crafts" at the 2003 Annual Meeting of the Society for American Archaeology in Milwaukee. This volume is an outgrowth of the symposium and has two interrelated aims.

First, it aims to illustrate the scope and depth of understanding and insights to be gained from properly contextualized excavations of workshops and other craft production loci in regard to the behavioral, material, and social components of craft production and the role of the producer and the consumer in production and their social identity. The intent of the volume is to demonstrate the productiveness of a workshop-based approach to reconstructing and understanding craft production and to complement what has been achieved by other means.

Subsumed within this aim is the aforementioned interest in redressing the skewed producer–consumer balance in our study of craft production by paying long-overdue attention to a producer or "bottom-up" perspective. Also related to the first aim is the goal to continue earlier advances in our understanding of the social identity of the artisans (e.g., Brumfiel 1991, 1998; Costin 1996, 1998; McGraw 1996; Wright 1991, 1996). The volume *Craft and Social Identity* (Costin and Wright 1998) has effectively illuminated the intimate linkage between social identity and craft production. The location, scale, intensity, and organization of production as well as the use and value of its products are inextricably tied to social identity and relations between both the producer and the consumer. M. Helms (1993), for example, brings attention to the symbiotic relationship between skilled artisans and their elite sponsors in their mutual quest for power and prestige and how artisans may have attained higher social status in the process. In the Maya area, there are various lines of evidence that suggest that the artisans who created monumental sculpture for the purpose of perpetuating elite propaganda were esteemed members of Maya nobility who were attached to royal courts as highly trained and skilled full-time specialists (e.g., Inomata 2001; Inomata and Triadan 2000; also see Inomata, this volume). Ames (1995) presents a similar argument for the important roles of social elites in craft production on the Northwest Coast. Much the same has been suggested for skilled Moche or Mochica potters in a workshop situated close to the ceremonial center of Huaca de la Luna at Moche (Uceda and Armas 1998). Is the binary vision of sponsors and producers as elites and nonelites, respectively, that emerges from hierarchical models of the organization of production valid or insightful? How can we address these and other related questions in regard to research methods? With what research methods can we address these and other related questions?

The second major aim of this book is to bring attention back to multicrafting, multicraft production, multicraft interaction, and related phenomena and rectify the perception of craft production that is skewed by emphasis on discrete, single-medium cases. We challenge the widespread tendency to divide cultural remains by material type rather than integrate them into a larger understanding of systems of production and use. Both of our two basic aims are underlain by an emphasis on data and the perspective from excavations of workshops and other production loci and the vision of craft production and associated technology as a material and social process. In respect to multicraft production, we are interested not only in issues of technological transfer, coevolution, and synergetic creativity but also in the organization, social identity, and interrelationships among producers, including competition for supplies, new members, and social prestige.

By *multicrafting*, I mean the same or closely related artisans performing a major part or the entirety of two or more distinct crafts involving different media or raw materials in close proximity to each other to produce items primarily for suprahousehold consumption, for example, the same group of women weaving cloth and making pottery (Mills, this volume) or making pottery and beer (Isbell, this volume). Although the simultaneous engagement in multiple crafts would not be physically feasible, a single individual or group could effectively apportion time and energy to perform facets or phases of multiple crafts in the same area or nearby locations. Multicrafting can be seasonal or year-round. Given the inherent difficulty in acquiring new motor skills and acquiring and maintaining diverse tools and raw materials (Arnold 1999), we would expect multicrafting to favor crafts based on similar technologies, for example, shell and stone bead making (Feinman and Nicholas, this volume; Stark, this volume) or bronze and iron smithing (Wells, this volume).

On the other hand, *multicraft production* describes the concurrent practice of multiple crafts by different individuals or groups, each specialized in one or more crafts, in the same space or in a series of adjacent spaces. An example would be a pairing of pottery making and metalworking (Goldstein and Shimada, this volume; cf. Kenoyer and Miller, this volume) that are both based on material transformation through heat application. In the case that Goldstein and I (this volume) discuss, charcoal generated by pottery firing appears to have been recycled as fuel for metalworking in another part of the same workshop. Potters also provided the tall urns used to build updraft furnaces. An additional factor that brought them to the same area may have been the need for ample space to store fuels and cope with the resultant detritus and environmental pollutions (also see Li, this volume; Kenoyer and Miller, this volume).

The term *coproduction* emphasizes a form of multicraft production in which artisans specializing in different crafts (maybe close to each other) collaborate in the design and manufacture of products (i.e., a feedback relationship). As Stark (this volume) notes, the resultant products are composites of parts typically of different media. Li (this volume) effectively illustrates this process in regard to Late Shang section-mold bronze casting. In this case, the artisans who prepared clay models and their section molds worked in tandem with those responsible for subsequent casting that took place at a distinct location, but both groups shared a good deal of knowledge of the overall production process and final products.

Although multicraft production implies the concurrent practice of multiple crafts, parts of the activities may be better seen as *sequential*, as in the aforementioned case of charcoal from pottery firing serving as metalworking fuel. Much the same can be said about coproduction. The temporal gap between the two succeeding stages was likely to be determined by multiple factors such as the nature of the products and the intensity and scale of production

As seen from the above discussion, multicraft production and multicrafting subsume diverse production organizations and producers. The above distinctions are, however, preliminary formulations. It is hoped that this publication will spur their testing and refinement.

CONTENTS AND ORGANIZATION OF THE VOLUME

This volume constitutes a collective effort to attain the two aims and answer some of the questions posed above. In addition to an emphasis on contextualized excavations of workshops or other production loci and a dyadic vision of craft production and related technology as a material and social process, the book attempts a cross-cultural examination of the organization of preindustrial multicraft production and the social identities of the artisans involved. The comparative perspective is critical to the clarification of contextual variability in issues raised earlier. The book as a whole is comparative in multiple senses. It compares cases involving different sets of crafts practiced in varied environmental, temporal, and cultural contexts. Although the pre-Hispanic New World case studies predominate, the book contains comparative studies covering Western Europe (Wells), the Indus Valley (Kenoyer and Miller), and Central China (Li). The case studies represent settlements and cultures of different scale and sociopolitical integration, ranging from the relatively small, nucleated Mogollon settlement of a "middle-range" society in Arizona (Mills) to the vast city of Harappa in the Indus Valley (Kenoyer and Miller). As reflected in the book title, however, cases from complex societies predominate.

These cases vary in the theoretical approaches and analytical methods utilized. Though no particular approach or theoretical perspective was imposed or predominates, the popularity and productivity of the combined use of archaeometric, experimental, ethnohistoric, and ethnographic data are apparent in various chapters. In regard to data sources, although the book emphasizes data from excavated workshops or other production loci, contributors have utilized various combinations of other sources. Regardless of differences in data source, all contributors recognize the critical importance of elucidating the social, historical, and natural contexts in which craft production and usage of its products occurred.

This volume is organized thematically and

begins with case studies by Mills, Goldstein and me, Isbell, Feinman and Nicholas, and Inomata of multicrafting and multicraft production based on excavations of production loci at a single place or a few nearby sites. The following three chapters, by Wells, Kenoyer and Miller, and Li, are more regional and diachronic syntheses but also discuss a variety of production organizations, including coproduction and sequential production. The final part of the book offers two chapters by Stark and Ramírez, who draw information from intensive regional survey and surface collections and sixteenth- and seventeenth-century documents, respectively. Their contributions are intended to show the diversity of effective avenues for reconstructing craft production, particularly in regard to the identity of producers and multicraft production that do not rely on workshop excavations.

What follows is a discussion of how different chapters relate to each other, to the issues raised above, and to the two principal aims of the book. The idea of elucidating the social, historical, and natural contexts of multicraft production is at the core of Barbara J. Mills's chapter as she examines "how multicraft production, migration, and identity intersect" during the period spanning A.D. 1000–1325 among seven sites pertaining to "middle-range societies" in the Mogollon Rim region of east-central Arizona. This was a period of extensive rural–rural migration and a sudden intensification in the production of multiple crafts in the region. In the context of such social fluidity, where and by whom were these crafts produced, and how did the production and products reflect this dynamic situation? Taking advantage of the fine and reliable chronological control afforded by tree ring dating, Mills tracked how crafting, migration, and other social changes intersected. With the premises that migration (1) involves changes in the way that social identity is constituted and (2) creates economic, social, and political asymmetries between hosts and guests, she argues that the intensification of craft production was part of the construction of new identities and the negotiation of social inequalities that followed the migration of new households into established communities, that craft production afforded opportunities for the expression and reformulation of identities.

Mills "emphasizes how the household and suprahousehold production of goods emerged as two social contexts for multicraft production in the Southwest" and "shows how identity and craft production were interlinked." In regard to the latter, she suspects that in market economy–oriented production, identity and crafting are more separated than in smaller-scale nonmarket production in the Southwest, where "the social contexts of production and consumption were probably all well known by most community members." It is a hypothesis that will interest anyone examining the linkage between crafting and identity in complex societies.

Mills also brings out the situational concept of workshop, that we cannot assume contexts devoted exclusively to craft production and that the degree of exclusivity and intensity of use of a given space for craft activities was dependent on the situation. Mills finds that during the late eleventh and early twelfth centuries when small groups migrated to an essentially uninhabited area, the great kiva (communal ritual structure) served as the integrative mechanism and pottery and steatite bead making were conducted evenly and "probably practiced by most households on a relatively equal basis." In contrast, during the late thirteenth century, households were migrating into areas with a much larger host population, resulting in unprecedented population nucleation. Concurrently, dramatic changes in the intensity and concentration of craft production by some households can be documented, including the appearance of intensive multicraft areas (ceramics, bead making, shell working, and probably basketry and weaving), specialized production tools, and an uneven distribution of production-related tools and debris. Not only did multicrafting by some households occur within the context of increased production specialization, but intensification of one craft was not accompanied by the de-emphasis or abandonment of other craft production activities. Further, there were two distinct contexts of craft production: intensive multicrafting by some households and the crafting of textiles in kivas. Mills's study "emphasizes how workshops...were only a small part of multicraft specialist production." Overall, together with Feinman and Nichols (this volume), Mills shows that household production is found in societies of varying scale and complexity and that production is quite flexible in scale and intensity, ranging from that for their own consumption to that for community use.

The contribution by David J. Goldstein and me offers an effective methodology for elucidating multicraft production within a single workshop while underscoring the importance of examining the entire sequence of operation. Our study focuses on fine pottery production using molds and copper and gold alloy smithing at a 1,000-year-old workshop on the north coast of Peru. We show that what may have been of little value for one group of artisans may have been valuable resources for another group of artisans working nearby. This is akin to the "chained" relationship of two groups of obsidian workers at Otumba, a Late Postclassic city-state in the Basin of Mexico, that Otis Charlton (1993, 1994) has documented. There, exhausted prismatic cores from blade making were used by lapidaries as tools and as raw material for ear spools and other ground and polished products.

This pairing that Goldstein and I demonstrate is not a surprise given that both are based on the principle of material transformation through the controlled use of heat and atmospheric conditions, as well as attendant shared needs for refractory firing structures (kilns and furnaces) and quality fuels (Matson 1989; Wertime 1964). I hypothesized that, because metals had a higher social and symbolic importance than pottery in the Sicán culture, smiths would have enjoyed preferential allocation of limited hardwood fuels over potters in a multicraft production context. We tested this through a comparative analysis of carbonized fuel remains from excavated pottery kilns, metalworking furnaces, and domestic hearths (cf. Johannessen and Hastorf 1990; Lennstrom and Hastorf 1995).

Paleoethnobotanical analysis of fuel remains was an integral part of a multifaceted analytical program that included (1) ethnoarchaeology of fuel gathering and consumption as well as charcoal making, (2) field firing experiments employing a replica kiln and a replica furnace, and (3) chemical and physical determination of the material components and performance quality of both Sicán and modern replica kilns and furnaces as well as their fired products and by-products. The analytical program dovetailed with a broader and longer-term regional study of Sicán cultural developments (i.e., appropriate contextualization).

Analytical results largely contradicted the hypothesis, showing instead various alliances and continuities between pottery making and metalworking. Although there are discernable qualitative differences between the fuel residues associated with these two crafts, charcoal for metalworking furnaces appears to have been generated as a by-product of the hardwood fuels used for the reduction firing of ceramics, together with possible onsite charcoal production. This conclusion, together with other lines of evidence, argues that the "choice" and apportionment of different fuels were left in the hands of local metalworkers and potters, who were likely to have directly negotiated with each other for mutually acceptable arrangements. The good deal of technical variation in pottery production documented at the workshop further points to considerable autonomy in spite of their prestigious products.

Overall, our findings are much closer to postulates of the agency-based and technological choice approaches than to those of the political economic model. Our study also illustrates that even a detailed social and symbolic analysis of finished products is inadequate in elucidating "technological choices" and interaction among artisans in the production process.

William H. Isbell presents a stimulating agency-based vision of the role and social identity of the inferred wives and concubines of local elites in the urban "palace community" (cf. Kolata 1983) of Conchopata in the central highlands of Peru during the first half of its occupation, dated to circa A.D. 550–850. Isbell makes a cogent argument that Conchopata offers a unique opportunity to explore relations among craft specializations, identity construction by artisans, and changing images of the city as its social and economic structure evolved through time. His emphasis is on defining the social and political contexts of craft production and the identity of craftspersons, more than on the technology, organization, or functional interrelationships among different activities.

More specifically, Isbell suggests that these women were engaged in multicrafting and services—including making ornate, oversized ceramic urns and jars, brewing beer, hosting the competitive and status-building feasts that their lords sponsored, and perhaps assisting in ritual smashings/offerings of the fancy vessels they produced. Based on the tomb of a male adult surrounded by at least six women and juveniles, as well as a clear predominance of females among the other human remains recovered, Isbell argues for polygamy among elites and thus a pool of women available for crafts and remarks that these women "employed the products of their arts to project feminine agency into dynamic collective events where important social and political negotiations took place." Isbell, along with Mills, argues that feasts served to not only shape and validate identity but also intensify craft production (also Spielmann 1998).

The multicrafting and services in which these women were engaged can be seen as a case of sequential production, starting with pottery production, progressing to use (beer making and feast hosting), and ending with their ritual "sacrifice." Perhaps reflecting the unique character of the excavated areas (palaces, temples, and immediate surroundings), various findings do not fit in conventional models of ceramic production. Cook and Benco (2001) report six firing areas and four concentrations of ceramic manufacturing tools distributed across the excavated areas of the site and smaller quantities of potting tools scattered in many more locations. They conclude that the spatial patterns of production at Conchopata do not fit with any of the types of pottery specialization identified by Costin (1991) or Sinopoli (1988).

A set of essays from Mesoamerica illuminates the contextual and organizational variability in multicraft production and the social identities of associated artisans as well as the importance of household production. Based on their multiyear investigation of both modern and pre-Hispanic household-based production in the Oaxaca region of south-central Mexico, Gary M. Feinman and Linda M. Nichols examine specific organizational features of multicrafting and multicraft production, factors that may account for variability, and their implications for understanding the regional political economy. They show that, similar to today, most craft production in the Classic period Valley of Oaxaca took place in domestic settings. At the excavated Classic settlements of Ejutla and El Palmillo, they found large quantities of debris attesting multicraft production (e.g., ceramics, ground- and chipped-stone tools, lapidary objects, and marine shell ornaments, as well as remains from the processing of maguey for fiber) at individual household and community levels. It is understandable that shell and lapidary works were found in close association in some Ejutla households as both employed chert microdrills and similar motor skills. Similar household-level multicrafting (faience manufacture, seal making, shell working) is documented by Kenoyer and Miller (this volume) for the third millennium B.C. Indus city of Harappa.

It is apparent that, at Ejutla and El Palmillo, households engaged to varying degrees in specialized crafts and that the crafts emphasized were markedly different at the two sites. At the same time, no single craft industry prevailed at either site, and most crafts appear to have been practiced only on a part-time basis. Thus, Feinman and Nichols suggest that "within individual households, some production could be for exchange beyond the community…, whereas other goods were produced largely for local consumption.… In Classic period Oaxaca, it is evident that domestic production could produce craft goods at very different volumes or intensities."

The considerable variability in multicraft

production organizational form and its productive output and intensity is indeed one of the main findings that emerges from various chapters in this book based on the regionally contextualized producer-workshop perspective. We simply cannot assume a given organizational form and its intensity or even degree of specialization. The main conclusion to be drawn from the decentralized, largely autonomous household-based craft production that Feinman and Nichols document, together with the conspicuous absence of large-scale storage facilities, is that the Classic regional economy in Oaxaca was based on some kind of marketplace exchange similar to that of today and at the time of Spanish conquest.

At the hastily abandoned Classic Maya site of Aguateca in Guatemala, Takeshi Inomata takes advantage of uniquely well-preserved remains (e.g., tools and component parts) found in residences and other elite contexts to examine craft production by the Maya elite from the perspectives of multicraft production and performance theory. His study is built on the premise that craft production is "not simply an economic activity but often has a quality of performance." He defines *performance* as "communicative physical acts that involve an actor (or actors) and an audience." This performance perspective "mediates the top-down [consumer/sponsor perspective] and bottom-up [producer perspective] approaches in the study of craft production," allowing an in-depth examination of how producer and consumer perspectives are intertwined and shape each other. These are among the main concerns of the book.

Inomata presents a persuasive case for elites performing actual craft (in many ways, multicrafting) work as creative, economic, and, more importantly, political and ideological acts with profound consequences for the competition and alliance or collaboration among them as well as to distinguish them from their subordinates. He argues that craft production in elite residences occurred in visible areas intended to be seen and impress fellow elites—performance in the true sense of the word.

Mirroring one of the volume's concerns, Inomata cautions against the imposition of classificatory schemes created by archaeologists, which may not fit the reality of past craft practices; he notes that, "for Maya craftspersons, the socially constructed values and meanings of objects may have shaped more important categorizations of products than those determined by material types." Taking advantage of insights from written sources, Inomata also illuminates how the esoteric knowledge of Maya elite culture was given material expression in multicraft goods; he argues that it is incumbent on us to pay attention "not only to the visible and tangible domain of materiality but also to the abstract and cognitive domains of knowledge and belief."

As in the contribution by Goldstein and me, Inomata probes deeply into facets of multicraft production: competition and cooperation, together with associated synergetic constraints and potential. Inomata argues that Maya elite craft activities involved collaboration and competition paralleling similar processes that took place in the royal court and that the act of production itself, beyond the economic and political gains expected from the exchange of products, conferred much prestige and power to the elite artisans (Helms 1993). Further, his case study, like other contributions in the volume, challenges the clear-cut producer–consumer and attached–independent production dichotomies.

As expected, case studies from the Old World present us with both similarities and differences in approaches to and opinions concerning various issues on multicraft production and others of interest to this volume. Peter S. Wells examines changes and stability in technology, organization, and the nature and identity of workers engaged in ceramic and metal ornament production against the background of major social, economic, and political transformations in temperate Europe between the Late Bronze Age (1200–800 B.C.) and the end of the Early Roman period (A.D. 300). This macroregional and diachronic study is conducted from the vantage of access to both textual and material evidence, which in turn affords insights into both producers and consumers/sponsors. For example, Wells documents impressive variation in the social identity and character of producers, ranging from specialized craft workers of prestigious elite items such as polychrome pottery and gold ornaments to entrepreneurial (independent) workers peddling their products or providing them on a made-to-order basis to soldiers manning the frontier along the Danube River.

Wells argues that multicrafting was limited or absent in Late Bronze Age temperate Europe, that more and more studies are pointing to a pervasive pattern of small-scale, dispersed production as opposed to the earlier conception of centralized production in urban settings. He also suggests that some craft workers, especially metalworkers, enjoyed special status. Their craft used intense fire to fundamentally transform seemingly mundane ores into valuable new substances, metals, a metamorphosis perhaps underscored by the association of magic with their processes and results. The association of the transformative aspects of smelting with supernatural forces or magic has been documented in other areas of both the Old and New Worlds (e.g., David and Kramer 2001; Schmidt 1996; Shimada and Merkel 1991). Wells suggests that these powerful associations may have played a role in helping some of these metalworkers maintain a certain level of independence within the social and political context in which they worked.

Of much interest to cross-cultural comparison (including sociopolitical complexity and scale) of production organization is Wells's finding that the small-scale and dispersed organization of manufacturing persisted throughout later prehistory and the Early Roman period. This continuity is remarkable given the major changes in the social and political landscapes during this fifteen-century-long time span, including the establishment of the large *oppida*, fortified settlements, at the end of the Late Iron Age and invasion by the Roman Empire. Although aspects of manufacturing became more specialized with greater social differentiation, the persistence of dispersed patterns of manufacturing is quite evident. For example, the growth of new, larger political units such as urbanized oppida in the later part of the Early Iron Age was accompanied by an increase in the number of workshops and the development of series production in pottery and in ornaments. However, people built more time-tested small kilns in many locations within the same settlement rather than building and concentrating large kilns in a few locations. In other words, increased production output was achieved by multiplying the basic traditional production units, not by transforming them into highly concentrated or tightly controlled forms. In the case of pottery production, Wells concludes that "the fundamentally dispersed character of craft production [in temperate Europe during 1200 B.C.–A.D. 300] reflects [a] pattern of segmentary, noncentralized political organization."

A similar solution to the demand for increased production output is seen in the New World. Dispersed, small-scale "modular" production organization seen in the pre-Hispanic Moche and Sicán cultures on the north coast of Peru allows easy recruitment of workers along extant kinship and economic specialization lines and preserves time-tested technology as well as a sense of group identity, unity, and continuity, while promoting the integration of different ethnic and social groups (Shimada 2001; Shimada and Wagner 2007). It appears that the Roman Empire also adopted this "indirect administrative" setup in the frontier regions of temperate Europe, maintaining the integrity of traditional local patterns of craft production and supply, together with associated social and political systems.

In general, as with the case of Chinese archaeology (Li, this volume), Wells notes that "close attention to patterns of workshop distribution and activity is still relatively new [in the archaeology of temperate Europe], and much remains to be done before we can say that we have a good understanding of the production side of the equation." However, his chapter effectively shows that efforts to redress this situation have already resulted "in the realization that workshop activity was much more widespread than earlier investigators had thought and that a much wider range of individuals was crafting than imagined."

Based primarily on data culled from 16

seasons of fieldwork by the Harappa Archaeological Research Project within the extensive city of Harappa and its surrounding communities in the Indus Valley, Jonathan Mark Kenoyer and Heather M.-L. Miller offer a valuable synthesis of a long-term trajectory (Upper Paleolithic through the second millennium B.C.) of how different crafts developed and interacted and the factors and processes that shaped and brought them together (and, in some cases, kept them apart). They argue that the continuity and change seen in one craft cannot be generalized to others and that "each occurrence of a craft must be studied carefully to understand its relationship to other crafts at the site," underscoring the importance of excavating production loci.

Of particular interest to this volume is Kenoyer and Miller's explication of the factors and conditions that promote multicraft production. They note, for example, that crafts based on "extractive" or "reductive technologies" such as shell and ivory working, steatite and agate bead making, and stone tool production appear to have been practiced together in distinct areas of Harappa (Harappa Phase, 2600–1900 B.C.). The authors suggest factors such as the sharing of menial labor and raw materials and the need to combine many of the different materials into more complex objects as contributing to the observed clustering. In contrast, "transformative" crafts that involve the transformation of raw materials by various pyrotechnological (application of heat) or chemical processes such as copper working, pottery firing, and the manufacture of faience and steatite seals appear to have been undertaken in relative isolation, not only from reductive crafts but also from each other. The physical separation of transformative crafts from habitational areas and other crafts, however, is expected given that they produce nuisance smoke, ash, and heat and require much space for kilns or furnaces and the storage of fuels and other items. In the case of the segregation of copper working, Miller (2007) suggests that elite interest in monopolization may have been more important than pollution concerns. This situation contrasts with the coexistence of pottery making and metalworking and the absence of readily recognizable on-site supervision at Huaca Sialupe discussed by Goldstein and me.

Kenoyer and Miller conclude that the association of crafts is defined by a combination of both (1) the similarities or complementarities of the manufacturing processes and (2) the relative values of the final products. The relative value is itself based on complex factors relating to the availability of raw materials and the technology used for production. In regard to the second factor, they suggest that the crafts that became most important for reinforcing social and ritual status (e.g., faience molded tablets or steatite inscribed tablets) were ones that could be efficiently controlled by new elites and powerful merchants of the Indus cities. They note that, although there is little evidence for direct control of most crafts at Harappa, the fact that all of the crafts were taking place inside the city wall may indicate a degree of indirect control through the regulation of movement into and out of the craft area. Further, they suggest that "the presence of chert weights and weight manufacturing debris in the area just inside the gateway may indicate the taxation of goods coming into or leaving this walled sector of the city." In general, Kenoyer and Miller appropriately relate the development and specialization of craft production to "fundamental social processes of crafting power... that were going on during the initial phase of urban development."

They are also concerned with the complex network of trade and craft relationships between intraurban crafts and those in specialized craft villages in surrounding areas. A similar center–hinterland or regional perspective characterizes various case studies from both the New and Old Worlds, particularly those of Stark (this volume) and Feinman and Nicholas (this volume).

Issues surrounding the social identities of the artisans are difficult to resolve. The Harappa project faces such daunting challenges as 15 m of superimposed deposits and the cultural practice of dumping refuse in open drains. From an ethnoarchaeological perspective, however, Kenoyer and Miller argue for the importance of itinerant craft workers (particularly bead makers) who carried their entire stock and tools with them and moved from one town to the next, following major markets or seeking festivals and work along the edge of the street or in any available open area. Ramírez (this volume) and Wells (this volume) echo the same concern. Related to the question of the artisans' role is the need to recognize the distinct possibility that craftspersons producing items of a given style may not belong to the same ethnic or social group that the style symbolizes. In essence, it highlights the potential difference between consumer-based versus producer-based perspectives. Methodologically speaking, Kenoyer and Miller, along with Goldstein and me (this volume) and Li (this volume), shed light on the importance of (1) situating workshop excavations within a comprehensive regional study and (2) self-corrective use of ethnoarchaeology, experiments, and archaeometry.

Yung-ti Li presents a valuable and insightful synthesis of a large corpus of data on the production of complex ritual bronze vessels in and around the last Shang capital of Anyang (ca. 1200–1000 B.C.) in China. His study differs from other chapters of this book in its adoption of the operational sequence approach. Anthropologically oriented, detailed studies of ancient Chinese production and technology in English are scarce (e.g., Bagley 1993, 1995), and Li's contribution goes a long way in filling this notable gap. He first details the history of research on the relevant materials and technology and the emerging operational sequence. What emerges is an understanding of the close relationship between bronze casting and pottery technology, that is, "a co-craft based on the mastery of two very different materials, metal and clay."

Li refines Ursula Franklin's (1983) conception of early Chinese bronze casting as an example of what she terms "prescriptive technology": that Chinese casting involved skilled clay workers and metal casters, with each performing specific "prescribed" tasks (i.e., specialized) in a long, predetermined sequence of operations commensurate with individual expertise. Franklin contrasts this ancient version of "assembly-line" production organization with "holistic technology," in which the entire manufacturing process is performed by the same artisans. Production organizations approximating the latter are well documented in this volume, particularly by Goldstein and me.

Li's detailed review of the available data shows that bronze casting organizations at Anyang were much more complex than Franklin theorizes, with prescriptive and holistic technologies coexisting (albeit at different locations). This finding also illustrates the limitations of reconstructing the organization of craft production from finished products, as Franklin has done. Li shows that the changes in production organization were more closely related to increased demand stemming from the expanding population (including elites) and a greater investment in political economy than to shifts in managerial control. Given the dispersed locations of the six excavated bronze foundries within the extensive capital region of Anyang, Li suggests correspondingly different groups in charge, such as the ruling elite and local lower elite lineages.

Overall, Li's review of workshop excavation data suggests that the production organization of valuable bronzes at Anyang may have been a compromise between exercising control and meeting the logistics of workshop operations. His insightful conclusions largely derive from carefully considering workshop data within the nested sociopolitical contexts in and around the royal capital. The data in turn are supplemented by information from modern replication bronze casting experiments conducted in conjunction with archaeometric analyses of relevant archaeological remains, raw materials, and experimental products and ethnographic and ethnohistorical descriptions of Chinese pottery production.

Unlike most of the chapters, which emphasize fine-grained data from excavated workshops, Barbara L. Stark derives her data from full-coverage regional surveys and attendant systematic collections in the western Lower Papaloapan Basin of south-central Veracruz, Mexico, conducted in 1984–2002. Her study focuses on the factors and processes underlying the nucleation and intensification of craft production during the Classic (A.D. 300–900) and Postclassic (A.D. 900–1521) periods.

Although surface evidence has various limitations, such as the underrepresentation of crafts that produce relatively little material debris (e.g., spinning and lapidary work), Stark illustrates how such evidence illuminates the spatial patterning of craft production, that is, dispersed versus concentrated and stable versus shifting locations of production, and even the production intensity in some cases. Her measures of intensity of production based on surface collection would interest others planning regional full-coverage surveys.

In her study area, evidence of more spatially concentrated production increases in the Late Classic period, especially in the vicinity of major centers, and becomes even more marked in the Postclassic period. Some of the nucleated crafts represent more intensive production as well. Concomitantly, there are increasing signs of possible multicrafting. Stark concludes that "multicrafting may have been part of a wider escalation of craft specialization and one form of craft intensification."

Stark further argues that the origins and spread of specialization and marketing in Mesoamerica are closely linked to a series of changes resulting from the collapse of large pan-regional states. More specifically, of the factors considered for the nucleation and intensification of craft production—local competition, population growth, increased marketing, economies of scale, the transportability of products, labor costs, economies of place, and political strategies—she sees a combination of the last two, which crosscut polities, as best accounting for the observed changes. She notes that economies of place "made craft nucleation at centers efficient in terms of transport and access to consumers, and labor demands help make sense of the lack of nucleation of some crafts." The multiple factors that Stark evaluates are well worth attention by those investigating craft production in market economies.

Akin to what Mills (this volume)and Feinman and Nichols (this volume) find for eleventh- to fourteenth-century Arizona and the Classic period Oaxaca, respectively, Stark reports for the period under study that "all the evidence suggests production in household contexts, with differences in the intensity of production from a level ancillary to household activities to workshop levels" in regard to pottery making, cotton spinning, and prismatic obsidian blade making. Stark offers a critical discussion of the concept of multicrafting (cf. Mills, this volume). She applies "*multicrafting* to a range of situations involving marked variation in the products or technologies of specialized producers who shared a social unit that had direct bearing on their craft work." Further, multicrafting also occurs when "quite functionally distinct products are fabricated within the same broad technological procedures," as in the case of domestic ceramic vessels and ritual ceramic figurines; that is, multicrafting does not necessarily involve different media as I defined above. It is evident that our current definitions of multicrafting and other related terms should be considered "works in progress" to be refined as more relevant studies are conducted. Stark's usage of "scales of difference" in distinguishing multicraft situations, however, is an idea worth pursuing. Thus, the same artisan engaged in pottery making and the lapidary arts would represent one extreme within this scale of difference. Perhaps the pottery making and metalworking that Goldstein and I (this volume) discuss would correspond to an intermediate point in the scale given that, in spite of the very different materials involved, they share the same thermodynamic and chemical principles and fuels.

The ethnohistorical contribution by Susan E. Ramírez constitutes a notable departure from the book's stated emphasis on the primary data garnered from excavations of production loci. Instead, Ramírez critically reviews early Spanish colonial documents for indications of occupational specialization, social and productive relationships among craft specialists on the late pre-Hispanic north coast of Peru (ca. fifteenth–sixteenth centuries following successive Chimú and Inca domination), and their social identities. Her review is intended to serve as a comparative yardstick, with which to measure the relative effectiveness of archaeological studies with similar aims but without the aid of textual sources. In effect, her chapter highlights the importance of supplementing the workshop-centered approach with as many other approaches as possible.

What emerges from her documentary analysis are revealing insights, both discrepancies and concordance (though rare) with archaeological findings, and strong caveats to be heeded by archaeologists who rely on documentary evidence to buttress or base their interpretations. Ramírez shows how a lengthy list of occupational specialists (over 20 types, including many craftspersons) from colonial days may be misleading as a reflection of the late pre-Hispanic situation. More specifically, some may not be pre-Hispanic in origin but, rather, a mistranslation of native terms or an imposition of European occupational categories motivated by the need to identity professions for tributary collection. It is apparent, though, that there were many craft specialists who are rarely studied by archaeologists (e.g., itinerant cotton-cloth painters). How we may identify and study them, however, is left to the archaeologist.

One of Ramírez's main concerns is the social identity of craftspersons and their productive roles. She reminds us that, although skilled full-time craft specialists receive much archaeological attention, their production was inextricably based on the support of numerous farmers and fisherfolk as well as family members, who, for example, were often called on to acquire or produce needed supplies as tribute for their ethnic lords (and, by extension, for the state). Although this specialist–commoner relationship is an integral part of the prestige–staple goods production model of Inca state political economy (e.g., D'Altroy and Earle 1985), it is well worth remembering that in the late pre-Hispanic Andes, what archaeologists often attribute to specialist production represented in part "less skilled" commoners' tributary compliances to the state. Differences in skill level between specialized and nonspecialized production (see Costin and Hagstrum 1995; Ingold 2000) is not always apparent and should not be generalized across different crafts. Metalsmithing, for example, often has a master–apprentice organization in which less skilled apprentices perform simpler, repetitious, but time-consuming tasks (e.g., Shimada and Griffin 2005).

It is also evident that, although archaeologists often see production output as a simple binary distinction between the producer's own household versus extrahousehold consumption, commoners' products were also intended for their own ethnic lords and deceased ancestors. Ramírez reminds us that commoners sometimes needed to change location temporarily to comply with their tributary duties, which may well have resulted in multiethnic production settings. In regard to the oft-debated question concerning the "intensity of specialization" (cf. Costin 1991; Longacre 1999; Morrison 1994; Rice 1996), she regards references to access rights to or ownership of agricultural lands as the telling indicator that metalsmiths and potters were full-time and part-time specialists, respectively.

Conclusion

The ten chapters representing case studies and regional syntheses from diverse geographical areas, time periods, and sociopolitical complexities break important new ground in the anthropological study of the creative role and social identity of the producer and multicraft production. This volume redresses our skewed conception and approach to craft production that has been predominantly shaped by studies focused on discrete single-medium crafts, finished products, and the consumer. It reminds us that many crafts that are analytically treated as discrete were practiced concurrently in close proximity to each other, maintaining creative, social, and material interaction to varying degrees. Just as we need a balanced evaluation of the roles of producer and consumer in craft production, we need to pay as much attention to multicraft situations as to those characterized by crafts practiced singly. Last but by no means least, this volume illustrates the importance of properly contextualized workshop excavations and the definition of the entire sequence of operation in documenting craft production as both a social and a material process.

References Cited

Ames, Kenneth M.

1995 Chiefly Power and Household Production on the Northwest Coast. In *Foundations of Social Inequality*, edited by T. Douglas Price and Gary M. Feinman, pp. 155–188. Plenum Press, New York.

Arnold, B.

1999 "Drinking the Feast": Alcohol and the Legitimation of Power in Celtic Europe. *Cambridge Archaeological Journal* 9:71–93.

Arnold, Dean E.

1985 *Ceramic Theory and Cultural Process*. Cambridge University Press, New York.

Bagley, Robert W.

1993 Replication Techniques in Eastern Zhou Bronze Casting. In *History from Things: Essays on Material Culture*, edited by S. Lubar and W. D. Kingery, pp. 231–241. Smithsonian Institution, Washington, D.C.

1995 What the Bronzes from Hunyuan Tell Us about the Foundry at Houma. *Orientations*, January: 46–54.

Bey, G. J., III, and C. A. Pool (editors)

1992 *Ceramic Production and Distribution: An Integrated Approach*. Westview Press, Boulder.

Bijker, W., T. Hughes, and T. Pinch (editors)

1987 *The Social Construction of Technological Systems: New Directions in the Sociology and History of Technology*. MIT Press, Cambridge.

Bronitsky, Gordon

1986 The Use of Materials Science Techniques in the Study of Pottery Construction and Use. In *Advances in Archaeological Method and Theory*, edited by M. B. Schiffer, pp. 209–273. Academic Press, New York.

Brumfiel, Elizabeth

1991 Weaving and Cooking: Women's Production in Aztec Mexico. In *Engendering Archaeology*, edited by J. Gero and M. Conkey, pp. 224–255. Blackwell, Oxford.

1998 The Multiple Identities of Aztec Craft Specialists. In *Craft and Social Identity*, edited by C. Costin and R. Wright, pp. 145–152. Archaeological Papers No. 8. American Anthropological Association, Washington, D.C.

Brumfiel, Elizabeth, and Timothy Earle (editors)

1987 *Specialization, Exchange, and Complex Societies*. Cambridge University Press, Cambridge.

Childs, S. Terry

1994 Native Copper Technology and Society in Eastern North America. In *Archaeometry of Pre-Columbian Sites and Artifacts*, edited by D. A. Scott and P. Meyers, pp. 229–253. Getty Conservation Institute, Marina del Rey.

Clark, John E.

2001 Workshops. In *The Oxford Encyclopedia of Mesoamerican Cultures: The Civilizations of Mexico and Central America*, edited by D. Carrasco, Vol. 3:336–338. Oxford University Press, London.

Connell, Samuel V.

2002 Getting Closer to the Source: Using Ethnoarchaeology to Find Ancient Pottery Making in the Naco Valley, Honduras. *Latin American Antiquity* 13(4):401–417.

Cook, Anita G., and N. Benco

2001 Vasijas para la Fiesta y la Fama: Producción Artesanal en un Centro Urbano Huari. In *Huari y Tiwanaku: Modelos vs. Evidencias*, edited by P. Kaulicke and W. H. Isbell, pp. 489–504. Boletín de Arqueología PUCP, Vol. 4. Fondo Editorial, Pontificia Universidad Católica del Perú, Lima.

Costin, Cathy L.

1991 Craft Specialization: Issues in Defining, Documenting, and Explaining the Organization of Production. In *Archaeological Method and Theory*, edited by Michael B. Schiffer, Vol. 3:1–56. University of Arizona Press, Tucson.

1996 Craft Production and Mobilization Strategies in the Inka Empire. In *Craft*

Specialization and Social Evolution: In Memory of V. Gordon Childe, edited by B. Wailes, pp. 211–225. Symposium Series, Vol. 6. University Museum of Archaeology and Anthropology, University of Pennsylvania, Philadelphia.

1998 Introduction: Craft and Social Identity. In *Craft and Social Identity*, edited by Cathy L. Costin and Rita P. Wright, pp. 3–18. Archaeological Papers of the American Anthropological Association. Arlington, Virginia.

2000 The Use of Ethnoarchaeology for the Archaeological Study of Ceramic Production. *Journal of Archaeological Method and Theory* 7(4):377–403.

2001 Craft Production Systems. In *Archaeology at the Millennium: A Sourcebook*, edited by Gary M. Feinman and T. Douglas Price, pp. 273–327. Plenum, New York.

Costin, Cathy L., and Melissa Hagstrum

1995 Standardization, Labor Investment, Skill, and the Organization of Ceramic Production in Late Prehispanic Peru. *American Antiquity* 60:619–639.

Costin, Cathy L., and Rita Wright (editors)

1998 *Craft and Social Identity*. Archaeological Papers No. 8. American Anthropological Association. Washington, D.C.

D'Altroy, Terence N., and Timothy K. Earle

1985 Staple Finance, Wealth Finance, and Storage in the Inka Political Economy (with comment). *Current Anthropology* 26:187–206.

David, Nicholas, and Carol Kramer

2001 *Ethnoarchaeology*. Cambridge University Press, Cambridge.

Dobres, M.-A.

2000 *Technology and Social Agency*. Blackwell, Oxford.

Dobres, M.-A., and C. Hoffman

1994 Social Agency and the Dynamics of Prehistoric Technology. *Journal of Archaeological Method and Theory* 3:211–258.

Dobres, M.-A., and C. Hoffman (editors)

1999 *The Social Dynamics of Technology: Practice, Politics, and World Views*. Smithsonian Institution Press, Washington, D.C.

Feinman, Gary M.

1999 Rethinking Our Assumptions: Economic Specialization at the Household Scale in Ancient Ejutla, Oaxaca, Mexico. In *Pottery and People: A Dynamic Interactions*, edited by James M. Skibo and Gary M. Feinman, pp. 81–98. University of Utah Press, Salt Lake City.

Feinman, Gary M., Linda M. Nicholas, and William D. Middleton

1993 Craft Activities at the Prehispanic Ejutla Site, Oaxaca, Mexico. *Mexicon* 15:33–41.

Franklin, Ursula M.

1983 On Bronze and Other Metals in Early China. In *The Origins of Chinese Civilization*, edited by D. N. Keightley, pp. 279–296. University of California Press, Berkeley.

Gosden, Chris, and Yvonne Marshall

1999 The Cultural Biography of Objects. *World Archaeology* 31(2):169–178.

Hayashida, Frances M.

1995 *State Pottery Production in the Inka Provinces*. Ph.D. dissertation, University of Michigan. University Microfilms, Ann Arbor.

1999 Style, Technology, and State Production: Inka Pottery Manufacture in the Leche Valley, Peru. *Latin American Antiquity* 10(4):337–352.

Helms, Mary

1993 *Craft and the Kingly Ideal: Art, Trade, and Power*. University of Texas Press, Austin.

Hosler, Dorothy

1994 *The Sounds and Colors of Power: The Sacred Metallurgical Technology of Ancient West Mexico*. MIT Press, Cambridge.

Ingold, Tim

2000 *The Perception of the Environment: Essays on Livelihood, Dwelling and Skill*. Routledge, London.

Inomata, Takeshi

2001 The Power and Ideology of Artistic Creation: Elite Craft Specialists in Classic Maya Society. *Current Anthropology* 42(3):321–349.

Inomata, Takeshi, and Daniela Triadan

2000 Craft Production by Classic Maya Elites in Domestic Settings: Data from Rapidly Abandoned Structures at Aguateca, Guatemala. *Mayab* 13:57–66.

Janusek, John W.

1999 Craft and Local Power: Embedded Specialization in Tiwanaku Cities. *Latin American Antiquity* 10(2):107–131.

Johannessen, Sissel, and Christine A. Hastorf

1990 A History of Fuel Management (A.D.

500 to the Present) in the Mantaro Valley, Peru. *Journal of Ethnobiology* 10(1): 61–90.

Jones, Andrew
2002 *Archaeological Theory and Scientific Practice.* Cambridge University Press, Cambridge.

Kingery, W. David
1993 Technological Systems and Some Implications with Regard to Continuity and Change. In *History from Things: Essays on Material Culture*, edited by Steven Lubar and W. David Kingery, pp. 215–230. Smithsonian Institution Press, Washington, D.C.

Kolata, Alan L.
1983 Chan Chan and Cuzco: On the Nature of the Ancient Andean City. In *Civilization in the Ancient Americas: Essays in Honor of Gordon R. Wiley*, edited by R. M. Leventhal and A. L. Kolata, pp. 345–371. University of New Mexico, Albuquerque.

Kolb, Charles C. (editor)
1988 *Ceramic Ecology Revisited.* British Archaeological Reports, International Series 436(ii). Oxford.

Lechtman, Heather N.
1977 Style in Technology—Some Early Thoughts. In *Material Culture: Styles, Organization, and Dynamics of Technology*, edited by H. N. Lechtman and R. S. Merrill, pp. 3–20. West Publishing Co., St. Paul.

Lechtman, Heather, and Arthur Steinberg
1979 The History of Technology: An Anthropological Point of View. In *The History and Philosophy of Technology*, edited by G. Bugliarello and D. B. Doner, pp. 135–160. University of Illinois Press, Champaign-Urbana.

Lemonnier, Pierre
1992 *Elements for an Anthropology of Technology.* Anthropological Papers, No. 88. Museum of Anthropology, University of Michigan, Ann Arbor.
1993 Introduction. In *Technological Choices: Transformation in Material Cultures since the Neolithic*, edited by Pierre Lemonnier, pp. 1–35. Routledge, London.

Lennstrom, Heidi A., and Christine Hastorf
1995 Interpretation in Context: Sampling and Analysis in Paleoethnobotany. *American Antiquity* 60(4):701–721.

Longacre, William A.
1999 Standardization and Specialization: What's the Link? In *Pottery and People: A Dynamic Interaction*, edited by J. M. Skibo and G. M. Feinman, pp. 44–58. University of Utah Press, Salt Lake City.

Lyon, Patricia J.
1991 Andean Art and Its Cultural Implications. In *Ancient Art of the Andean World*, edited by Shozo Masuda and Izumi Shimada, pp. 27–45. Iwanami Shoten, Publishers, Tokyo.

Matson, Frederick R.
1989 Ceramics: The Hub of Ancient Craft Interplay. In *Cross-Craft and Cross-Cultural Interactions in Ceramics*, edited by P. E. McGovern, pp. 13–28. Ceramics and Civilization Vol. 4. American Ceramic Society, Westerville, Ohio.

McGovern, Patrick E.
1989 Ceramics and Craft Interaction: A Theoretical Framework, with Prefactory Remarks. In *Cross-Craft and Cross-Cultural Interactions in Ceramics*, edited by P. E. McGovern, pp. 1–11. American Ceramic Society, Westerville, Ohio.

McGraw, Judith A.
1996 Reconceiving Technology: Why Feminine Technologies Matter. In *Gender and Archaeology*, edited by Rita P. Wright, pp. 52–78. University of Pennsylvania Press, Philadelphia.

Miller, H. M.-L.
2007 Associations and Ideologies in the Locations of Urban Craft Production at Harappa, Pakistan (Indus Civilization). In *Rethinking Specialization in Complex Societies: Archaeological Analysis of the Social Meaning of Production*, edited by Z. X. Hruby and R. Flad, pp. 37–51. Archaeological Papers of the American Anthropological Association, Vol. 17. University of California Press and the American Anthropological Association, Berkeley.

Moholy-Nagy, Hattula
1990 The Misidentification of Mesoamerican Lithic Workshops. *Latin American Antiquity* 1:268–279.

Morrison, K. D.
1994 The Intensification of Production: Archaeological Approaches. *Journal of Archaeological Method and Theory* 1:111–159.

Otis Charlton, Cynthia

1993 Obsidian as Jewelry: Lapidary Production in Aztec Otumba, Mexico. *Ancient Mesoamerica* 4:231–243.

1994 Plebians and Patricians: Contrasting Patterns of Production and Distribution in the Aztec Figurine and Lapidary Industries. In *Economies and Polities in the Aztec Realm*, edited by Mary G. Hodge and Michael E. Smith, pp. 195–219. Studies in Culture and Society Vol. 6. Institute for Mesoamerican Studies, University at Albany, State University of New York, Albany.

Peacock, D. P. S.

1981 Archaeology, Ethnology, and Ceramic Production. In *Production and Distribution: A Ceramic Viewpoint*, edited by H. Howard and E. Morris, pp. 187–194. International Series 120. British Archaeological Reports, Oxford.

Pfaffenberger, Bryan

1992 Social Anthropology of Technology. *Annual Review of Anthropology* 21:491–516.

Rice, Prudence M.

1987 *Pottery Analysis: A Source Book*. University of Chicago Press, Chicago.

1996 Recent Ceramic Analysis: 2. Composition, Production and Theory. *Journal of Archaeological Research* 4:165–202.

Santley, R. S., and K. G. Hirth (editors)

1993 *Prehispanic Domestic Units in Western Mesoamerica: Studies of the Household, Compound, and Residence*. CRC Press, Boca Raton.

Schiffer, Michael B.

1976 *Behavioral Archaeology*. Academic Press, New York.

1992 *Technological Perspectives on Behavioral Change*. University of Arizona Press, Tucson.

Schiffer, Michael B., and James M. Skibo

1987 Theory and Experiment in the Study of Technological Change. *Current Anthropology* 28:595–622.

Schmidt, Peter R.

1996 *The Culture and Technology of African Iron Production*. University of Florida Press, Gainesville.

Sellet, F.

1993 *Chaîne Opératoire*: The Concept and Its Application. *Lithic Technology* 18:106–112.

Shimada, Izumi

1978 Economy of Prehistoric Urban Context: Commodity and Labor Flow in Moche V Pampa Grande, Peru. *American Antiquity* 43:569–592.

1985 Perception, Procurement and Management of Resources: Archaeological Perspective. In *Andean Ecology and Civilization*, edited by Shozo Masuda, I. Shimada, and Craig Morris, pp. 357–399. University of Tokyo Press, Tokyo.

1994a *Pampa Grande and the Mochica Culture*. University of Texas Press, Austin.

1994b Prehispanic Metallurgy and Mining in the Andes: Recent Advances and Future Tasks. In *In Quest of Mineral Wealth: Aboriginal and Colonial Mining and Metallurgy in Spanish America*, edited by Alan K. Craig and Robert West, pp. 37–73. *Geoscience and Man* Vol. 53. Louisiana State University, Baton Rouge.

1996 Sicán Metallurgy and Its Cross-Craft Relationships. *Boletín del Museo del Oro* 41:27–61.

2001 Late Moche Urban Craft Production: A First Approximation. In *Moche Art and Archaeology in Ancient Peru*, edited by Joanne Pillsbury, pp. 177–205. National Gallery of Art, Studies in History of Art 63. Washington, D.C.

2005 Experimental Archaeology. In *Handbook of Theories and Methods in Archaeology*, edited by Herbert D. G. Maschner and Christopher Chippindale, pp. 603–642. AltaMira Press, Walnut Creek, California.

Shimada, Izumi (editor)

1998 *Andean Ceramics: Technology, Organization, and Approaches.* MASCA Research Papers Vol. 15. MASCA, University of Pennsylvania Museum for Archaeology and Anthropology, Philadelphia.

Shimada, Izumi, Adon Gordus, Jo Ann Griffin, and John Merkel

1999 Sicán Alloying, Working and Use of Precious Metals: An Interdisciplinary Perspective. In *Metal in Antiquity*, edited by S. M. M. Young, A. M. Pollard, P. Budd, and R. A. Ixer, pp. 301–309. BAR International Series 792. Archaeopress, Oxford.

Shimada, Izumi, and Jo Ann Griffin

2005 Precious Metal Objects of the Middle

Sicán. In "Mysteries of the Ancient Ones," special edition, *Scientific American* 15(1):80–89.

Shimada, Izumi, and John Merkel
1991 Copper Alloy Metallurgy in Ancient Peru. *Scientific American* 265(1):80–86.

Shimada, Izumi, and Ursel Wagner
2007 Craft Production on the Pre-Hispanic North Coast of Peru: A Holistic Approach and Its Results. In *Archaeology as Anthropology: Theoretical and Methodological Approaches*, edited by James Skibo, Michael Grave, and Meriam Stark, pp. 163–197. University of Arizona Press, Tucson.
In press Tecnología y Organización de la Producción de Cerámica Negra en la Costa Norte del Perú. *Corriente Arqueológico* 2.

Sillar, Bill, and Michael Tite (editors)
2000 Technological Choices in Ceramic Production. Theme issue, *Archaeometry* 42(1).

Sinopoli, Carla
1988 The Organization of Craft Production in Vijayanagara, South India. *American Anthropologist* 90:580–597.

Skibo, James
1992 *Pottery Function: A Use-Alteration Perspective*. Plenum, New York.

Smith, Cyril S.
1965 Materials and the Development of Civilization and Science. *Science* 148:908–917.
1970 Art, Technology and Science: Notes of Their Historical Interaction. *Technology and Culture* 11:493–549.

Spence, Michael W., J. Kimberlin, and Garman Harbottle
1984 State-Controlled Procurement and the Workshops of Teotihuacán, Mexico. In *Prehistoric Quarries and Lithic Production: New Directions in Archaeology*, edited by J. E. Ericson, and B. A. Purdy, pp. 97–105. Cambridge University Press, Cambridge.

Spielmann, Katherine A.
1998 Ritual Influences on the Development of Rio Grande Glaze A Ceramics. In *Migration and Reorganization: The Pueblo IV Period in the American Southwest*, edited by Katherine A. Spielmann, pp. 253–279. Arizona State University Anthropological Research Papers 51. Tempe.

Stark, Barbara L.
1985 Archaeological Identification of Pottery Production Locations: Ethnoarchaeological and Archaeological Data in Mesoamerica. In *Decoding Prehistoric Ceramics*, edited by Ben A. Nelson, pp. 158–194. Southern Illinois University Press, Carbondale.

Stark, Miriam
2003 Current Issues in Ceramic Ethnoarchaeology. *Journal of Archaeological Research* 11:193–242.

Topic, John R., Jr.
1982 Lower-Class Social and Economic Organization at Chan Chan. In *Chan Chan: Andean Desert City*, edited by M. E. Moseley and K. C. Day, pp. 145–175. University of New Mexico Press, Albuquerque.
1990 Craft Production in the Kingdom of Chimor. In *The Northern Dynasties: Kingship and Statecraft in Chimor*, edited by Michael E. Moseley and Alana Cordy-Collins, pp. 145–176. Dumbarton Oaks Research Library and Collection, Washington, D.C.

Tringham, Ruth
1978 Experimentation, Ethnoarchaeology, and Leapfrogs in Archaeological Methodology. In *Explorations in Ethnoarchaeology*, edited by Richard A. Gould, pp. 169–199. University of New Mexico Press, Albuquerque.

Turner, Margaret H.
1988 The Lapidary Industry of Teotihuacán, México. Unpublished Ph.D. dissertation, Department of Anthropology, University of Rochester.

Uceda, Santiago, and José Armas
1998 An Urban Pottery Workshop at the Site of Moche, North Coast of Peru. In *Andean Ceramics: Technology, Organization and Approaches*, edited by Izumi Shimada, pp. 63–89. MASCA Research Papers Vol. 15. MASCA, University of Pennsylvania Museum for Archaeology and Anthropology, Philadelphia.

Urban, Patricia A., E. Christian Wells, and Marne Ausec
1997 The Fires without and the Fires Within: Evidence for Ceramic Production Facilities at the Late Classic Sites of La Sierra, Northwestern Honduras, and in Its Environs. In *The Prehistory and History of*

Ceramic Kilns, edited by Prudence Rice, pp. 173–194. American Ceramic Society, Columbus, Ohio.

van der Leeuw, S. E.

1977 Towards a Study of the Economics of Pottery Making. *Ex Horreo* 4:68–76.

1984 Pottery Manufacture: Some Complications for the Study of Trade. In *Pots and Potters: Current Approaches in Ceramic Archaeology*, edited by Prudence M. Rice, pp. 55–69. UCLA Institute of Archaeology Monograph 24. University of California Press, Los Angeles.

Wagner, Ursel, R. Gebhart, E. Murad, Josef Riederer, Izumi Shimada, C. Ulbert, and Fritz E. Wagner

1998 Production of Formative Ceramics: Assessment by Physical Methods. In *Andean Ceramics: Technology, Organization, and Approaches*, edited by Izumi Shimada, pp. 173–198. Supplement to MASCA Vol. 15. Museum Applied Science Center for Archaeology, University of Pennsylvania Museum for Archaeology and Anthropology, Philadelphia.

Wertime, Theodore A.

1964 Man's First Encounters with Metallurgy. *Science* 146(3649):1257–1267.

Wright, Rita P.

1991 Women's Labor and Pottery Production in Prehistory. In *Engendering Archaeology: Women and Prehistory*, edited by Joan M. Gero and Margaret W. Conkey, pp. 194–223. Basil Blackwell, Oxford.

1996 Technology, Gender, and Class: Worlds of Difference in Ur III Mesopotamia. In *Gender and Archaeology: Essays in Research and Practice*, edited by R. Wright, pp. 79–110. University of Pennsylvania Press, Philadelphia.

I

Case Studies in Multicraft and Related Production

1

Multicrafting, Migration, and Identity in the American Southwest

BARBARA J. MILLS

The dramatic restructuring of craft production in the late prehistoric period of the American Southwest offers the opportunity to investigate several themes of interest to this book. In particular, I look at the sudden intensification in the production of multiple crafts, who produced those crafts, and where those objects were produced. I contextualize this discussion by looking at how this intensification was related to other social changes—especially changes that involved the restructuring of social identities and the establishment of new suprahousehold ritual institutions following an intensive period of in-migration in the Mogollon Rim region of east-central Arizona.

I argue that the intensification of craft production was part of the construction of new identities and the negotiation of social inequalities that followed the migration of new households into established communities. Individual, household, and community forms of migration were all part of the social landscape of Southwestern communities. However, the structure of these migrations—the distances moved, the social scales of migration, and the durations that migrants stayed in guest communities—differed widely across the region and at different times. Because of the fine-grained chronological control and a prolonged interest in social and ritual reorganization on the part of archaeologists working in the project area, we are able to more precisely track how crafting, migration, and other social changes intersected.

This case study provides two important points of comparison for other studies in this volume. First, it emphasizes how the household and suprahousehold production of goods emerged as two social contexts for multicraft production in the Southwest. During the late thirteenth and fourteenth centuries, both of these scales of production were the locus of production, coinciding with a major period of population migration and aggregation. Although both of these contexts were important, this study especially underscores the flexibility and importance of the household in the intensification of the production of multiple crafts (see also Feinman 1999; Feinman and Nicholas, this volume). Labor was reorganized to produce a variety of goods more intensively but primarily within the social unit of the household. Thus, this study emphasizes how workshops—production locations separate from the household—were only a small part of multicraft specialist production.

Second, the Southwest example also shows how identity and craft production were interlinked. This has been brought home in a number of papers but especially in a volume edited by Costin and Wright (1998; see Inomata, this volume), in which several of the chapters discuss how producer and consumer identities were important to what was made and by whom. In the case of the Southwest, prestige was based on activities that enhanced participation in ritual activities, and many of the objects that were produced by multicrafters were used in activities associated with ritual obligations and feasting, including ornaments, ceramics, and textiles. As Spielmann (2002) and others have argued, ritual production is an important part of specialization in "middle-range" societies such as the Southwest. In the case study that I discuss in this chapter, it is clear that this is an important and understudied aspect of the political economy and cannot be untangled from the social transformations that took place in the post-migration communities of the late thirteenth through fourteenth centuries, as new identities were constructed through ritual participation.

MULTICRAFTING: DEFINITIONS, SCALES, AND COMPLEMENTARITY

In this chapter, I use a definition of *multicrafting* as the production of multiple crafts at the same social scale for consumption beyond the unit of production. I use the term *social scale* following Costin (1991, see also 2001), who defines it as the unit of production. Thus, *multicrafting* as used in this chapter refers to a case of specialized production in which different kinds of products were being made within a specific social setting. That setting may range from one individual within a household to the community, depending on the way production was organized. Although the production of multiple crafts may be conducted by the same individual, it is clear that there are cases in which multiple hands were involved in the production of a single object. These "hidden" producers have been recognized in the literature on craft production for some time (e.g., Kramer 1985; Wright 1991). These multiple hands could also have been involved in the production of multiple objects. What links them, and what makes them important and salient for comparison with the other cases in this volume, is that they were producing in the same social setting and were producing for consumption outside that social setting.

The specific identification of who was producing what is very difficult to ascertain archaeologically. It depends on the scale of the analysis being undertaken as well as the quality of the archaeological data being collected. At the largest scale of the region we may have little information about who was actually producing—or for whom—but we may know that there was a co-occurrence of specialized craft production at that particular spatial scale. At the most local scale, given high-resolution data, we may be able to talk about the production of multiple crafts by the same producer. Archaeologists rarely are able to identify the products of individual producers—even when production was organized at this social scale. More often we work with the scale of the household, clusters of households, suprahousehold organizations, and communities. These different scales have architectural analogues, and these analogues provide the spatial basis for looking at both direct and indirect indicators of production.

In the Southwest, like many other areas of the world, multicrafting was conducted at each of these spatial scales: households, household clusters, suprahousehold organizations, and communities. And as in other areas, it is clear that specialization at the household level was not only frequent but highly variable. Individuals within households might have specialized in a particular craft, but it is also evident that the entire household was involved in the production of one object or a class of objects (Mills 2000). This flexibility of the household in both the number of producers and the intensity of their production has been brought out many times in the Southwest literature (e.g., Hagstrum 1995; see also other chapters in Mills and Crown 1995) but is certainly not unique to this area. For example, Feinman (1999; see also Feinman and Nicholas, this volume) clearly shows that this is the pervasive pattern in Mesoamerica. One need only look at contemporary ceramic production in many parts of Mesoamerica and the Southwest to see how flexible household craft specialization really is; many of the objects that are entrenched within the tourist market are made by specialized households, not by producers in workshops separate from the household scale. These may be concentrated in villages within the region, but the social scale of production was and still is the household.

An important part of multicrafting is the complementarity among different crafts. When tools, materials, and transformative technologies are common to the production of different crafts, the movement of labor from one kind of object to another is easier to accomplish. McGovern (1989) calls this "cross-craft interaction," and Goldstein and Shimada (this volume) illustrate how shared material use resulted in the multicrafting of ceramics and metallurgy in the Andean area. In the Southwest during the nineteenth century, blacksmiths became jewelers at least in part because of the common tools, skills, and facilities that the production of the two materials shared (Mills 1995). Other forms of complementarity in craft specialization include the production of chipped-stone tools for hunting and lapidary work for ornaments. In the case that I discuss below, complementarity may have been present in the production of ornaments using minerals such as turquoise and the production of ceramics decorated with glaze paints that used mineral fluxes that were found in similar geological contexts and were quarried at the same time.

MIGRATION, IDENTITY, AND CRAFT PRODUCTION

The case study that I discuss comes from excavations at a set of Ancestral Pueblo settlements in the Mogollon Rim area of east-central Arizona. Migration, which occurred at different times and at different social scales (Herr 2001; Mills 1998), is pivotal to the history of this entire region. Contemporary social science literature is replete with case studies of migration that illustrate how migrants are incorporated into host communities and the effects of migration on both hosts and guests. Because most of these studies are based on colonial and postcolonial societies, and because they usually involve migration from rural to urban communities and the shift from agricultural to wage labor, they are often difficult to apply to archaeological cases in which rural to rural migration is more common, such as the Mogollon Rim region. Nonetheless, anthropological, sociological, and other studies underscore two very important general observations: (1) that migration involves changes in the way that social identity is constituted—whether looking at gender, age, class, and so on (Anthias 1998); and (2) that migration creates economic, social, and political asymmetries between hosts and guests (e.g., Esses et al. 1998).

Both of these observations directly relate to craft production and to the archaeology of the Mogollon Rim region. First, they are related because the act of producing is also an act of identity formation. What one makes, how well one makes it, who one produces with, who the recipient of the object is, and what social contexts the object is used in are all opportunities for the expression and reformulation of identity (e.g., Costin and Wright 1998). Second, economic, social, and political asymmetries are enhanced by the production and use of certain objects that have symbolic significance (e.g., Robb 1999). In recent years there has been an expansion of models from those that consider items to have been highly valued for the rarity of their materials or the fineness of their workmanship to a consideration of objects within their social contexts of consumption. Although the distance from which source materials were acquired is one way of determining value (Helms 1988, 1993), it is the conclusion of many studying craft production, especially in middle-range societies, that the social context of use and how the knowledge of production was transmitted—whether ritual or some other basis—best reflect differences in the value of objects (Mills 2000, 2004; Spielmann 1998, 2002).

This consideration of materiality has enhanced many recent discussions of the biographies of objects (e.g., Marshall and Gosden 1999), but those discussions have, surprisingly, veered away from that part of object biographies that begins with production to an almost exclusive discussion of the many contexts of consumption in which objects were used. A reintegration of production with consumption provides the framework for acknowledging the importance of objects in the

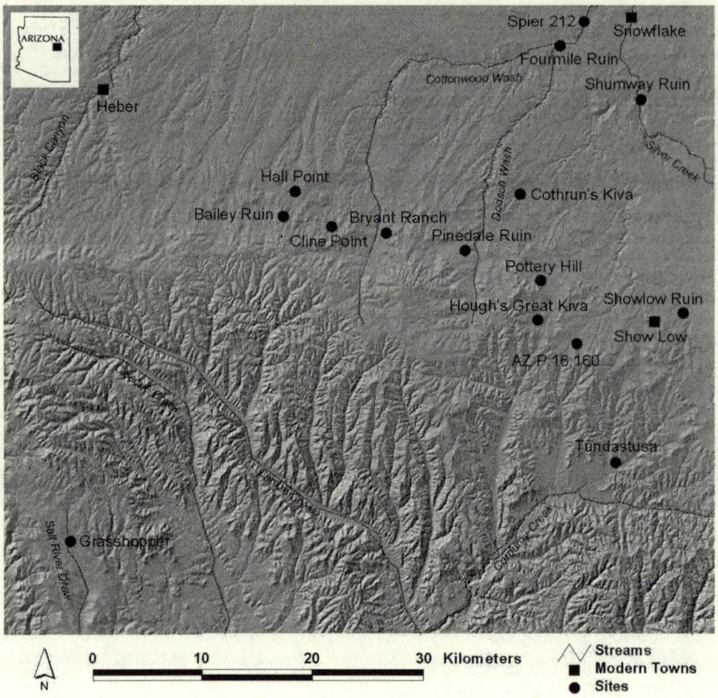

FIGURE 1.1. Location of the Silver Creek area and sites referred to in the text. Tla Kii and Broken K are situated outside the boundaries of this map.

expression of identity and the use of those objects in the negotiation of social asymmetries.

MULTICRAFTING AND MIGRATION IN THE SILVER CREEK AREA

To look at how multicraft production, migration, and identity intersect I focus on the Silver Creek area of east-central Arizona (Figure 1.1). Located at the edge of the Colorado Plateau, the Silver Creek sites are considered to be Ancestral Pueblo settlements. Since 1993 the Silver Creek Archaeological Research Project (SCARP) has conducted excavation and survey in the area. Seven sites have been excavated that date to the period between A.D. 1000 and 1325 (Fenn et al. 2006; Herr 2001; Kaldahl et al. 2004; Mills 1998; Mills et al. 1999b). Collections were made using a consistent set of excavation protocols, which allows comparisons to be made across this set of sites. Previous excavations at other nearby Ancestral Pueblo sites augment the data collected by SCARP. These include the work of Jesse Walter Fewkes (1904) at Fourmile and Pinedale Ruins; Emil Haury at Tla Kii and Pinedale (Haury 1985) and Showlow Ruins (Haury and Hargrave 1931); and the long-term excavations at Grasshopper Pueblo by the University of Arizona (Longacre et al. 1982; Reid and Whittlesey 1999).

Although I focus primarily on the data collected by SCARP, together these excavations reveal a picture of social fluidity and change that is useful for thinking about how craft production can be recontextualized (Costin 2001). The reconstruction of migration as one of the most important social processes of the period is a major part of this social context. Migration was a constant part of the social landscape in this area, as with many areas of the Southwest, but the scale, context, and effects of migration between the earlier and later sites were quite different. The contrast between them provides a backdrop for an interpretation of changes in the social contexts of craft production, multicrafting, and specialization.

Three of the SCARP excavated sites are

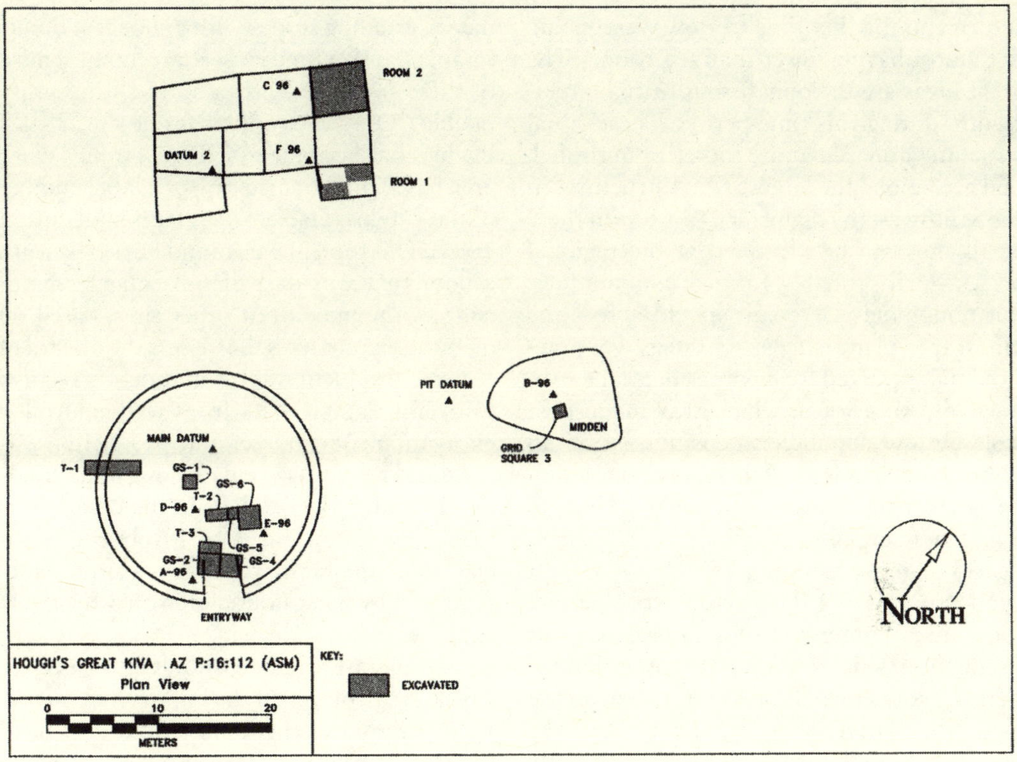

Figure 1.2. Hough's Great Kiva site showing small room block and associated circular great kiva.

roughly contemporaneous, dating to the A.D. 1000–1150 period (Hough's Great Kiva, Cothrun's Kiva, and AZ P:16:160[ASM]), and the other three are chronologically sequential within the period A.D. 1200–1325 (Pottery Hill, Bryant Ranch, and Bailey Ruin; Table 1.1). The chronological placement of these sites is possible through tree ring–dated ceramics, cross-dating, and frequency seriations of decorated ceramics (Mills and Herr 1999; Scholnick 2003). The latter three sites may overlap with each other, but their assemblages are distinct enough to provide a well-controlled data set for looking at relatively fine-grained changes, especially during the pivotal late thirteenth century.

Table 1.1. Date Ranges for the Occupation of Excavated Silver Creek Sites

Site	Date Range (A.D.)
Cothrun's Kiva	1000–1050
Hough's Great Kiva	1075–1175
AZ P:16:160	1075–1175
Pottery Hill	1200–1275
Bryant Ranch Pueblo	1260–1280
Bailey Ruin	1275–1325

The three earliest sites are associated with the migrations of the late eleventh and early twelfth centuries into the Mogollon Rim region. This period of migration in the area involved small groups moving into what was a largely unoccupied area with all of the social, political, and ecological characteristics of a frontier (Herr 2001). Although there are earlier sites in this area, the density was very low, and occupation may even have been seasonal. The migrants who came in the late eleventh and early twelfth centuries constructed masonry pueblos and were more residentially stable. They occupied dispersed households that were organized into communities with a single focal form of ritual architecture—the circular great kiva (Figure 1.2). Like earlier sites, settlements of this time period were small, with most having fewer than ten rooms. The great kivas are the only form of ritual architecture during this time period. These ritual structures are circular, are roofed or unroofed, and are among the largest ritual structures in the Southwest (by diameter). Clusters of these small sites can be identified at the regional level, which probably formed communities, but households were still spread across the landscape. A land extensive strategy was practiced that involved frequent residential movement of households, which was made possible by low population densities. As Sarah Herr (2001) has argued, these low population densities made "wealth-in-people" (Nyerges 1992) more important than wealth-in-things.

Although we have ample evidence of craft production during this early Pueblo period, there are no contexts of suprahousehold craft production separate from the household—that is, workshops. Production debris is not associated with the ritual architecture of the great kivas, most of which were apparently unroofed and served more as community arenas or gathering places. Other forms of suprahousehold architecture are not present during this time period. Unlike the case at other Ancestral Pueblo sites of this time period, such as Chaco, there are no room-block kivas. Instead, all production debris is from the floors of domestic structures or mixed with trash from domestic activities.

Neither is there much evidence for uneven production among households during this period. Household specialization has been documented in other areas of the Southwest, some of it dating even earlier than the Mogollon Rim great kiva sites (Hagstrum 1995; Hegmon et al. 1995). However, there is no evidence for this form of specialization in the Mogollon Rim area. Steatite beads and ceramics are the two best-documented crafts that were produced during the eleventh and twelfth centuries in the Mogollon Rim area (Haury 1985; Herr 2001), but most of the production evidence relates to ceramics. For ceramics, all of the indicators of production that we find in the Mogollon Rim area during later periods, such as prepared clays, pigments, grinding stones with pigment staining, scrapers, and unfired vessels, are found at most of these small eleventh- and twelfth-century pueblos. Only one context at one site (Tla Kii Pueblo) has been identified as a potter's work area based on the clustering of a large number of these items (Haury 1985; Herr and Stinson 2005). This tool kit was found buried beneath a floor that was part of subsequent remodeling of the pueblo. At other sites, based on quantitative analyses that assess the discard of ceramic production tools and debris relative to overall sample sizes from well-controlled excavation contexts, ceramic production was practiced relatively evenly among households (Mills et al. 1999a). Thus, in the case of this earlier migration into a relatively empty social landscape, craft production was probably practiced by most households on a relatively equal basis.

The migration of households into the Silver Creek area continued during the A.D. 1200–1350 period and was especially prevalent in the later half of the thirteenth century, when large areas of the Four Corners were depopulated (Lyons 2003; Mills 1998). In contrast to the case during the earlier period of migration, households moving into the Silver Creek area at this time were moving into an area with a much larger host population, including the descendants of the earlier migrants. Within a few short generations, a major aggregation of population occurred, from sites that were maximally 50 rooms, such as Pottery Hill, to sites that were maximally 250–500 rooms, such as Bailey Ruin (Figure 1.3).

In between sites such as Pottery Hill and Bailey Ruin are a few short-lived settlements that show clear-cut evidence of burning at the end of their occupations—an intentional method of ritual closure that contrasts with other abandonment behaviors practiced at earlier and later sites. These sites show marked differences in ceramic assemblages from other sites (Crown 1981; Fenn et al. 2006; Montgomery and Reid 1990; Scholnick 2003), and the one site with extensive excavation of extramural areas, Bryant Ranch Pueblo, shows clear evidence of suprahousehold feasting from the faunal assemblage and the cluster of

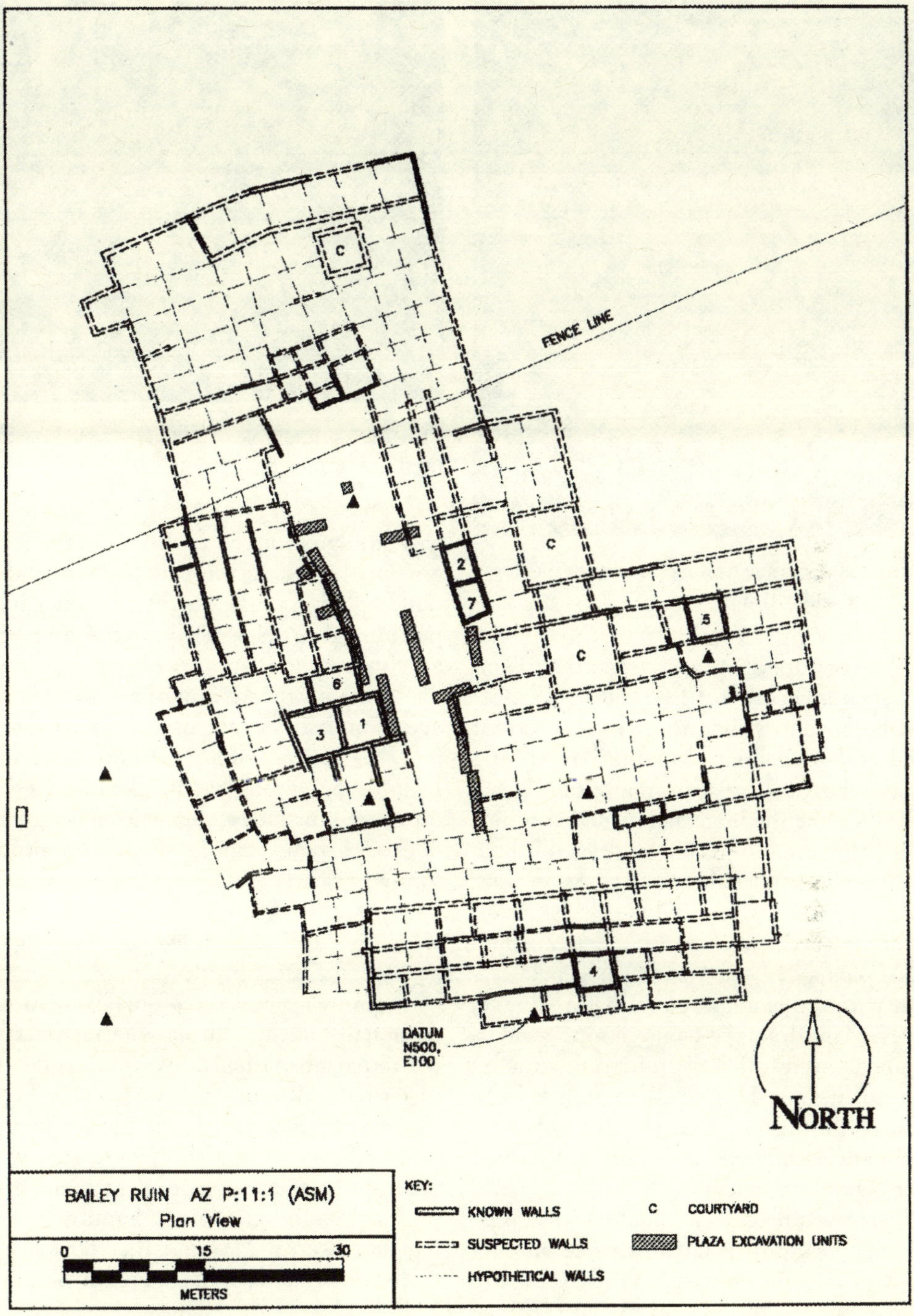

FIGURE 1.3. Plan of Bailey Ruin.

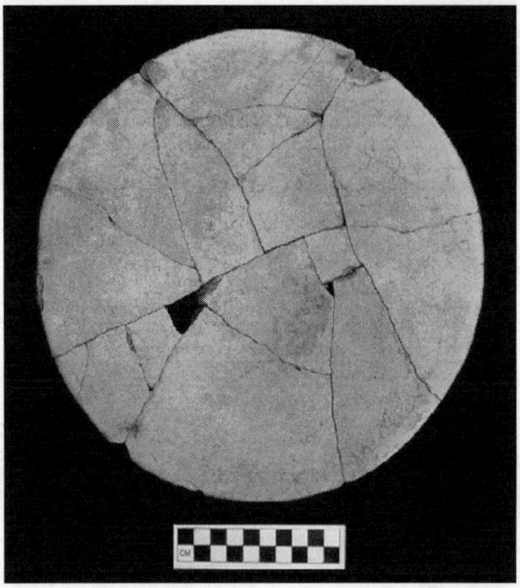

Figure 1.4. Ceramic turning plate from the floor of Room 7, Bailey Ruin.

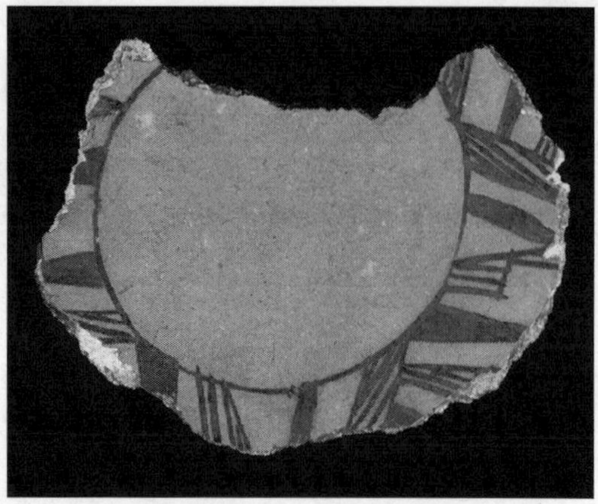

Figure 1.5. Example of a *puki* (recycled vessel) used as a turning plate from Room 3, Pottery Hill.

large, extramural features (Dean 2001; Fenn et al. 2006). These short-lived sites are significant because they indicate the instability of social and residential configurations during the period of most likely in-migration.

By the time the large aggregated communities of hundreds of rooms were established, the differences in ceramic *consumption* were downplayed. Thus, a paradox of this period is that we specifically cannot identify which households were hosts and which were guests in the parlance of migration. What we do see, however, are dramatic changes in the intensity and concentration of craft production by some households. This is evident from de facto refuse on room floors, as well as from the production debris in middens within abandoned rooms within the pueblos.

Floor assemblages with multiple craft production items, tools, and debris are commonly found in these later sites. Some of these are in old habitation rooms converted for storage and the intensive production of multiple crafts (Mills et al. 1999a; Triadan 1989). Several of these have been well documented at Grasshopper Pueblo, south of the Silver Creek cluster of sites. In the Silver Creek area, Room 5 at Bailey Ruin exemplifies this type of deposit. It is a former habitation room that was converted to a storage room by partitioning the southern one-third of the room off from the rest and building a bin in the northeast corner. These partitioned areas were plastered and probably used as granaries. A hearth in the middle of the room was still open, and the storage room had one unclosed door connecting it to the room next door. The floor assemblage included numerous fragments of worked and unworked turquoise, pigments, a large cache of ground stone, and clay. Ornament production was clearly conducted in this room, and based on the high amounts of pigments and clay, it was also the storage room for materials used in pottery production.

In another case, a ceremonial room was intentionally filled, and one-quarter of it was converted into a small domestic and manufacturing room (Room 7) that was used in the production of several craft items. On the floor was a complete ceramic plate (Figure 1.4), which was used as a base or turning plate in pottery production. In much of the Southwest, turning plates were made out of the bottoms of recycled bowls and jars (e.g., Figure 1.5). In the Mogollon Rim area, plates were also made. The association of these plates with pottery production tools in several areas of the Southwest confirms their use in pottery production. Specific varieties of plates (e.g., those with perforated rims) have even been identified as indicators of potter identities because of their time-transgressive spatial distribution paralleling routes of migration out of northeastern Arizona into the Mogollon Rim area and farther into southern Arizona (Christenson 1994; Lyons and Lindsay 2006; Mills 1998). In addition to the plate on the floor of Room 7 was a large cache of sherd-tempered clay, pigments of different kinds, worked and unworked *Olivella* sp. shells, and several quartz crystals that were probably used as cutting tools. Again, as in Room 5, both ornament and ceramic production activities were practiced in this room.

Ceramic tools and production debris in middens also provide evidence for the uneven and intensive production of ceramic vessels by some households at Bailey Ruin versus other households at the same site. In addition, these data can be compared with household discard assemblages from other sites to look at changes through time in production intensity, an important dimension of craft specialization (Costin 1991). In this stage of the analysis, we are looking at discard assemblages rather than de facto refuse from production activities on floors. This is an important difference because the trash thrown inside of neighboring rooms is slightly later than that from the same room's floor assemblages and should be kept analytically distinct. The items discarded provide a way to consistently index the intensity of production in all assemblages. By looking at the frequency of production tools and debris relative to the total ceramic assemblage, we are able to see what areas of the site contributed trash that is production intensive versus other areas of the site where production was less intensive.

Pottery production tools include *pukis* (recycled vessels), finished plates, scrapers with distinctive beveled edges, and pottery polishing stones. The frequency of all ceramic production items relative to the number of sherds in the discard assemblage shows that some households discarded many more items related to ceramic production than did others (Figure 1.6; Mills et al. 1999a). These households are mostly at Bailey Ruin.

Another indicator of the intensification of ceramic production is the replacement of pukis with a form that was specifically made as turning plates. As noted above, the Mogollon area has both pukis and plates. Project analysts consistently recorded the incidence of pukis versus plates at all of the sites excavated by SCARP. At the sites excavated by SCARP it is clear that plates became more popular through time, and by the time that Bailey was occupied, potters no longer recycled the bottoms of vessels and almost exclusively made new artifacts for supporting and turning vessels during the building process (Table 1.2). Among these plates are dozens of examples of perforated plates, although not all plates were perforated. The manufacture of a special form for pottery production rather than relying on broken vessels suggests a greater investment in the tools of production. Bailey Ruin had numerous plate fragments, with the largest number coming from only a few trash deposits such as the fill of Room 1 (Figure 1.7).

Table 1.2. Percentage of Pukis (Recycled Vessels) and Plates in Mogollon Rim Assemblages

Site	Form	Percent
Great Kiva sites	*Pukis*	33.0
Great Kiva sites	Plates	67.0
Pottery Hill	*Pukis*	25.0
Pottery Hill	Plates	75.0
Bailey Ruin	*Pukis*	1.2
Bailey Ruin	Plates	98.8

Pottery polishing stones also illustrate an uneven distribution across the sample of sites, with Bailey Ruin showing the greatest heterogeneity. Some households at Bailey Ruin were using and discarding more pottery polishing stones than other households were. Not only were these polishing stones more frequent (given sample size), but they were also more intensively used. The analysis of the number of used surfaces revealed a shift from more expediently used polishing stones at the earlier Great Kiva sites to those with multiple surfaces at Bailey Ruin (Valado 1999).

It is clear from the ceramic production data that some households were crafting more pots than others at Bailey Ruin. Based on chemical compositional analyses showing exact

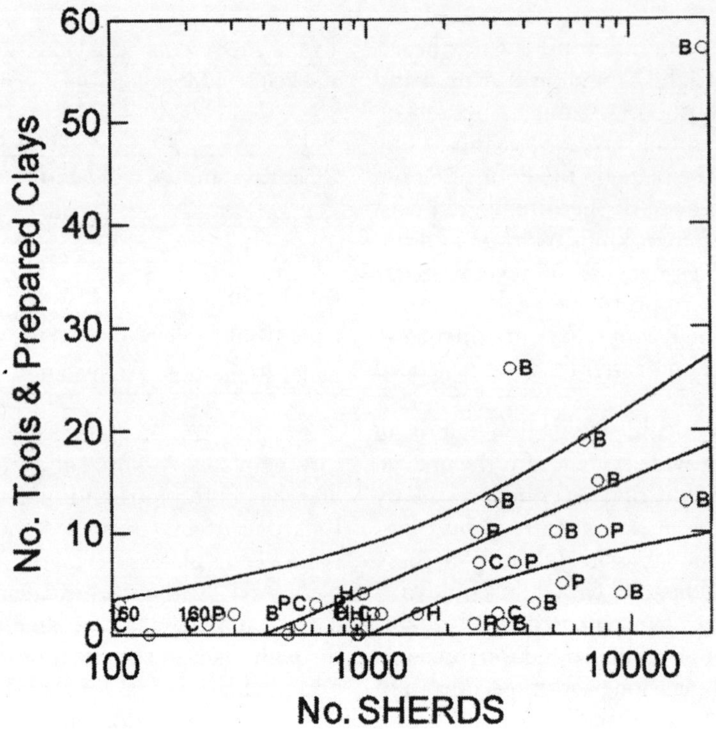

FIGURE 1.6. Frequency of ceramic production tools (scrapers, *pukis*, plates, and polishing stones) plotted against ceramic sample sizes (log10) for excavated proveniences at Cothrun's Kiva (C), Hough's Great Kiva (H), AZ P:16:160(ASM) (160), Pottery Hill (P), and Bailey (B), showing 95 percent confidence interval.

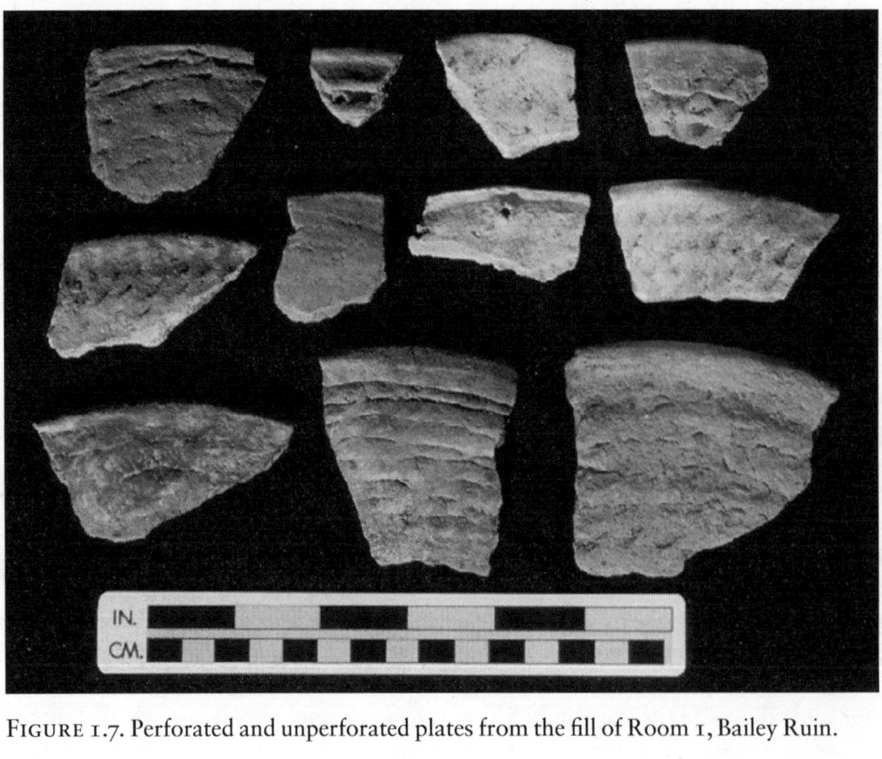

FIGURE 1.7. Perforated and unperforated plates from the fill of Room 1, Bailey Ruin.

matches between prepared clays from Room 1 and Room 1 fill sherds, these vessels minimally included Cibola White Ware and White Mountain Red Ware (Mills et al. 1999a). Some of the vessels from Bailey Ruin are quite large and were most likely to have been used in suprahousehold feasting (Figure 1.8). These special consumption events may have even helped to create a demand for increased ceramic production at this and other sites of the period, when feasting appears to have ratcheted up in scale (Isbell, this volume).

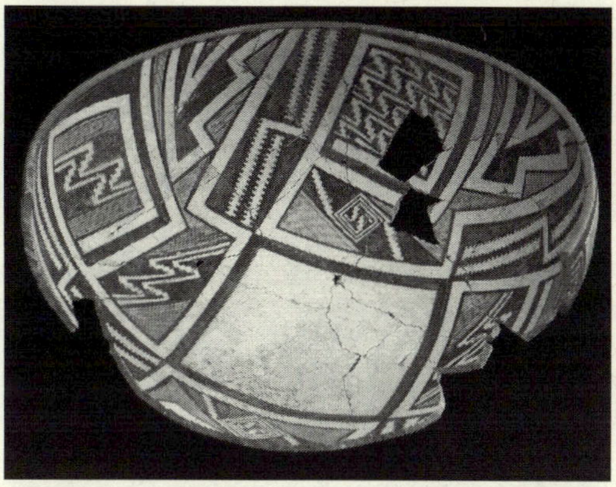

FIGURE 1.8. Interior of a White Mountain Red Ware vessel from Courtyard 2, Bailey Ruin.

The production specialization in ceramics is paralleled by data on ornament production out of shell, bone, and stone materials. For example, Room 1's trash fill contained shell ornaments, unworked shell, shell debris (Figure 1.9; Vokes 1999), and bone ring blanks (Figure 1.10; Zack-Horner 1999). As Zack-Horner (1999) has discussed, the specific bone tools in this same trash assemblage were probably used in basketry and other textile production based on polishing patterns and form, adding still another craft to the multicrafting that took place in these households.

Although more than one household was probably depositing trash in an individual room, the spatial association of different kinds of production tools and debris, the close packing of rooms, and the presence of extramural communal middens that ring the site suggests that only a limited number of households would in fact have been disposing of their trash in abandoned room spaces in the interior of the pueblo. Thus, differences in the trash assemblages that are identified

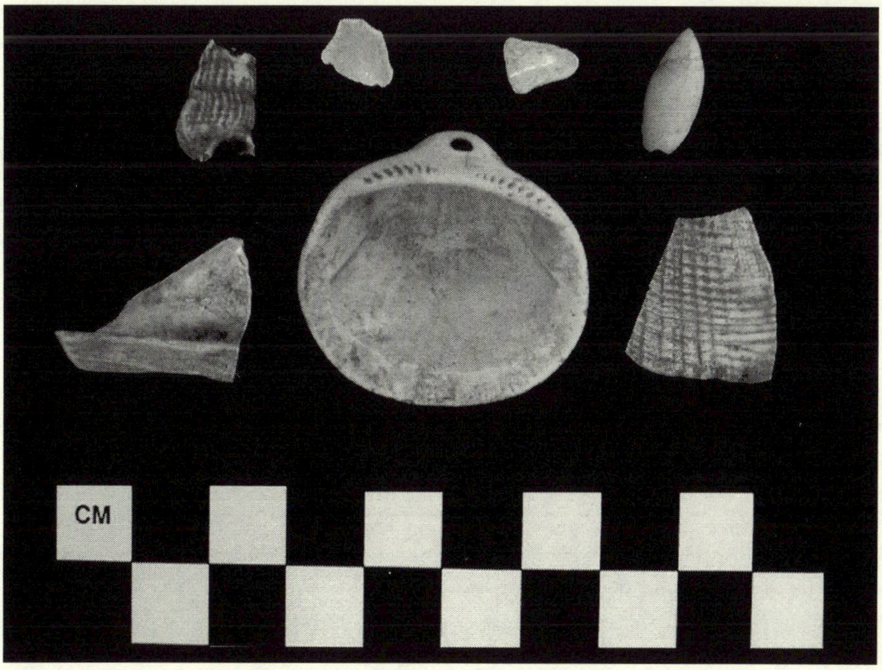

FIGURE 1.9. Shell debris and finished products from ornament production from the fill of Room 1, Bailey Ruin.

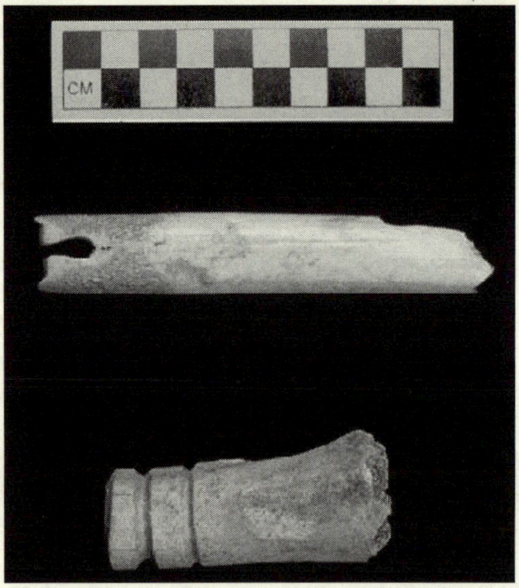

Figure 1.10. Bone tool and ornament production debris from the fill of Room 1, Bailey Ruin.

among the SCARP sites can be used to index the intensity of multicraft production by some households over others. The evidence from Bailey Ruin, based on the association of different craft production items on floors and in trash fill, shows that multicrafting by some households occurred within the context of increased production specialization. Although we would expect nearly all households in all time periods to participate in multicrafting, it is significant that when some households intensified production of one craft, such as ceramics, they did not do this at the expense of other craft production activities, especially the production of shell, bone, and mineral ornaments.

Households were not the only scale of craft production at Pueblo IV period sites. At the same time that Bailey Ruin was founded, we have the first evidence for the spatially and socially restricted production of textiles and other objects used in ritual performance. Loom anchors arrayed in pairs along the floors of suprahousehold ritual structures or kivas are found in many Pueblo IV period sites throughout the northern Southwest (Figure 1.11; Mills 2000: tab. 9.1; Webster 1997). In the Mogollon Rim area, these include the thirteenth-century site of Broken K Pueblo and the late-thirteenth- and early-fourteenth-century sites of Pinedale Ruin and Fourmile Ruin (Fewkes 1904; Haury and Hargrave 1931; Martin et al. 1967). These looms were used to make broad cloths, rather than the narrower textiles produced with the backstrap loom. In the Pinedale kiva shown in Figure 1.11, three pairs of loom holes are shown.

The production of these textiles in suprahousehold structures continued into the historic period and is well documented in photographs and ethnohistoric accounts (Brooks 1944; Stephen 1936). These lines of historical evidence indicate that weaving was done by men in these structures, along with the repair (if not production) of other objects used in ritual performances, including moccasins. One visitor to Hopi described these kivas as "workshops" because of the intensity of production and, presumably, their spatial segregation from domestic producers.

MULTICRAFTING, MIGRATION, AND IDENTITY

In the new social landscape of the late thirteenth century, there were at least two contexts of intensive multicraft production that corresponded with the depopulation of the Four Corners area and the migration of new people into the Silver Creek area. One was the intensive multicrafting by some households, which included ceramic containers as well as shell, bone, and turquoise ornaments. The other context of production was the crafting of textiles used in suprahousehold ritual structures. Based on continuities in production technology through the historic period, these textiles were most likely produced by men in the suprahousehold kiva settings, while women were the producers of ceramics and men were the producers of ornaments within the household. Nonetheless, these attributions, like all based on ethnographic cases, should be treated with caution.

Based on their usage, however, the production of the above objects would have promoted the status of their producers. Ceramics, especially polychrome bowls, were used in ritual feasting, as several analyses of bowl sizes have demonstrated (Mills in press; Spielmann

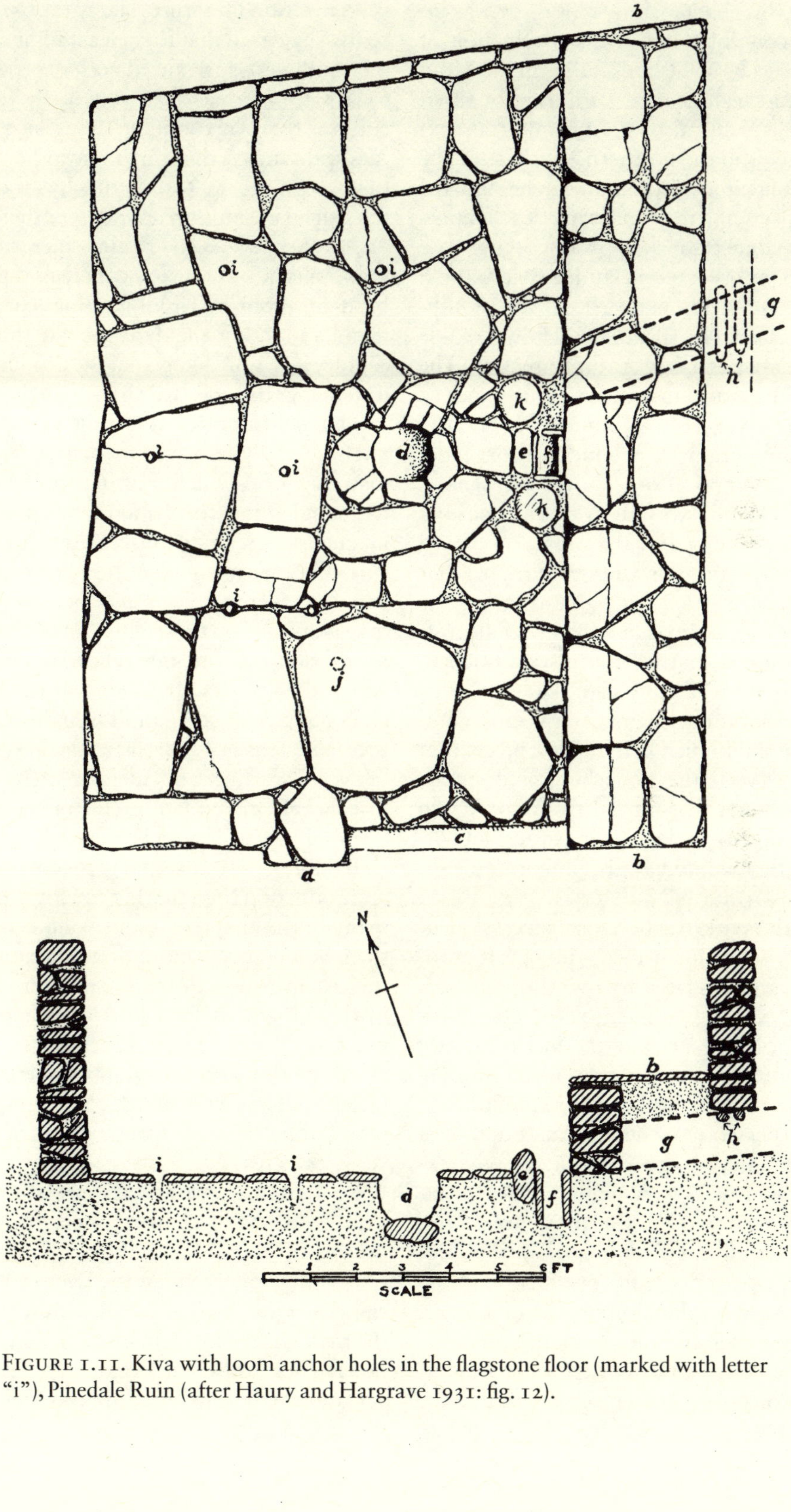

FIGURE 1.11. Kiva with loom anchor holes in the flagstone floor (marked with letter "i"), Pinedale Ruin (after Haury and Hargrave 1931: fig. 12).

1998; Van Keuren 2004). The analyses discussed in this chapter show that there was increasing concentration in the production of these objects by some households in the Mogollon Rim sites. Textiles were used in ritual performances and also became important exchange items in the northern Southwest. The weaving of textiles also enhanced the status of the producers and the ritual societies that provided the context for their production.

Like many other recent studies (Ames 1995; Feinman 1999; Inomata 2001), this Southwestern case study also shows how flexible the household unit is in craft production. The household can produce at different scales of intensity, ranging from production for household consumption to production for widespread consumption outside the household. When production was at its most intense in this area, multifunctional storage rooms attached to living rooms and other rooms that were no longer used for habitation are the most frequent contexts in which we find craft production raw materials, tools, and debris. In those households in which the multicrafting of ceramics and ornaments was most common, household members made objects that moved between the household and public spheres—bowls that moved from household consumption to public feasting events and ornaments that would have been worn by individuals in private and public.

In both contexts, the close succession of migration and intensified multicraft production can be related to two different scenarios of why intensification of craft production occurs, especially the intensification that results in multicraft specialized production. One of these is the risk-reduction scenario, discussed elsewhere in the Southwest by Rautman (1993) to account for why exchange frequencies were higher in agriculturally marginal areas than in those areas with greater agricultural potential. In some ethnographic models of ceramic production, those with access to less land often take up potting or other forms of craft production, sometimes called the "disenfranchised peasant model" (e.g., Arnold 1985; Foster 1973). Despite the prevalence of risk-reduction scenarios in the specialization literature, Harry (2005) has recently suggested that for at least some areas of the Southwest, specialized pottery production took place in areas that were better for agricultural than surrounding areas, not worse. As she discusses, there are many other factors that were more important for why production intensified in some areas and not others.

Another scenario relates increased ceramic specialization to the increased prestige brought about by ritual feasting (e.g., Spielmann 1998). For the dynamic late thirteenth century it is more likely that the production of different crafts was tied to the social contexts in which those crafts were used, including but not necessarily restricted to feasting. There are many ways in which crafting can be related to changing identities and the negotiation of inequalities created out of the migration process. Occupational identities are related to who crafts what objects and the social values that are placed on those different occupations. In addition, identities are related to the way in which the crafts were used, that is, their social contexts of consumption. I suspect that these two ways of looking at the intersection between crafting and identity were largely inseparable in the late prehistoric Puebloan communities discussed here.

If the items produced were part of commodity production for a market economy (such as many of the chapters in this volume discuss), then we might see more of a separation, but there is no evidence of this scale of production in the Mogollon Rim area. Who produced what and for whom and the social contexts of production and consumption were probably all well known by most community members within the small-scale societies of the Silver Creek area.

Some of the crafts were probably made as expressions of gender; some, of age; and others, of suprahousehold ritual participation within the community. Even within the class of ornaments, there were differences in the materials used, including shell, stone, and bone. These different ornaments have been suggested to be outward displays of identity at various levels, including gender, age, and ritual status. For example, women at Grasshopper Pueblo, located just south of the Mogollon Rim and approximately 30 air miles from Bailey Ruin, were more likely to be buried with bone and shell rings, whereas men were buried with shell pendants and bone hairpins (Reid and Whittlesey 1999:145).

The crafting of ceramic serving vessels that moved from domestic to ritual feasting contexts also would have provided a way in which different identities could be expressed. In the Southwest, polychrome ceramics were closely tied to the emerging ritual institutions of the late thirteenth and fourteenth centuries and the associated communal feasting that accompanied this ritual transformation (Crown 1994; Mills in press; Spielmann 1998; Van Keuren 2004). Polychrome ceramics moved between the domestic and ritual spheres and were used to serve foods in multiple contexts. Feasting was a context in which identities were expressed and where commensal politics were practiced. The production of these ceramics would have been an ideal way in which identity could be constructed. Some households (and most likely the women in those households) clearly made more vessels than others, and the participation of these households in ritual feasting would have established and reinforced differences in status and household identity.

Whether the households that engaged in multicrafting were migrant or first-comer households is difficult to determine. We have no architectural means of differentiating among households that could be related to the order in which families arrived in the Silver Creek area. That they were so quickly incorporated into the social life of the pueblos that they joined is significant for understanding the dynamics during this period. Maintaining distinctive identities based on where one came from is a part of migration scenarios in some areas of the world, but in this case, enclaves or what archaeologists in the Southwest have called "site unit intrusions" were not present in the Mogollon Rim area. Instead, migration resulted in the forging of new identities—occupational, ritual, and social. The proliferation of ritual structures during and following the migrations of the thirteenth century indicates that new institutions were introduced. As Hopi oral history recounts, migrants were often expected to bring a new song, a new dance, and even a new ritual society to the village to which they moved.

Ritual societies were organizations that brought migrants and hosts together at the same time that they created new divisions and social identities within Pueblo IV period communities (Kaldahl et al. 2004). They were also workshops where ritual knowledge was transmitted, including knowledge about how to produce objects important for ceremonial participation. In historic times, these workshops were places where other crafts were made besides cloth—such as moccasins—that were also used in ritual performances. Participation in the production of ritual clothing would have reinforced the overall participation in these suprahousehold ritual societies. Each step—the production of clothing, the practice of songs and dances, and the final performance itself—was a way in which identities were constructed and maintained.

In conclusion, the intensification of multicraft production documented for the late thirteenth and early fourteenth centuries in the Southwest was closely tied to factors relating to the aftermath of migration. Unlike the case in the earlier period of migration when the Mogollon Rim region was a frontier, the migrants of the thirteenth and fourteenth centuries had to negotiate an already populated social landscape. Craft production was intimately tied to the reformulation of social identities and the diversification of ritual institutions as newcomers and first comers interacted in postmigration aggregated villages. Multiple crafts were produced in households and suprahousehold settings, linked by their use in contexts of identity formation. This case study shows not only how multicrafting in a nonmarket economic setting can be present and practiced in multiple production contexts but also how multicrafting must be looked at within the broader social setting.

References Cited

Ames, Kenneth M.

1995 Chiefly Power and Household Production on the Northwest Coast. In *Foundations of Social Inequality*, edited by T. Douglas Price and Gary M. Feinman, pp. 155–188. Plenum Press, New York.

Anthias, Floya

1998 Evaluating "Diaspora": Beyond Ethnicity. *Sociology* 32(2):557–580.

Arnold, Dean

1985 *Ceramic Theory and Cultural Process*. Cambridge University Press, New York.

Brooks, J.

1944 Journal of Thales H. Haskell. *Utah Historical Quarterly* 12:69–95.

Christenson, Andrew L.

1994 Perforated and Unperforated Plates as Tools for Pottery Manufacture. In *Function and Technology of Anasazi Ceramics from Black Mesa, Arizona*, edited by Marion F. Smith Jr., pp. 55–65. Center for Archaeological Investigations Occasional Paper 15. Southern Illinois University, Carbondale.

Costin, Cathy L.

1991 Craft Specialization: Issues in Defining, Documenting, and Explaining the Organization of Production. In *Archaeological Method and Theory*, edited by Michael B. Schiffer, Vol. 3:1–56. University of Arizona Press, Tucson.

2001 Craft Production Systems. In *Archaeology at the Millennium: A Sourcebook*, edited by Gary M. Feinman and T. Douglas Price, pp. 273–327. Plenum, New York.

Costin, Cathy L., and Rita Wright (editors)

1998 *Craft and Social Identity*. Archaeological Papers No. 8. American Anthropological Association. Washington, D.C.

Crown, Patricia L.

1981 *Variability in Ceramic Manufacture at the Chodistaas Site, East-Central Arizona.* Ph.D. dissertation, University of Arizona, Tucson. University Microfilms, Ann Arbor.

1994 *Ceramics and Ideology: Salado Polychrome Ceramics*. University of New Mexico Press, Albuquerque.

Dean, Rebecca M.

2001 Social Change and Hunting during the Pueblo III to Pueblo IV Transition, East-Central Arizona. *Journal of Field Archaeology* 28(3–4):271–285.

Esses, Victoria M., Lynne M. Jackson, and Tamara L. Armstrong

1998 Intergroup Competition and Attitudes toward Immigrants and Immigration: An Instrumental Model of Group Conflict. *Journal of Social Issues* 54(4):699–724.

Feinman, Gary M.

1999 Rethinking Our Assumptions: Economic Specialization at the Household Scale in Ancient Ejutla, Oaxaca, Mexico. In *Pottery and People: Dynamic Interactions*, edited by James M. Skibo and Gary M. Feinman, pp. 81–98. University of Utah Press, Salt Lake City.

Fenn, Thomas, Barbara J. Mills, and Maren Hopkins

2006 The Social Contexts of Glaze Paint Ceramic Production and Consumption in the Silver Creek Area. In *The Social Life of Pots: Glaze Wares and Cultural Dynamics in the Southwest, A.D. 1250–1680*, edited by Judith A. Habicht-Mauche, Suzanne L. Eckert, and Deborah L. Huntley, pp. 60–85. University of Arizona Press, Tucson.

Fewkes, Jesse Walter

1904 *Two Summer's Work in Pueblo Ruins*. Bureau of American Ethnology 22nd Annual Report, 1900–01. Washington, D.C.

Foster, George

1973 [1948] *Empire's Children: The People of Tzintzuntzan*. Greenwood Press, Westport, Connecticut. (Original: Smithsonian Institution.)

Hagstrum, Melissa B.

1995 Creativity and Craft: Household Pottery Traditions in the Southwest. In *Ceramic Production in the American Southwest*, edited by Barbara J. Mills and Patricia L. Crown, pp. 281–300. University of Arizona Press, Tucson.

Harry, Karen G.

2005 Ceramic Specialization and Agricultural Marginality: Do Ethnographic Models Explain the Development of Specialized Pottery Production in the Prehistoric American Southwest? *American Antiquity* 70(2):295–319.

Haury, Emil W.

1985 *Mogollon Culture in the Forestdale Valley, East-Central Arizona*. University of Arizona Press, Tucson.

Haury, Emil W., and Lyndon L. Hargrave

1931 *Recently Dated Pueblo Ruins in Arizona.*

Smithsonian Miscellaneous Collections, Vol. 82, No. 11. Smithsonian Institution, Washington, D.C.

Hegmon, Michelle, Winston Hurst, and James R. Allison

1995 Production for Local Consumption and Exchange: Comparisons of Early Red and White Ware Ceramics in the San Juan Region. In *Ceramic Production in the American Southwest*, edited by Barbara J. Mills and Patricia L. Crown, pp. 30–62. University of Arizona Press, Tucson.

Helms, Mary

1988 *Ulysses' Sail: An Ethnographic Odyssey of Power, Knowledge, and Geographical Distance*. Princeton University Press, Princeton.

1993 *Craft and the Kingly Ideal: Art, Trade, and Power*. University of Texas Press, Austin.

Herr, Sarah A.

2001 *Beyond Chaco: Great Kiva Communities on the Mogollon Rim Frontier*. Anthropological Papers of the University of Arizona No. 66. University of Arizona Press, Tucson.

Herr, Sarah A., and Susan L. Stinson

2005 A Potter's Assemblage from Tla Kii Pueblo, Arizona. *Kiva* 71(1):57–78.

Inomata, Takeshi

2001 Power and Ideology of Artistic Creation: Elite Craft Specialists in Classic Maya Society. *Current Anthropology* 42(3):321–349.

Kaldahl, Eric, Scott Van Keuren, and Barbara J. Mills

2004 Migration, Factionalism, and the Trajectories of Pueblo IV Period Clusters in the Mogollon Rim Region. In *The Protohistoric Pueblo World, a.d. 1275–1600*, edited by E. Charles Adams and Andrew I. Duff, pp. 85–94. University of Arizona Press, Tucson.

Kramer, Carol

1985 Ceramic Ethnoarchaeology. *Annual Review of Anthropology* 14:77–102.

Longacre, William A., Sally J. Holbrook, and Michael W. Graves

1982 *Multidisciplinary Research at Grasshopper Pueblo, Arizona*. Anthropological Papers of the University of Arizona No. 40. University of Arizona Press, Tucson.

Lyons, Patrick D.

2003 *Ancestral Hopi Migrations*. University of Arizona Anthropological Papers No. 68. University of Arizona Press, Tucson.

Lyons, Patrick D., and Alexander J. Lindsay

2006 Perforated Plates and the Salado Phenomenon. *Kiva* 72(1):5–54.

Marshall, Yvonne, and Chris Gosden (editors)

1999 The Cultural Biography of Objects. Theme issue, *World Archaeology* 31(2).

Martin, Paul S., William A. Longacre, and James N. Hill

1967 *Chapters in the Prehistory of Eastern Arizona, 3*. Fieldiana: Anthropology 57. Field Museum of Natural History, Chicago.

McGovern, Patrick E. (editor)

1989 *Cross-Craft and Cross-Cultural Interactions in Ceramics*. Ceramics and Civilization, Vol. 1. American Ceramic Society, Westerville, Ohio.

Mills, Barbara J.

1995 Gender and the Reorganization of Zuni Craft Production: Implications for Archaeology. *Journal of Anthropological Research* 51:149–172.

1998 Migration and PIV Community Reorganization in the Silver Creek Area, East-Central Arizona. In *Migration and Reorganization: The Pueblo IV Period in the American Southwest*, edited by Katherine A. Spielmann, pp. 65–80. Arizona State University Anthropological Research Papers 51. Tempe.

2000 Gender, Craft Production, and Inequality. In *Women and Men in the Prehispanic Southwest: Labor, Power, and Prestige*, edited by Patricia L. Crown, pp. 301–343. School of American Research Press, Santa Fe.

2004 The Establishment and Defeat of Hierarchy: Inalienable Possessions and the History of Collective Prestige Structures in the Puebloan Southwest. *American Anthropologist* 106(2):238–251.

In press Performing the Feast: Visual Display and Suprahousehold Commensalism in the Puebloan Southwest. *American Antiquity*.

Mills, Barbara J., and Patricia A. Crown (editors)

1995 *Ceramic Production in the American Southwest*. University of Arizona Press, Tucson.

Mills, Barbara J., and Sarah A. Herr

1999 Chronology of the Mogollon Rim Region. In *Living on the Edge of the Rim: Excavations and Analyses by the Silver*

Creek Archaeological Research Project, 1993–1998, edited by Barbara J. Mills, Sarah A. Herr, and Scott Van Keuren, pp. 269–293. Arizona State Museum Archaeological Series No. 192. University of Arizona, Tucson.

Mills, Barbara J., Sarah A. Herr, Susan L. Stinson, and Daniela Triadan

1999a Ceramic Production and Distribution in the Silver Creek Area. In *Living on the Edge of the Rim: Excavations and Analyses by the Silver Creek Archaeological Research Project, 1993–1998*, edited by Barbara J. Mills, Sarah A. Herr, and Scott Van Keuren, pp. 295–324. Arizona State Museum Archaeological Series No. 192. University of Arizona, Tucson.

Mills, Barbara J., Sarah A. Herr, and Scott Van Keuren (editors)

1999b *Living on the Edge of the Rim: Excavations and Analyses by the Silver Creek Archaeological Research Project, 1993–1998*. Arizona State Museum Archaeological Research Series No. 192. University of Arizona, Tucson.

Montgomery, Barbara K., and J. Jefferson Reid

1990 An Instance of Rapid Ceramic Change in the American Southwest. *American Antiquity* 55(1):88–97.

Nyerges, A. Endre

1992 Ecology of Wealth-in-People: Agriculture, Settlement, and Society on the Perpetual Frontier. *American Anthropologist* 94(4):860–881.

Rautman, Alison E.

1993 Resource Variability, Risk, and the Structure of Social Networks: An Example from the Prehistoric Southwest. *American Antiquity* 58(3):403–425.

Reid, Jefferson, and Stephanie Whittlesey

1999 *Grasshopper Pueblo: A Story of Archaeology and Ancient Life*. University of Arizona Press, Tucson.

Robb, John E. (editor)

1999 *Material Symbols: Culture and Economy in Prehistory*. Center for Archaeological Investigations Occasional Paper No. 26. Southern Illinois University Press, Carbondale.

Scholnick, Jonathan

2003 Village Formation during the Pueblo III to Pueblo IV Period Transition: Contextualizing Bryant Ranch Pueblo, Arizona. Unpublished M.A. thesis, Department of Anthropology, University of Arizona.

Spielmann, Katherine A.

1998 Ritual Influences on the Development of Rio Grande Glaze A Ceramics. In *Migration and Reorganization: The Pueblo IV Period in the American Southwest*, edited by Katherine A. Spielmann, pp. 253–279. Arizona State University Anthropological Research Papers 51. Tempe.

2002 Feasting, Craft Specialization, and the Ritual Mode of Production in Small-Scale Societies. *American Anthropologist* 104(1):195–207.

Stephen, Alexander M.

1936 *Hopi Journal of Alexander M. Stephen*, vols. 1–2. Edited by Elsie Clews Parsons. Columbia University Press, New York.

Triadan, Daniela

1989 Defining Local Ceramic Production at Grasshopper Pueblo, Arizona. Unpublished M.A. thesis, Lateinamerikainstitut, Freie Universität, Berlin.

Valado, Trenna

1999 Ground Stone Technology in the Silver Creek Area. In *Living on the Edge of the Rim: Excavations and Analyses by the Silver Creek Archaeological Research Project, 1993–1998*, edited by Barbara J. Mills, Sarah A. Herr, and Scott Van Keuren, pp. 373–403. Arizona State Museum Archaeological Series No. 192. University of Arizona, Tucson.

Van Keuren, Scott

2004 Crafting Feasts in the Prehispanic Southwest. In *Identity, Feasting, and the Archaeology of the Greater Southwest*, edited by Barbara J. Mills, pp. 192–209. University Press of Colorado, Boulder.

Vokes, Arthur W.

1999 Shell Analysis. In *Living on the Edge of the Rim: Excavations and Analyses by the Silver Creek Archaeological Research Project, 1993–1998*, edited by Barbara J. Mills, Sarah A. Herr, and Scott Van Keuren, pp. 405–432. Arizona State Museum Archaeological Series No. 192. University of Arizona, Tucson.

Webster, Laurie D.

1997 Effects of European Contact on Textile Production and Exchange in the North American Southwest: A Pueblo Case Study. Unpublished Ph.D. dissertation, University of Arizona, Tucson.

Wright, Rita P.

1991 Women's Labor and Pottery Production in Prehistory. In *Engendering Archaeol-*

ogy: *Women and Prehistory*, edited by Joan M. Gero and Margaret W. Conkey, pp. 194–223. Basil Blackwell, Oxford.

Zack-Horner, Jennifer

1999 Aggregation and the Faunal Record. In *Living on the Edge of the Rim: Excavations and Analyses by the Silver Creek Archaeological Research Project, 1993–1998*, edited by Barbara J. Mills, Sarah A. Herr, and Scott Van Keuren, pp. 433–457. Arizona State Museum Archaeological Series No. 192. University of Arizona, Tucson.

2

Middle Sicán Multicraft Production: Resource Management and Labor Organization

DAVID J. GOLDSTEIN AND IZUMI SHIMADA

The succession of complex cultures (e.g., Cupisnique, Mochica [aka Moche], Sicán [aka Lambayeque], and Chimú) that flourished from the late second millennium B.C. to the fifteenth century A.D. on what is now the north coast of Peru produced a wide array of artistically and technologically superior craft products deserving of the characterization "master craftsmanship" (Bennett and Bird 1949; Sawyer 1968). The intense search for and focused excavation of the workshops that made these products began in the mid-1970s and has met with a good deal of success (e.g., Hayashida 1999; Lechtman 1976; Merkel et al. 1994; Shimada 1978, 1982, 1994a, 1994d, 1998; Shimada et al. 1982; Tschauner 2001). Since 1978, the Sicán Archaeological Project (SAP) under Shimada's direction has dedicated much of its attention and effort to "holistic studies" of the Middle Sicán (A.D. 900–1100) crafts (Figure 2.1). The Sicán distinguished themselves with their unprecedented production of black pottery (Shimada and Wagner 2001, 2007, in press), paddle-decorated pottery (Cleland and Shimada 1994, 1998), arsenical copper (Epstein 1993; Epstein and Shimada 1983; Merkel et al. 1994; Shimada and Merkel 1991; Shimada et al. 1982), and precious metals (Shimada and Griffin 1994; Shimada et al. 1999; Shimada et al. 2000). By *holistic study* we mean the interdisciplinary investigation of workshops and associated technology, production processes, and producers in their regional social and environmental contexts to elucidate the human (behavioral, social, and symbolic) and material dimensions of pre-Hispanic craft production and products (Shimada 1985; Shimada and Wagner 2001, 2007). Although these studies concentrated on single crafts (i.e., ceramics or smelting/metalworking), varied issues concerning possible relationships between or among different crafts and artisans came to the fore with the discovery of diverse luxury goods in an elite shaft tomb excavated in 1991–1992 (e.g., Shimada 1995; Shimada and Griffin 1994). Many of these artifacts, for example, a gold mask and headdresses, were assembled with components from crafts conventionally regarded as distinct, such as goldsmithing, lapidary, and feather working (Shimada 1996). Who designed and manufactured these composite items? How was their production supervised? If multiple artisans were involved, how were they functionally, spatially, and temporally organized? Were their productive activities coterminous or sequential? What were their social relations? Why are some crafts found together and not others? These are the questions that came to challenge us.

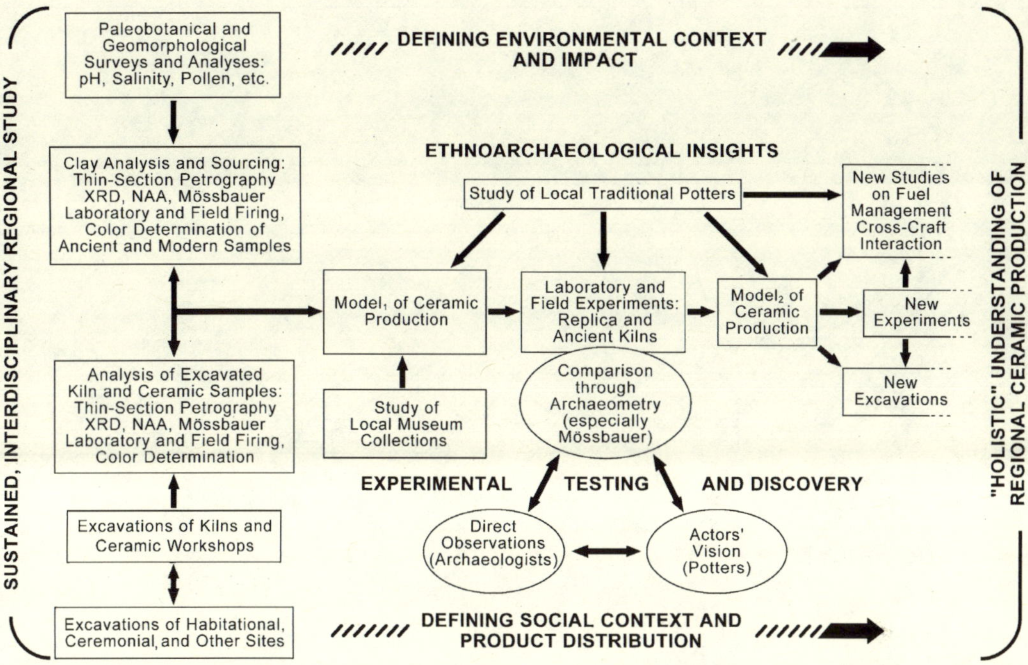

FIGURE 2.1. Graphic representation of the holistic approach to our investigation of Middle Sicán ceramic production and its multicraft relationships. Graphics by S. Mueller and I. Shimada. Reproduced with permission of the University of Arizona Press.

The 1996 discovery of a Middle Sicán multicraft workshop at Huaca Sialupe in the lower La Leche Valley on the northern north coast (Figure 2.2) provided us with an excellent opportunity to explore in depth some of the issues raised above. Our fieldwork (1999–2001) and related analysis have provided new insight into the technology and organization of craft production that spanned from about A.D. 950 to 1050. Our excavations and experimental work focused on production debris from the workshop, and here we look at the fuel remains to provide a basis for examining the issues of resource management and labor organization. Specifically, we employ a paleoethnobotanical approach to understand fuel use by Middle Sicán metalsmiths and potters who shared the workshop space 1,000 years ago. Generally, we wish to address the following questions: To what extent did ceramic production and metalworking complement or compete with each other in regard to the critical fuel supply and use? Who were the artisans involved, and how were they organized? From looking at the organic remains from fuel use areas, we reconstruct the use and view of organic fuel resources by ancient artisans at the site while answering these questions.

RESEARCH ISSUES AND CONSIDERATIONS

During its estimated 100-year existence, the workshop at Huaca Sialupe manufactured a variety of products in at least two media. In terms of ceramics, artisans relied on molds and small, manageable kilns to produce both oxidized and reduced finewares. Though decorated black pottery predominated the wasters recovered in our excavations, it is difficult to establish the actual proportions of oxidized versus reduced wares produced (see Taylor 2000). Metalsmiths in adjacent workspaces fabricated, among other products, thin sheets of arsenical copper and gold alloys using simple but efficient furnaces. Metalsmiths cut, shaped, and joined together the resultant sheet metal to produce predominantly small utilitarian objects and ornaments. There is no

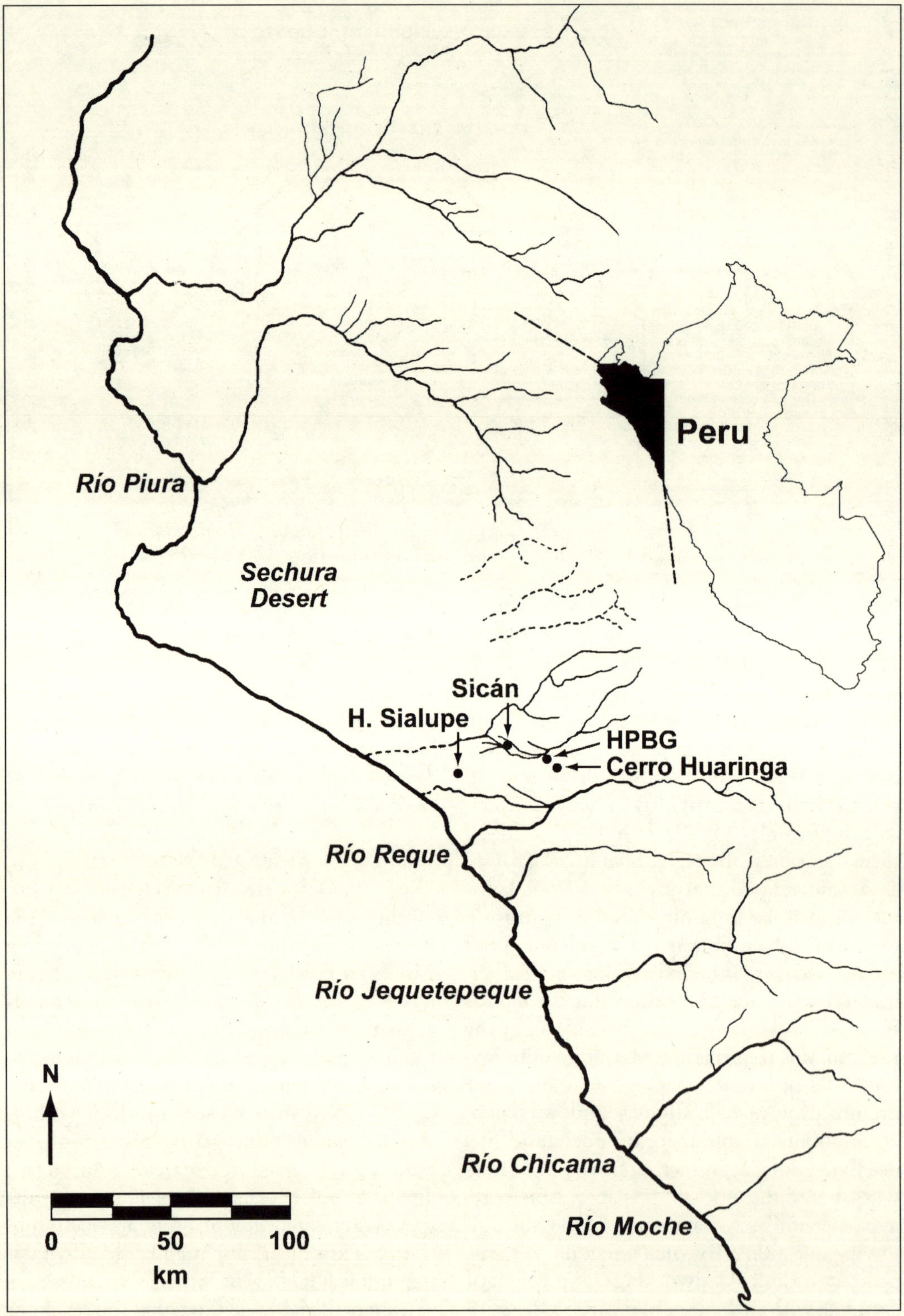

FIGURE 2.2. The Middle Sicán site of Huaca Sialupe and other major contemporaneous sites in the Lambayeque region on the north coast of Peru. H and HPBG stand for Huaca and Huaca del Pueblo Batán Grande. Graphics by S. Mueller and I. Shimada.

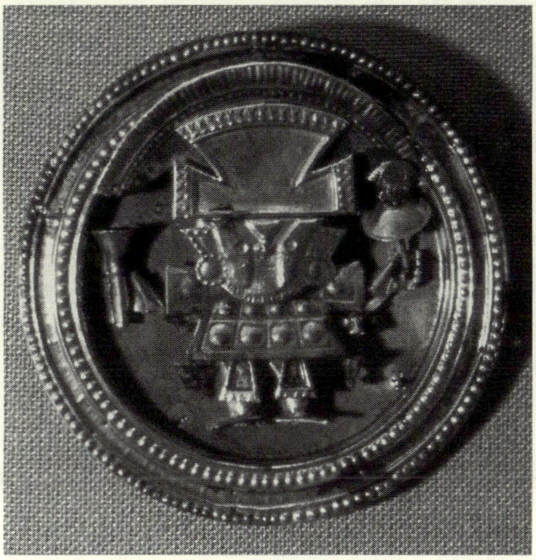

Figure 2.3. Middle Sicán gold ear spool with a representation of the standing Sicán Lord, earthly alter ego of the supernatural Sicán Deity. Hamburg Ethnological Museum, Hamburg. Photo by I. Shimada.

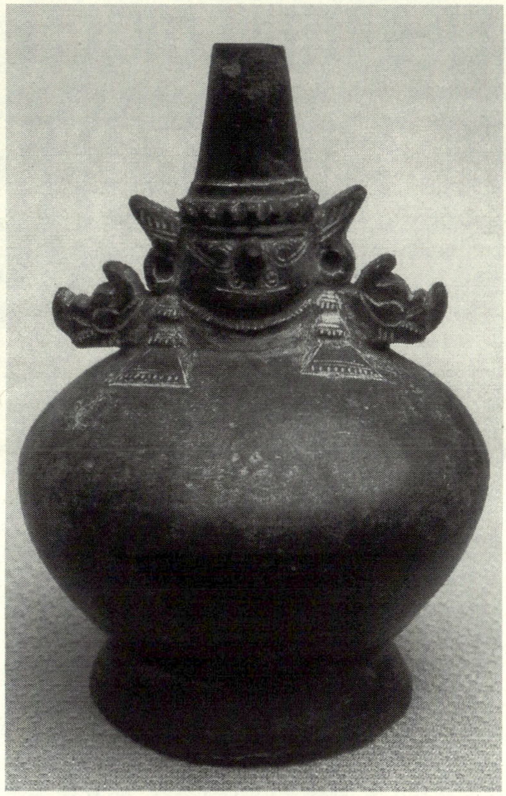

Figure 2.4. Example of a Middle Sicán black bottle with the head of the Sicán Deity decorating the base of the spout. Museo Amano, Lima. Photo by I. Shimada.

evidence of any on-site smelting that would have required specialized equipment, a reliable supply of ores, and labor (Shimada and Merkel 1991). Our reconstruction of on-site activities is based on excavation of the site in 1999 and 2001 and the experimental reconstruction of the kilns and furnaces, described below, during July 2000.

Our research centers on concerns previously highlighted in the literature on ancient craft production. First, Matson (1989:23–24; also Wertime 1964) has argued that the pyrotechnologies used in ceramic and metal production developed through a series of synergistic processes. He argues that metal furnaces and ceramic kilns developed parallel to one another technologically and that, because of similar concerns over fuel exploitation, associated industries employed complementary fuel use strategies. Second, apart from these developmental concerns, previous work in Sicán period contexts indicates that ceramic and metal products had different roles and value in the Middle Sicán culture. Well-polished and black ceramic bottles bearing a similar image of the Sicán Deity, the principal deity of the Sicán religion, were a signature craft product for middle- and high-status burials of the culture. These items were clearly valued, exchanged, and even imitated in form and finish well outside of the Sicán territory (Shimada 1990; Shimada and Wagner 2001, 2007, in press). As noted above, the large-scale production of true black pottery is one of the major Sicán legacies. Yet the Middle Sicán artistic and technical standard-bearers are metal objects, particularly those of precious metals that were available only to the social elite (Figure 2.3; e.g., Cleland and Shimada 1998; Shimada 1994b, 1994c). The black pottery (Figure 2.4), in essence, is a simplified version of precious metal vessels in form, color, and iconographic content.

Given the higher social value metalwork held relative to pottery and the diverse roles metals played in the Middle Sicán, we have elsewhere expected that metal production received preferential material and labor support from the social elite (e.g., Cleland and Shimada 1998; Shimada 1994b, 1994c). More specifically, we have argued that Sicán metalworkers had greater access over potters to supplies of high-quality hardwood fuels, a potentially limited resource in the region's ecologically specialized dry tropical forests (Proyecto Algarrobo 1997; Shimada and Wagner 2001). Through the analysis of charcoal from production contexts, this study revisits these and related issues of multicraft interaction at Huaca Sialupe.

METHODOLOGY

Few studies of pre-Hispanic craft production have examined fuel use in any depth, with most projecting ethnographic or ethnoarchaeological observations into the past (e.g., Arnold 1985; Sillar 2000; Winterhalder et al. 1974). These studies do not benefit from the examination of excavated production contexts. Thus, the deliberate sampling and examination of fire use contexts excavated from Huaca Sialupe's production features are unique to our study. We have identified three production features at the site, ceramic kilns, metalworking furnaces, and hearths. We interpret the hearth features from their contexts as the remains of cooking fires, not related directly to ceramic or metalworking fire use. To complement our archaeological work, we conducted firing experiments in the field using a full-scale replica ceramic kiln and metalworking furnace (Shimada et al. 2003a; Shimada et al. in press; Shimada and Wagner 2001, 2007, in press). Additionally, we draw on our own ethnoarchaeological study of traditional charcoal making to understand the kinds of fuel remains that are likely to be encountered in the region (Goldstein 2002). Excavated fire use features ($n = 153$) from the 1999 and 2001 field seasons provide the materials analyzed in our discussion. The workshop areas discussed here consist of two conjoined mounds, I and II, within the five mounds that make up the site. The preserved remains of the workshop cover an estimated 1,750 m^2 (ca. 70 m N–S × 28 m E–W), approximately 40 percent (ca. 600 m^2) of which was excavated (Shimada and Wagner 2007, in press). Although production materials were scattered in much of the five-mound complex, the areas excavated exhibited by far the highest concentrations of fuel and manufacturing debris on the surface. Thus, it is likely that artifact production took place to varying degrees in much of the site but most intensely and on the largest scale at Mounds I and II.

We completely sampled features associated with production for examination. However, where this was not possible ($n = 32$), we removed a measured quantity relative to the total feature volume as recovered by excavation, 25 percent of the total volume. The 153 feature samples were dry sieved through a standard series of fine-grade geological sieves (4.0–0.425 mm). We sorted the materials under magnification using incident light and used standard anatomical techniques for wood (Wheeler and Baas 1998) and seed identification (Martin and Barkley 1961). This includes comparisons with modern thin sections from systematically identified wood voucher specimens collected in the field and samples of known charred material. Our identifications of the ancient charcoal specimens are based on magnified views (<180×) of the standard three dimensions (tangential, radial, and cross sections) of charcoal greater than 2 mm in size. Similarly, we compared the seeds and other plant materials with reference materials collected from the area. We examined the remains from the experimental firing of full-scale replica kilns and metalworking furnaces, produced in 2000, in tandem and the same manner as the archaeological materials. This process is ongoing, and the results presented here are the first in a series of evaluations of the fuel remains.

ENVIRONMENTAL BACKGROUND: SICÁN PERIOD IN NORTH COASTAL PERU

In the earliest phase of its research, the SAP focused some of its resources on reconstructing the paleoenvironment in the La Leche Valley and adjacent coastal area (Shimada 1981, 1982). The geomorphological work carried out by the SAP indicated that no agricultural activity had taken place in the immediate vicinity of the site of Sicán, the Middle Sicán capital, in what today is delimited by the National Historic and Ecological Sanctuary of the Poma Forest, Batán Grande, Peru (Craig and Shimada 1986). This work determined that the modern climate and environmental processes do not differ significantly from those of the Sicán period, save for the impact of 500 years of landscape modifications by humans, for example, an overall decrease in plants/animal populations and community biodiversity (Craig 1985; Shimada 1995). Work by Craig and Shimada (1986) indicates that periodic inundations, related to El Niño/Southern Oscillation (ENSO) rains, have been consistently episodic and greatly varied in their intensity during the more recent geological epoch. The sediment deposits appear erratic and catastrophic, indicating the recurrent drought-and-flood regime for at least the past 1,500 years (Craig and Shimada 1986). The work of Shimada et al. (1991), using ice cores from glaciers from the southern Peruvian Andes, has identified a series of geoclimatically forced severe drought events in the Common Era. This confirms Craig and Shimada's (1986) data for the broad regional trends of precipitation for the research area. Corroboration by radiocarbon-dated lake diatom strata from southern Ecuador indicates that these sequences are reliable (Shimada 1994a). Both climate and geomorphological data demonstrate that the present dominant climatic regime, including ENSO events, is similar to that during the Sicán period in the La Leche Valley, allowing for extrapolation of modern phytogeography onto the past. These studies set the stage for the analysis of these remains using modern flora from the area as comparative material for the identification and interpretation of the archaeobotanical samples.

The site of Huaca Sialupe is located just southeast of the modern pottery-making town of Mórrope, in the Department of Lambayeque, at the southern boundary of the Sechura desert (Figures 2.2, 2.5). Dry tropical forest dominates the desert's southern edge (Proyecto Algarrobo 1997; Tosi 1960). The regional woody flora consists of xerophytic trees, low shrubs, and cacti. The tree species are adapted to an extremely dry environment (precipitation <10 mm/year), and typically the biodiversity of the woody plants is manifest of two main families: Capparidaceae (caper family)—*Capparis* spp. (four species); and Fabaceae (bean family)—*Prosopis* spp. (three species) and *Acacia* spp. (two species [Proyecto Algarrobo 1997]). These three tree genera constitute the dominant mature-stature trees in the modern forest in the region. Presently, the cessation of modern charcoal production and agricultural clearing is at the heart of a regional campaign to stop deforestation in the area and preserve the remaining areas of dry coastal forest (Cuba Salerno et al. 1998; Risco 1998). Dry coastal forests are extremely vulnerable to deforestation by ecosystem disruption because of the slow annual growth of the dominant tree species, uncertain periodic water and nutrient recharge, and low biodiversity (Bullock et al. 1995; Cuba Salerno et al. 1998). Most trees in the area are long lived, at least 200 years, and are in their adolescence up to around 60 years of life. However, in most instances there are very few extant trees over 40 years of age (Ocampo 1991). Forest resources in the surrounding area are quickly disappearing as they are cut to create agricultural fields, supply fuel, and provide building materials (Risco 1998).

Previously published archaeological and experimental research in the area identifies only two species of primary wood fuel from select contexts, *Prosopis pallida* and *Capparis angulata* (Shimada and Merkel 1991; Shimada et al. 1998; Wagner et al. 1998). These studies recognize *Prosopis* spp., locally called *algarrobo*, as the primary high-quality fuel resource available in the area; *Capparis angulata*, locally *sapote*, though it does release energy as fuelwood comparable to *Prosopis* spp. (see Table 2.1), does not make good charcoal and is rapidly consumed in the fire (Shimada 1981, 1985).[1] Both were heavily used and traded in the colonial period as wood fuel and were used to produce goods, such as furniture or fences (Ramírez 1996). Rostworowski de Diez Canseco (1981) cites sixteenth-century

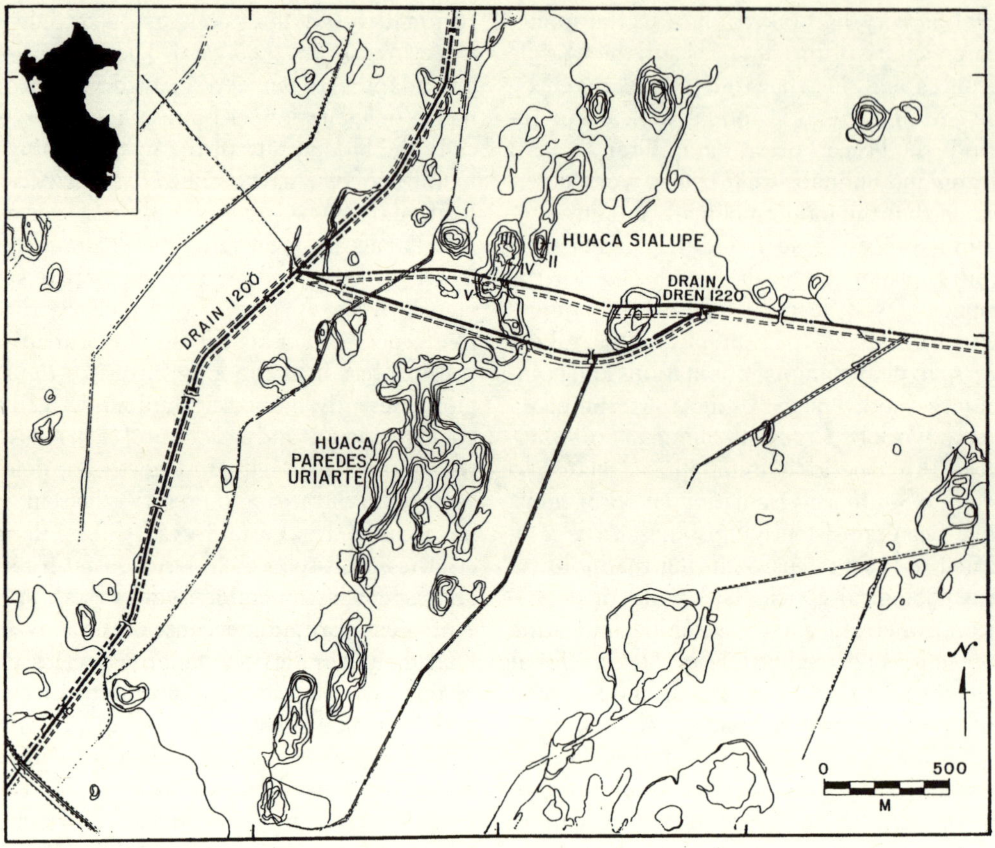

FIGURE 2.5. The five mounds that make up the site of Huaca Sialupe and its immediate context, including the site of Huaca Paredes-Uriarte, the inferred local Middle Sicán political and religious center. Drawing by C. Samillán. Reproduced with permission of the University of Arizona Press.

TABLE 2.1. Comparison of KJ/Kg Values of Select Wood Genera, Highlighting Woods from the Research Area

TREE SPECIES	CALORIC VALUE (DRY WOOD) KJ/KG
Acacia spp., corresponds to local *A. macracantha*	19,054
Capparis spp., corresponds to local *C. angulata* or *C. ovalifolia*	20,101
Cercidium spp., corresponds to local *C. praecox*	18,782
Prosopis spp., corresponds to three species of *Prosopis* spp. available locally	20,737
Quercus alba[a]	17,340
Eucalyptus globulous[b]	20,719

Source: Harker et al. 1982.

[a] White oak is not native to the area and is presented here as a temperate hardwood for comparison.

[b] Eucalyptus is not native to the area but is presently in common use in the area for reforestation and wood lot cultivation.

documents that mention 15 other economically useful trees for construction and fuel in the region, indicating that our understanding of ancient fuel resources is incomplete.[2] These two genera represent only seven of the woody plant species dominant in the region today (Cuba Salerno et al. 1998; Proyecto Algarrobo 1997). Because *Prosopis* spp. (Fabaceae) meets the physical demands of both sustained ceramic firing and metal smelting and smithing, the SAP has proposed that it was the cornerstone of Middle Sicán production (Shimada 1985, 1994b; Shimada and Merkel 1991; Shimada and Wagner 2001). If true, then from what we know about both dry tropical forests and the ecology of this species, the implication for forest destruction is profound.

The three *Prosopis* spp. species (*P. pallida*, *P. juliflora*, *P. affinis*) provide more than just high-grade wood and charcoal fuel; this genus is the backbone of the mature dry forest ecosystem (Delvaud 1984; Proyecto Algarrobo 1997). Dry tropical forests are highly specialized ecosystems that rely on a limited number of tree species to maintain soil structure and provide shelter to other plants and animals critical to forest regeneration and regrowth (Bullock et al. 1995; Cuba Salerno et al. 1998). *Prosopis* spp. both sustains soil fertility as a nitrogen fixer and helps maintain soil surface water through its lateral root system (Ocampo 1991; Proyecto Algarrobo 1997). This species plays a significant role in preventing soil erosion and desertification through both the shade and the soil cohesion offered by its branch and root systems, respectively (Luna-Suárez et al. 1998). Additionally, trees in this genus offer high-protein animal fodder in the form of their beanlike fruits, which are also used to create a sweet syrup, *algarrobina*, and a fermented beverage, *chicha de algarrobo* (Risco 1998). Local populations use the resin from the trunk as an adhesive and to fill dental caries (Risco 1998). These trees require nearly 100 years to grow to mature stature (Ocampo 1991). However, modern charcoal makers frequently cut them before they reach the age of 30 years (Risco 1998). The propensity of charcoal makers and agriculturalists to cut these trees has significantly affected forest regeneration throughout the colonial and modern era (Delvaud 1984; Risco 1998). As a result, *Prosopis* spp. charcoal production has become a critical deforestation problem over the past 200 years and has forced the regulation of forest products in the coastal areas of Peru (Ferreyra 1987). At present, rural forest residents produce *Prosopis* spp. charcoal illegally for additional household income to be sold in local and regional markets and for domestic use as fuel (Risco 1998).

The sustained clearance of *Prosopis* spp. in antiquity, or the concentrated removal of any dominant tree species, was likely deleterious to the forest ecosystem. The SAP's archaeological and experimental work suggests that metalworking and smelting alone may have adversely affected regional *Prosopis* spp. populations and regional forest structure. For instance, smelting experiments indicate that roughly 13,000 kg (13 metric tons) of charcoal would have been required to produce 100 kg of raw copper alloy ingots, an estimate that does not include the fuel required for subsequent ingot refinement, forging, and other metalworking tasks (Figure 2.6; Shimada and Merkel 1991).[3] Given that a commoner's burial invariably contained one to a few copper alloy objects and that many of the at least dozens of elite Middle Sicán shaft tombs in the area (e.g., the Huaca Loro East Tomb and the Huaca Menor [Pedersen 1976; Shimada 1995; Shimada et al. 2000]) contained up to three-quarters of a metric ton of metalwork, the draw on fuel resources for Sicán production would have been staggering (Shimada 1985). From an ecological understanding of *Prosopis* spp., dry tropical forests, and experimental production estimates, artisans likely used specific fuel use strategies to maintain production and mitigate the impact on fuel resources during the Middle Sicán.

MULTICRAFT INTERACTION: FEATURE RECOVERY AND DEFINITION

One category of features critical to our study is ceramic kilns. The remains of both oxidation and reduction kilns with distinct morphologies were found at the downwind, north end of the workshop (Shimada and Wagner

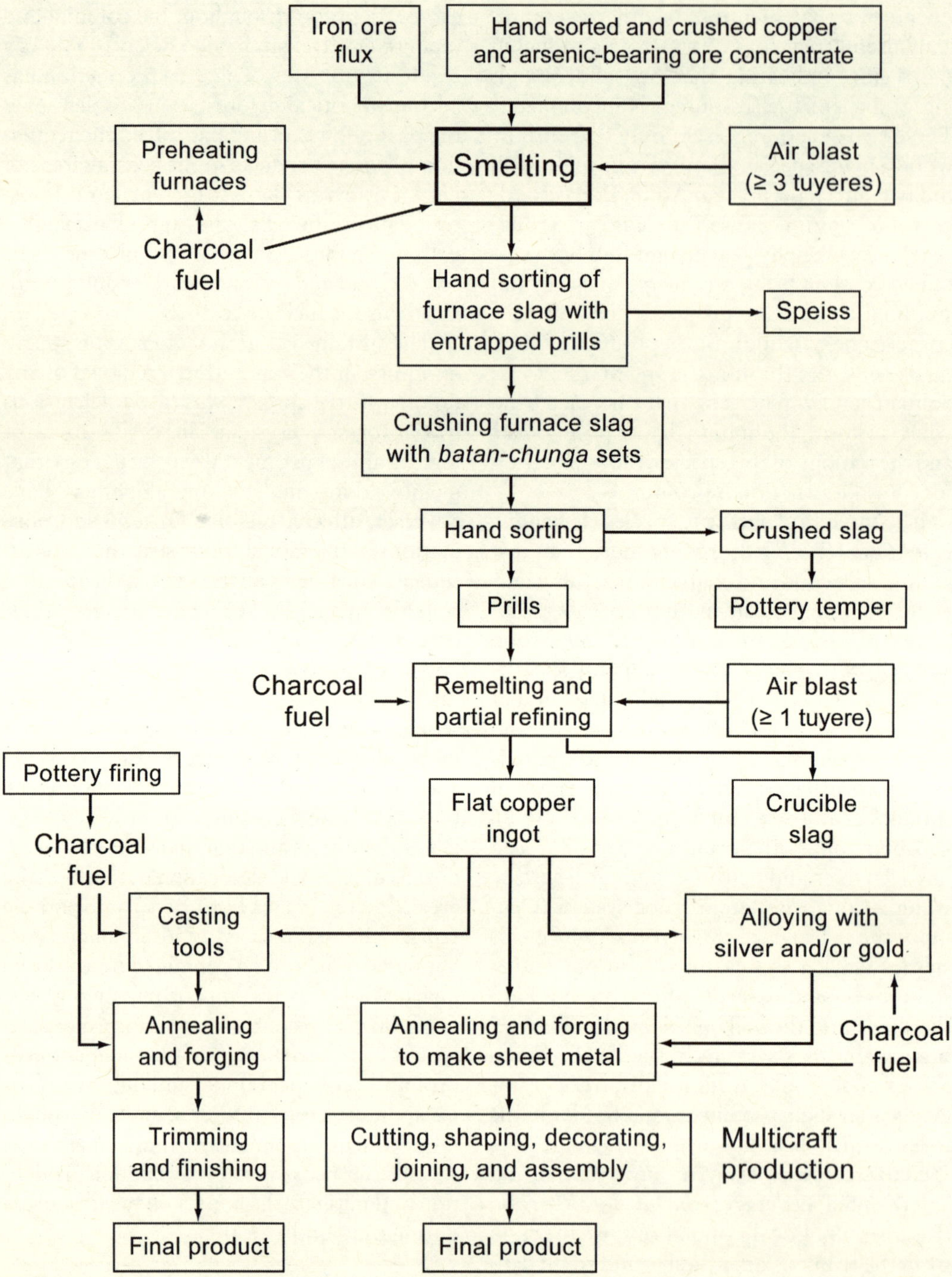

FIGURE 2.6. Flow chart of Middle Sicán arsenical copper production and its varied material and human correlates. Graphics by S. Mueller and I. Shimada. Reproduced with permission of the University of Arizona Press.

2001, 2007). This location is logical for removing the heat, smoke, and debris generated by kilns and furnaces.

Field and laboratory firing experiments (see below) as well as a comprehensive archaeometric analysis of samples of excavated and experimentally fired ceramics have shown that the widely valued glossy Middle Sicán black pottery owes its distinctive appearance to a number of factors. The vessel to be fired was first thoroughly burnished in a leather-hard state to create a charged surface, onto which a smooth layer of amorphous carbon from organic compounds (such as dried llama dung) formed during firing under strongly reducing conditions at temperatures of between 800 and 900°C (Shimada and Wagner 2001, in press; Shimada et al. 2003a; Shimada et al. 2003b; Shimada et al. 2003c; Shimada et al. in press). Thin-section microscopy and scanning electron microscopy showed that carbon penetrated a few millimeters into the body of the ceramics, contributing to the uniformly and truly black coloration. The metallic sheen found on many Middle Sicán black vessels is related to the formation of graphite crystals on the surface through prolonged exposure of the carbon to temperatures of at least 900°C.

Our excavations indicated that the sort of reduction firing described above was conducted in the evenly spaced kilns set into the floor of the large, walled, but unroofed area designated as Room 39 (Figure 2.7). The room is situated just south of a cluster of rooms that were engaged in formation and finishing (Figure 2.8). These kilns were elliptical or teardrop shaped and had their large firing chambers oriented downwind. Their slightly concave, sloping clay-lined bottoms were recessed circa 15 cm below floor level and measured around 100 × 45 cm wide, with some degree of variation. The excavated kilns were thoroughly blackened, including one with over 50 fine blackware sherds mixed among compacted charcoal powder (Figure 2.9). The associated floors were similarly blackened with charcoal. Fragments of collapsed linings indicate that the kilns originally had incurving sidewalls to create a dome-like firing chamber with a height of circa 30 cm above the floor.

Oxidation firing employed a similarly ovate kiln structure that was slightly larger in size (Figure 2.10). These were built of adobe bricks set in mud mortar and placed in pits circa 125 × 60 cm wide and 35 cm deep. Evidence from the workshop indicates that a wide range of relatively small, decorated vessels were mold-made at the site. Some vessels were embellished with appliqué or stamped decoration. The preserved portions and scattered structural remains suggest that the kilns used incurving walls to create semiclosed firing chambers. All but one of the kilns were oriented north–south, taking advantage of the prevailing winds for drafts and smoke removal. Associated with these kilns is a series of elongated beds of ash, charred twigs, and pulverized charcoal found nearby that we interpret as the remains of vessel preheating areas used before firing. Because of the overall good state of preservation of these features we can reconstruct a sizable portion of both oxidation and reduction firing processes.

We also recovered the remains of metalworking furnaces at the north end of the workshop (Mound I). The additional ones were set well into the earthen floors of a large room at the south end of the workshop (Mound II; Figure 2.3). These are made of large, inverted ceramic urns typically around 45–55 cm and 45 cm in diameter and height, respectively. The largest furnace documented thus far measures 85 cm in diameter and stands 73 cm high. Their bottoms were removed to create openings circa 20–25 cm in diameter at the top. The exteriors are covered with an insulating layer of clay and pieces of adobe block up to 12 cm in thickness (Figure 2.11). The furnaces contained at the bottom a jumble of carbonized organic fuel materials (ca. 8 liters). Each of these furnaces has a round 8–9 cm draft hole cut into the body near floor level facing the direction of the prevailing wind. Similar to the ceramic kilns, these furnaces were positioned to maximize the natural draft. Additionally, the best-preserved furnaces include two short walls flanking the draft hole to funnel wind

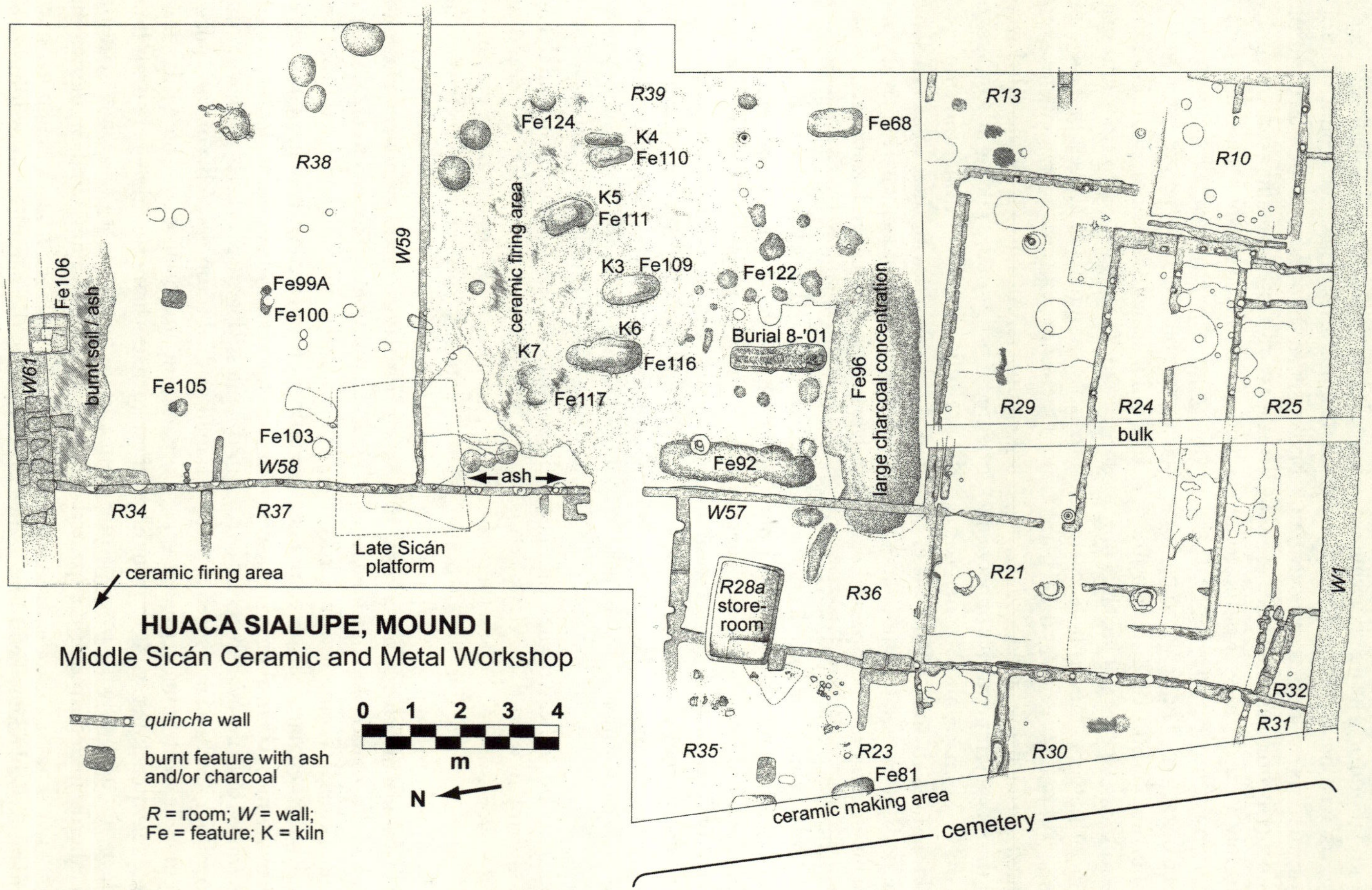

FIGURE 2.7. Ceramic formation and firing areas of the Huaca Sialupe workshop on Mound I. The large room in the middle of the plan, Room 39, contained small, evenly spaced kilns used for reduction firing. Drawing by C. Samillán, I. Shimada, and S. Mueller.

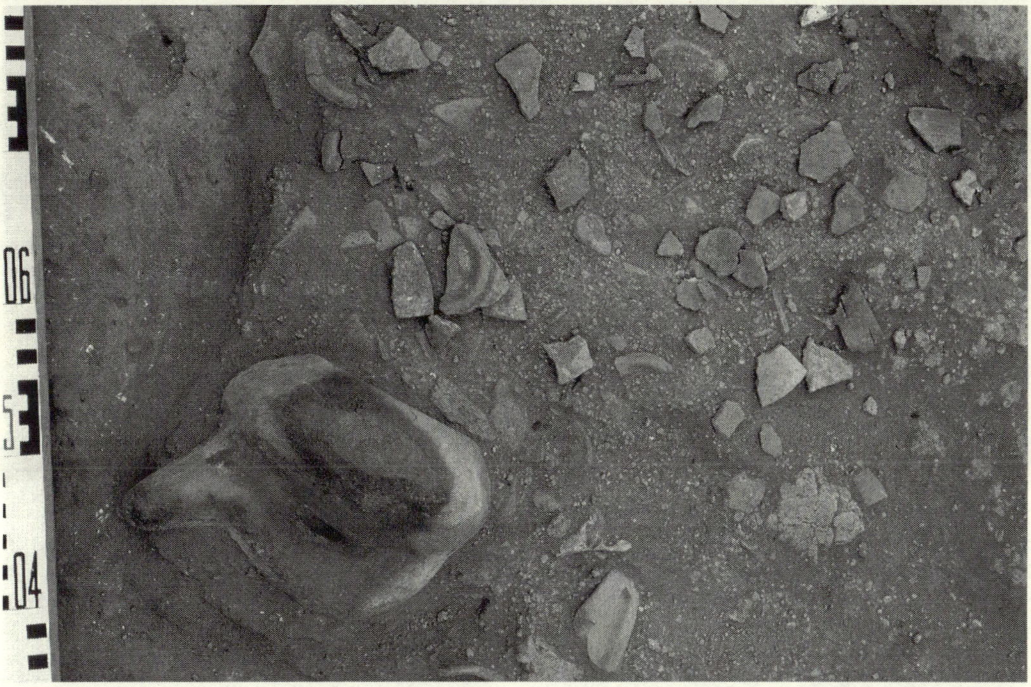

FIGURE 2.8. Floor of Room 23 with a nearly intact mold and fragments of unfired bowls, Mound I, Huaca Sialupe. The scale is in centimeters. Photo by I. Shimada.

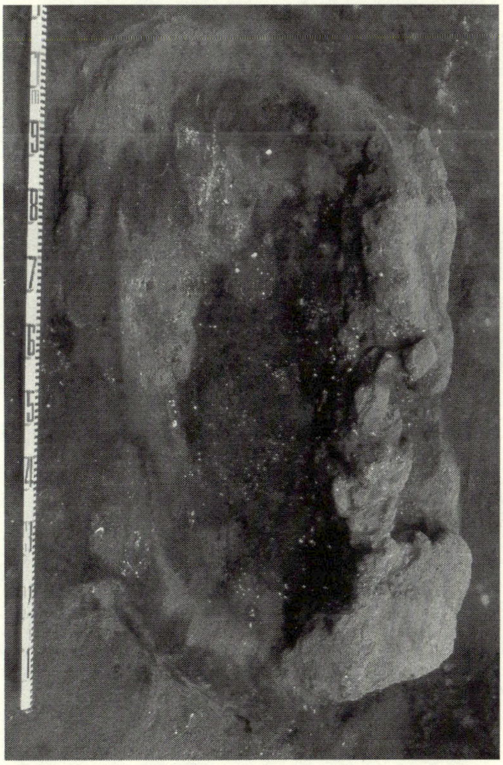

FIGURE 2.9. A reduction firing kiln in Room 39 at the Huaca Sialupe workshop. Photo by I. Shimada.

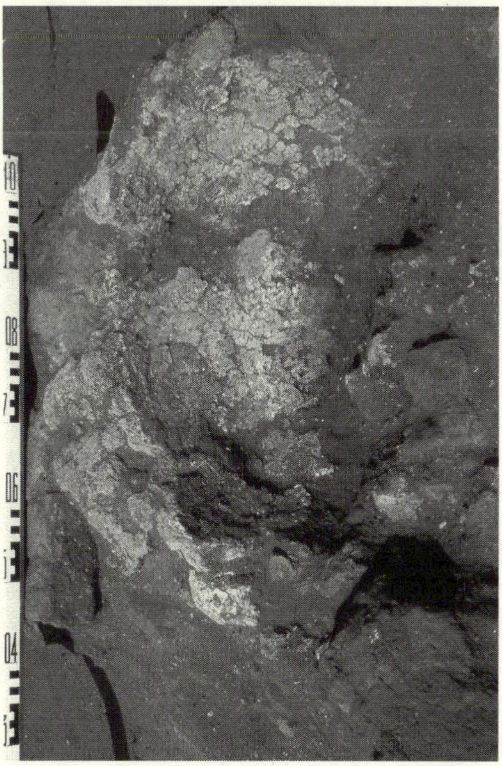

FIGURE 2.10. Example of the semiclosed kilns used for oxidization firing at the Huaca Sialupe workshop. Photo by I. Shimada.

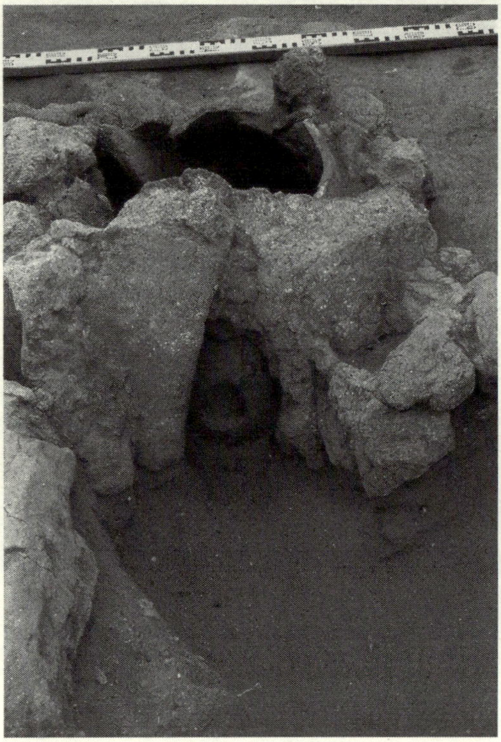

Figure 2.11. An excavated, well-preserved, metalworking furnace with its adobe–clay insulation largely intact from Feature 58, Mound II. Photo by I. Shimada.

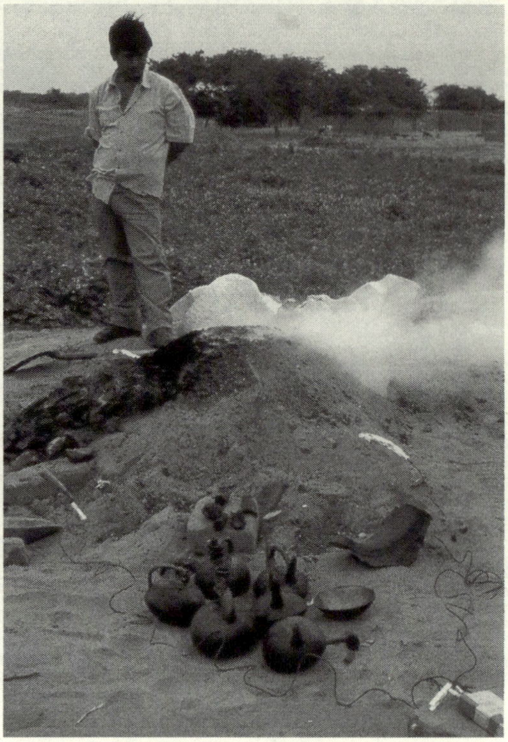

Figure 2.12. The 2000 ceramic firing experiment using a full-sized replica kiln. Ceramist, José Sosa, monitors the kiln. Wires connect Ni/Cd thermocouples placed in different locations within the kiln to digital thermometers. Photo by I. Shimada.

toward the fuel inside the furnace. The slag, spills, and scrap metal recovered in association with these furnaces indicate that the furnaces were used for melting ingots, as well as the alloying and annealing of hammered sheet metal (Shimada and Wagner 2001, 2007; Shimada et al. in press).

FIRING EXPERIMENTS

Complementing the archaeological, archaeometric, and paleoethnobotanical data and insights are those derived from the SAP's firing experiments (Figure 2.1). We conducted a series of firing experiments at the southern edge of Huaca Sialupe in July 2000 to gain comparative data for understanding the material and behavioral correlates of Middle Sicán ceramic firing and metalworking and the physical performance of the associated kilns and furnaces. Such experiments afford a unique opportunity to observe the decision-making processes and behaviors involved in successful firing, which is commonly regarded as the most challenging stage of pottery making (Shimada 2005). The 2000 experiments built on experience and knowledge gained from similarly conceived research conducted in 1992 using a 3,000-year-old kiln and its modern replica (e.g., Shimada et al. 1998; Shimada et al. 2003a; Wagner et al. 1998).

José Sosa prepared the full-scale replica kiln and pottery used in the 2000 experiments (Figure 2.12). He is a professional ceramist from Chulucanas, a well-known pottery-making town in the Department of Piura, with a good working knowledge of pre-Hispanic ceramic technology. The replicas approximated the measurements, materials, and techniques as determined by our prior excavations and attendant archaeometric analysis of relevant samples (Shimada and Wagner 2001; Shimada et al. 2003a; Shimada et al.

2003b; Shimada et al. 2003c). Fuels for the experimental firings consisted of dry, fallen hardwood branches (1–4 cm in diameter) collected from the forest floor. *Prosopis* spp. was used at a ratio of 5:1 by weight to *Capparis angulata*. Both species of wood exhibited bark beetle burrowing on the outermost wood surfaces of the collected materials. Our ceramist used leaves and small twigs as kindling and occasionally to boost combustion in the later stages of firing. He also charged the kiln with crushed dry llama dung (brought in from Callejón de Huáylas) and immediately and thoroughly sealed it with sand and mud to create a reducing atmosphere. The reducing conditions in the kilns not only produced the carbon/graphite layer on the surface of the ceramics but also reduced the hardwood fuels to charcoal. This charcoal remained in the kiln in significant amounts after firing was complete (ca. 6 kg hardwood charcoal). We left the kiln with its load of some dozen vessels to cool naturally for at least two hours (to a temperature of ca. 100°C) and, once, overnight. Sosa continuously monitored the kiln for the entirety of each firing. Independent of Sosa's own observations, we placed four Ni/Cd thermocouples, connected to digital thermometers, in various locations within the kiln to monitor its performance. Our experiments produced glossy black pottery that closely approximates that produced 1,000 years ago at the Huaca Sialupe workshop. They also yielded vessels with incomplete reduction and mottled appearance, just as we have documented among wasters recovered from Huaca Sialupe. We collected the fuel residues in the form of charcoal from the kilns for analysis.

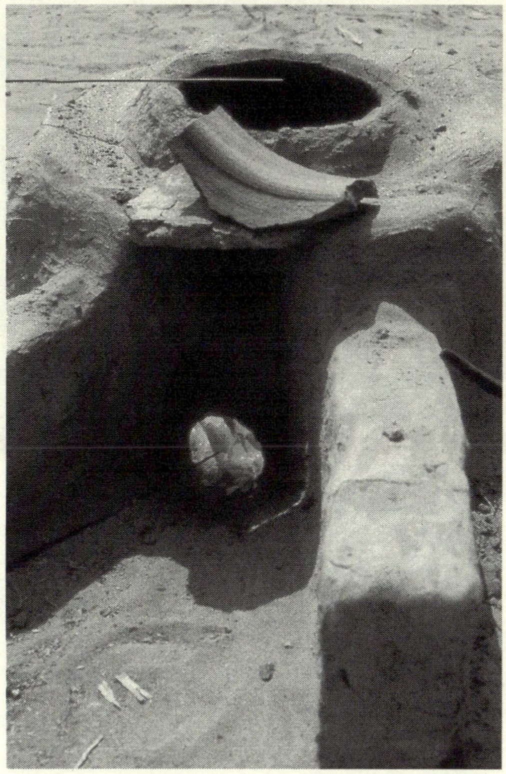

Figure 2.13. Full-sized replica metalworking furnace made for the 2000 experiments. Note the intense fire. Two wires belong to Ni/Cd thermocouples used to monitor temperatures at different locations in the furnace. The large sherd was used as a removable lid for air intake. Compare it with the 1,000-year-old furnace in Figure 2.11. Photo by I. Shimada.

We made our replica metalworking furnace with locally made adobe bricks, clays, and a *porrón* (large ceramic storage urn) that closely matched in size, shape, materials, and techniques the Middle Sicán originals previously documented at Huaca Sialupe (Figure 2.13). We charged the furnace with *Prosopis* spp. charcoal to the height just below the bottom of the circular draft hole that faced the prevailing wind. Our experiment showed that, with only natural wind fanning the charcoal bed, the furnace is capable of achieving an internal temperature between 1,000 and 1,100°C within 20 minutes from the time of ignition. Egressing air at the top of the furnace also reached 800°C. It was evident, additionally, that we could control easily the furnace temperature to suit a variety of metalworking tasks, such as melting ingots, alloying metals, and annealing metals, found in association with the original furnaces. We accomplished this level of control by adjusting the draft entering the furnace with a small potsherd. The experiment effectively showed that the Middle Sicán furnaces made with inverted urns were highly efficient and flexible devices well suited for a variety of metalworking tasks. Additional details of the experimental setups and the results of an archaeometric analysis of fired vessels and the kiln lining are described elsewhere (Shimada and Wagner 2001, 2007; Shimada et al. 2003a; Shimada et al. 2003b; Shimada et al. 2003c).

CHARCOAL FURNACES

Work completed by Goldstein (2002), observations by Shimada (1985), and recent work by Cayo (2003) demonstrate that charcoal can be produced in quantity at the household or small scale. Goldstein (2002) observed in the field that one person could collect easily 500 kg of *Prosopis pallida* and construct a small charcoal kiln to produce around 250 kg of hardwood charcoal. The labor investment to set the kiln and collect the wood requires a single day for one person, and the firing takes four days. The labor investment and productivity of these small-scale furnaces depend on the quantity of wood used, but this work demonstrates the ease with which charcoal can be prepared. The wood is arranged in a set of parallel lines on an area of ground cleared of loose soil and debris. Once arranged, a layer of *poña*, or kindling, collected from the surrounding area is laid on top of the logs. Poña consists of leaves, refuse, small twigs, and animal dung gathered from below the trees in the area surrounding the kiln. The entire pile is covered with compact peds of the soil dug from the perimeter of the piled logs and kindling. Hardpan clods are preferred, and loose dirt is saved for the last, uppermost covering of the kiln above the clods. Overall, the processing of 500 kg of wood into charcoal results in a pile 1.5 m high with a surface area that covers 4 × 6 m. A small area is left open at the downwind side where a small initial fire is started, called the *caja*. Once the caja catches fire, it too is covered, and the fire is left for several days to burn through the entire pile. During the second day, small vents are made at the end of the kiln opposite to the caja by inserting long straight sticks about 1 cm in diameter to let the smoke and gases escape. When the firing is complete and successful, the charcoal makers carefully excavate the pile, and the carbonized branches are removed intact, often in single pieces.

Processes such as these may have generated the features excavated at the very northern end of Mound I. In 1999, in a series of excavated test pits, we encountered at least two pits reaching a depth of 70 cm below the surface that contained an estimated 30–40 liters or more of seemingly fresh charcoal lumps (i.e., 1–2 cm in diameter). In light of the ethnoarchaeological correlate for small-scale charcoal making, these pits could be the remains of the perimeter trenches that correlate to dirt clod removal in modern charcoal kiln construction. Their contents may be explained by leveling of the area after the kiln was dismantled, during which the dregs from the charcoal kiln ended up as perimeter pit infill. There is no doubt that ash and burned soil mixed with some small charcoal lumps were discarded in this sector. In particular, we recovered numerous, superimposed lenses containing such mixtures from the NE corner of Mound I. The depth and quality of the charcoal recovered from these features may be the remains of local, small-scale charcoal furnaces associated with metal production at Huaca Sialupe.

PALEOETHNOBOTANICAL ANALYSIS AND INTERPRETATION

Many of the metalworking furnaces contained several liters of carbonized fuel remains. In some cases, the lack of ash and the relatively full furnaces indicate that these were charged for use just before abandonment (Ursel Wagner, personal communication regarding charcoal quality as revealed by scanning electron microscopy, Chiclayo, Lambayeque, July 2001). The furnaces provide a detailed record of collected fuel materials used for metal fabrication. Of the carbonized materials recovered, 80 percent is wood charcoal from hardwood trees from the Fabaceae family. The small and consistent diameter size, 2–6 cm, together with their anatomical characteristics, indicates that this sample represents wood from small branches. There is little or no indication that large pieces of trunk or primary branch wood were used for fuel.

In comparison, the charcoal recovered from the ceramic kilns contains notably smaller sized twigs than those seen in the metal furnaces, most representing branches around 0.5

Table 2.2. Summary of Contexts Sampled at Huaca Sialupe and Their Contents

Feature Type	Blackware Ceramic Kilns (*n*=13)	Preheating Pits (*n*=5)	Metalworking Furnaces (*n*=8)	Domestic Hearths (*n*=25)
Principal Fuels	Small Twigs, Household Garbage	Small Herbaceous Materials and Twigs	*Capparis* spp. and Legume Branchwood (e.g., *Prosopis* spp., *Acacia* spp., *Cercidium* spp.)	*Capparis* spp. and Legume Branchwood, Herbaceous Materials, and Household Garbage
Average Diameter of Carbonized Twig	0.5 cm< x < 2.0 cm	0.5 cm< x < 1.5 cm	2.0 cm< x < 6.0 cm	0.5 cm< x < 8.0cm
Degree of Combustion	Incomplete/ Cleaned with Small Carbonized Twigs Remaining (Except for #68, see discussion)	Incomplete/ Small Twig Charcoal Remains/Ash	Complete/ Some Charcoal Leftover	Near Complete/ Ash

cm and no larger than 2 cm in diameter (Table 2.2). Although our analysis at this point is more qualitative than quantitative, the majority of the remains from these features indicate a mixture of hardwoods from the family Fabaceae with additional other local woods. There are few remains of large pieces of hardwood in any of the kilns. The lack of large pieces of charcoal, ash, and overall limited remains, however, may well indicate that the kilns were cleaned prior to abandonment. In fact, one well-preserved kiln, Feature 68 in the floor of Room 39 (Figure 2.14), roughly contemporaneous with the aforementioned metalworking furnaces, is an exception in that the residual fuel load was present at excavation. This is the only kiln located during our excavations that had a remaining fuel load and may indicate that the Sialupe artisans did not clean this kiln following firing.

Hearths constitute the third set of fire use features systematically sampled at Sialupe. These are mostly ephemeral features without solid enclosing structures. The location and morphology of these features lead us to believe that they were ancillary to industrial production and related to the domestic economy of the site. Their insubstantial nature permits the following observations regarding domestic fire use relative to craft production. First, hearths are small, circa 15 cm in diameter and rarely larger than 30 cm. Burning tended to be complete, leaving a shallow ashy lens and a visibly smaller quantity of charcoal when compared with the production features. It is noteworthy that the quality of the wood charcoal is different from that seen in the furnaces and kilns. We see no apparent size grading as present in the other contexts. Instead, hearth charcoal spans a wide range of sizes representing the smallest twigs to remnants of large branch or trunk wood. Additionally, it appears that a wider range of taxa is represented to include families outside the Fabaceae.

The Leguminous hardwood category constitutes at least six different species from the family Fabaceae with extremely similar anatomies from three different genera: *Prosopis* spp., *Acacia* spp., and *Cercidium* sp. The most important of these genera are known today as algarrobo, *faique*, and *palo verde*, respectively, in the region and along the northern Peruvian coast. We present these three genera together as a combined category of taxa for this

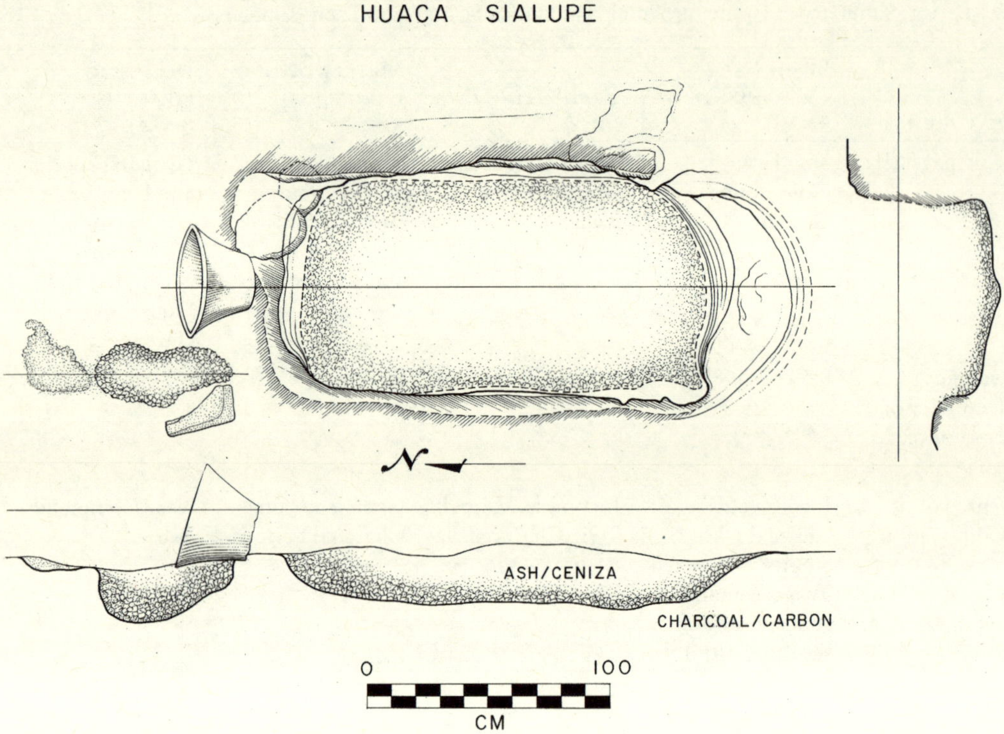

Figure 2.14. Kiln in the ceramic firing area that preserved its residual fuel load, Feature 68, Room 39, Mound I. Drawing by C. Samillán and I. Shimada.

analysis because of their extremely similar wood anatomies and a lacuna in the descriptions of local wood anatomies. Two species in the family Capparidaceae, *Capparis angulata* and *C. ovalifolia*, represent the two nonlegume species most relied on. All of these tree species are indigenous to the area, with no evidence for fuel importation or intensive cropping of fuelwoods.

Seeds are present in most of the deposits. We can classify these seed types into three complexes of herbaceous plants, crop plants, and potential fodder species. The herbaceous plants include seeds and leaves from the families Gramineae, Compositae, Verbenaceae, Chenopodiaceae, and Boraginaceae. Crop plant macroremains include *Zea mays*, maize or corn, and *Oxalis tuberosum*, locally called *oca*. The remains of animal fodders include leaves, pods, and seeds from local trees that are known in the colonial past and the present to be heavily relied on for grazing: *Prosopis* spp., *Acacia* spp., and *Parkinsonia aceulata*, known as *cabelludo*, *palo verde*, or *corona de cristo*. The presence of these herbaceous and food species in the hearth contexts is somewhat predictable; however, some of the fodder and crop macroremains also occur in the production contexts. These seed and fodder materials probably entered the industrial contexts through the secondary deposition of domestic garbage as fuel or the use of camelid dung as fuel. For instance, in both the replication experiments of Middle Sicán black pottery firing performed by José Sosa (Shimada and Wagner 2001, 2007, in press) and the Formative period ceramic firing by Victor Chang (Shimada et al. 1998), the potters used dry herbaceous plants and tree leaves as supplementary tinder to aid hardwood in catching fire.

The majority of the wood used in the kilns and furnaces presents signs of heavy insect activity. The channels that riddle the wood structure of even the smallest twigs are between 1 and 3 mm across and are full of pulverized sawdust-like frass. Exit holes and channeling below the bark are present. This damage is related to boring beetles. Members of the Burpestidae or Scylotidae beetle families are likely candidates based on the morphology of the damage (Borror and White 1970; Rogers 2000). These beetles feed on either sick, felled, or dead trees. We do not know if the wood gathered for use at Huaca Sialupe was alive during infestation. The thoroughness of consumption coupled with the absence of tool marks on the examined pieces, however, strongly argue for postmortem infestation. The patterning of the infestation is similar to that on modern samples of dead and fallen branch wood gathered in the field and used in the firing experiments discussed above. The nature of the insect activity and the apparently low representation of carbonized mature tree parts indicate that gathered fuelwood from the forest floor was likely the prime fuel source in all three types of fire use features.[4]

CONCLUSIONS AND IMPLICATIONS FOR FUTURE MULTICRAFT STUDIES

Smithing associated with furnaces requires a stable long-lasting heat source that yields consistent, high temperatures, making charcoal best suited for the purpose.[5] We assumed previously that Sicán artisans converted wood to charcoal elsewhere and transported it to Huaca Sialupe for use in the metal furnaces (Shimada and Merkel 1991; Shimada and Wagner 2001). Based on the present evidence we believe that metalworkers at the Sialupe workshop utilized at least two strategies to acquire charcoal for their metalworking furnaces. One may have been on-site small-scale charcoal production, and another may have been the complementary use of the charcoal remaining from the reduction firing involved in blackware production.

First, we find along with the wood charcoal a complex of other burned organic materials in the furnaces. Modern charcoal production studied in the Batán Grande region of the La Leche Valley by Goldstein indicates that the collection of poña, kindling materials used for the initial combustion of the charcoal furnace, involves gathering leaves and garbage scattered throughout the local settlement. This kindling incorporates leaves, textiles, sherds, the husks of fruits, small twigs, and in some cases animal dung. The Sialupe samples contain textile fragments, the fruits and seeds of fodder plants, grass leaves, tuber parts, corncobs, and animal dung. These materials are well mixed into the charcoal found within the archaeological furnaces. From these remains and the modern analogue it is conceivable that charcoal was locally produced at the Sialupe workshop to provide a steady supply of charcoal on an "as needed" basis. As our experimental work showed, charcoal production is not inherently difficult; nor does it require large labor or hardwood supplies (Food and Agricultural Organization 1983).

Second, our ceramic firing experiments and excavation data paint a more complex picture for charcoal procurement. Various lines of evidence taken together indicate that fuelwood was conserved for the firing of ceramics, and hardwood *charcoal* was apportioned more for metalworking. Our ceramic firing experiments described above demonstrated that the reduction firing of a single kiln could yield up to 25 liters of charcoal after each firing.

Yet the majority of excavated kilns were clean of ash or large pieces of charcoal. A metalworking furnace charged with 20 liters of charcoal could function at its full efficiency given our replication experimental results. The lack of large charcoal pieces in the archaeological kilns may indicate that the residual charcoal was removed from kilns after firing and used elsewhere. This inferred cleaning and reuse might account for the apparent discrepancy in the size grades of wood seen between the excavated kilns and furnaces. We interpret these observations as indications that a complementary technological innovation between metalworking and blackware ceramic production involved the reuse of carbonized branch wood from the kilns as a source for charcoal for the furnaces.

We began this essay by noting Matson's (1989) inference regarding ceramic and metal craft interaction and the parallel development of ceramic kilns and metallurgical furnaces with regard to fuel procurement. Independently, our earlier investigations of 3,000-year-old Cupisnique ceramic production in the La Leche Valley and 1,000-year-old Middle Sicán arsenical copper smelting alerted us to a series of intriguing material and organizational parallels and possible cross-fertilization between these two crafts (Shimada et al. 1998: 56–58). In addition to obvious similarities in regard to the basic designs, construction materials, and fuels of ceramic kilns and smelting furnaces, both crafts were characterized by what Shimada and Wagner (2007) call the modular organization of production, which involves largely self-contained artisans or groups of artisans working in close proximity yet largely independently to produce a similar or essentially identical array of products. The same artisans were involved in much, if not the entirety, of the production process. This approach is different from assembly-line production organization, typically manned by specialized workers under close supervision, each having a specific task and station along a lineally organized production sequence and layout (Shimada and Wagner 2007). This organization principle is in stark contrast to the craft production and public construction we document from the Cupisnique through the Moche, Sicán, and Chimú periods.

Building on the above understanding of the traditions and characteristics of regional craft production, our recent work at Huaca Sialupe illuminates the complex relationship between ceramic and metal production at that workshop during the Middle Sicán. Although we could assert that three distinct features, metal furnaces, ceramic kilns, and domestic fire areas, represent three classes of activity, it is evident that these three contexts for fire use had a well-developed interplay and are difficult to tease apart when we attempt to reconstruct fuel use by local artisans working at Sialupe. We suggest that these apparently distinct uses of fire (i.e., ceramic firing, metalworking, and domestic) were integrated at the artisan level through the horizontal interaction of technology and expertise.

To achieve and maintain the over 800°C temperatures necessary for the firing of black pottery, hardwood fuel was essential. On the other hand, the elegantly simple and efficient furnaces, together with strong natural drafts, gave metalworkers the flexibility to use charcoal fuels from a variety of sources. What we reflect on here is that fuel quality may not have been a limiting factor to Sicán artisans; instead, multi-crafting may have been lucrative and sustainable because of complementary use, where one person's residual was another's valuable fuel.

Hence, rather than viewing fuel use as being overdetermined by Middle Sicán sponsorship and the differential value of expected products, the selection and apportionment of different fuels were influenced by experience-based needs and left in the hands of local metalworkers and potters. These groups of artisans who labored close to each other are likely to have directly negotiated their fuel usage. The manner in which potters and metalworkers used fuels as reconstructed in the present study accords well with the picture that has emerged from independent analysis of styles and techniques of ceramic manufacture (Shimada and Wagner 2007). In essence, these studies serve as an invaluable indicator of innovation, autonomy, and economy in the Middle Sicán.

This essay demonstrates the importance of context in studying oft-ignored production debris as items worthy of investigation, particularly what remain of fuels. Such studies effectively illuminate the constraints and potential of a given environment on a wide variety of pyrotechnologies and the creative human solutions to them. Finally, our case study serves as a demonstration of the productivity of combining multiple analytical avenues with the excavation of production sites to achieve a comprehensive reconstruction of production process and technology, what we call a holistic understanding.

Acknowledgments.

The 1999 and 2001 fieldwork at Huaca Sialupe was supported by grants to Izumi Shimada from the H. John Heinz III Fund for Latin American Studies, the Committee for Research and Exploration of the National Geographic Society (Grant Nos. 6451-99 and 7039-01), and the Wenner-Gren Foundation for Anthropological Research (Grant No. 6877). The 2000 field experiments were made possible by a Southern Illinois University Faculty Research grant to Izumi Shimada. National Science Foundation Dissertation Improvement Grant #0211364 provided the support for David Goldstein's 2002 field research. His analysis was supported by a doctoral fellowship from the Graduate School at Southern Illinois University at Carbondale. Archaeometric analyses by Ursel Wagner were supported by a research grant from the German Research Council. Additional archaeometric analyses and consultation were offered by Aniko Bezúr, Department of Conservation, Buffalo State College. We are indebted to members of the 1999, 2000, and 2001 seasons of the Sicán Archaeological Project for their assistance in data and sample collection, as well as the staff at the Museo Nacíonal Sicán in Ferreñafe, Peru. The participation of César Samillán and the potter José Sosa in our 2000 field furnace and kiln-firing experiments was invaluable. Last, we thank Melody Shimada for her valuable editorial assistance. We are solely responsible for any errors or oversights that may have escaped our attention.

Notes

1. There are at least three genera of *Prosopis* spp. known in the research area: *P. pallida*, *P. juliflora*, and *P. affinis*. Present research worldwide continues to have difficulty classifying these species (Pasiecznik et al. 2001). Here we use *Prosopis pallida* as the most general species and *Prosopis* spp. as a generic term that encompasses the three species.
2. Other potential economic activities certainly competed for fuel resources, e.g., construction; this chapter does not address wood resource use apart from fuel.
3. Hardwood converts to charcoal at a weight ratio of approximately 6 to 1 (Horne 1982). Our data from the experimental kiln (2000) indicate a hardwood to charcoal ratio of 5.5:1. Thus, the 13,000 kg of charcoal that would be consumed to smelt 100 kg of arsenical copper would represent 71,500–78,000 kg (71.5–78 metric tons) of hardwood. This estimate effectively highlights the potential impact of intensive metallurgical production on local forests.
4. Gathered materials are the common source for traditional metal workshops in modern western Asia as documented in the ethnographic record (Matson 1966).
5. To date there is no evidence that Sicán pyrotechnologies at Sialupe or any other Sicán sites utilized lignite or other coal minerals.

References Cited

Arnold, Dean E.
1985 *Ceramic Theory and Cultural Process*. Cambridge University Press, New York.

Bennett, Wendell C., and Junius Bird
1949 *Andean Culture History*. American Museum of Natural History, Garden City, N.Y.

Borror, Daniel J., and Richard E. White
1970 *Peterson Field Guide to Insects*. Houghton Mifflin Co., Boston.

Bullock, S. H., H. A. Mooney, and E. Medina
1995 *Seasonally Dry Tropical Forests*. Cambridge University Press, Cambridge.

Cayo, Ludwing
2003 Informe: Proceso de Elaboracíon de Carbón en el Departmento de Potosí. Proyecto Arqueológico Porco-Potosí. Report on file at Colorado State University, Dr. Mary Van Buren, Colorado Springs.

Cleland, Kate M., and Izumi Shimada
1994 Cerámica Paleteada: Tecnología, Modos de Producción y Aspectos Sociales. In *La Tecnología y la Organización de Producción de las Cerámicas en los Andes Prehispánicos*, edited by Izumi Shimada, pp. 321–348. Fondo Editorial, Pontificia Universidad Católica del Perú, Lima.
1998 *Paleteada* Pottery: Technology, Chronology and Sub-Culture. In *Andean Ceramics: Technology, Organization and Approaches*, edited by I. Shimada, pp. 111–150. MASCA Research Papers Vol. 15. MASCA, University of Pennsylvania Museum for Archaeology and Anthropology, Philadelphia.

Craig, Alan K.
1985 Cis-Andean Environmental Transects: Late Quaternary Ecology of Northern and Southern Peru. In *Andean Ecology and Civilization*, edited by Shozo Masuda, Izumi Shimada, and Craig Morris, pp. 23–44. University of Tokyo Press, Tokyo.

Craig, Alan K., and Izumi Shimada
1986 El Niño Flood Deposits at Batán Grande, Northern Peru. *Geoarchaeology* 1(1):29–38.

Cuba Salerno, A. B., A. Silva Peralta, and C. Cornejo Flores (editors)
1998 *Bosques Secos y Desertificación*. Instituto Nacional de Recursos Naturales, Lima.

Delvaud, Claude C.
1984 *Las Regiones Costeñas del Perú Septentrional*. Centro de Investigación y Promoción del Campesinado/Pontificia Universidad Católica del Perú, Lima.

Epstein, Stephen M.
1993 Cultural Choice and Technological Consequences: Constraints of Innovation in the Late Prehistoric Copper Smelting Industry of Cerro Huaringa, Peru. Unpublished Ph.D. dissertation, Department of Anthropology, University of Pennsylvania, Philadelphia.

Epstein, Stephen M., and Izumi Shimada
1983 Metalurgia de Sicán: Una Reconstrucción de la Producción de la Aleación de Cobre en El Cerro de los Cementerios, Peru. *Beiträge zur Allgemeinen und Vergleichenden archäologie* 5:379–430.

Ferreyra, R.
1987 *Estudio Sistemático de los Algarrobos de la Costa Norte del Perú*. Ministerio de Agricultura, Lima.

Food and Agricultural Organization
1983 *Simple Technologies for Charcoal Making*. Food and Agricultural Organization of the United Nations, Rome.

Goldstein, David J.
2002 Small-Scale Charcoal Making in La Zaranda, Lambayeque, Peru. MS on file, Department of Anthropology, Southern Illinois University, Carbondale.

Harker, A. P., A. Sandels, and J. Burley
1982 *Calorific Values for Wood and Bark and a Bibliography for Fuel Wood*. Tropical Products Institute Technical Report G 162. Overseas Development Association, London.

Hayashida, Frances M.
1999 Style, Technology, and State Production: Inka Pottery Manufacture in the Leche Valley, Peru. *Latin American Antiquity* 10(4):337–352.

Horne, Lee
1982 Fuel for the Metal Worker. *Expedition* 3:6–13.

Lechtman, Heather N.
1976 A Metallurgical Site Survey in the Peruvian Andes. *Journal of Field Archaeology* 3:1–42.

Luna-Suárez, S., M. L. Luna-Guido, J. T. Frias-Hernández, V. Olalde-Portugal, and L. Dendooven
1998 Soil Processes as Affected by Replacement of Natural Mesquite Ecosystem with Maize Crop. *Biology of Fertilized Soils* 27:274–278.

Martin, Alexander C., and William D. Barkley
1961 *Seed Identification Manual*. University of California Press, Berkeley.

Matson, Frederick R.
1966 Power and Fuel Resources in the Ancient Near East. *The Advancement of Science* 23(10):146–153.
1989 Ceramics: The Hub of Ancient Craft Interplay. In *Cross-Craft and Cross-Cultural Interactions in Ceramics*, edited by P. E. McGovern, pp. 13–28. Ceramics and Civilization Vol. 4. American Ceramic Society, Westerville, Ohio.

Merkel, John, Izumi Shimada, C. P. Swann, and R. Doonan
1994 Pre-Hispanic Copper Alloy Production at Batán Grande, Peru: Interpretation of the Analytical Data for Ore Samples. In *Archaeometry of Pre-Columbian Sites and Artifacts*, edited by D. A. Scott and P. Meyers, pp. 199–227. Getty Conservation Institute, Marina del Rey, California.

Ocampo, M.
1991 *Metodología para la Determinaciín de la Edad y las Tasas de Crecimiento para las Especies Forestales de los Bosques Secos de la Costa Norte del Perú*. Proyecto Algarrobo con Fundación para el Desarrollo Nacional, Chiclayo.

Pasiecznik, N. M., P. Felker, P. J. C. Harris, L. N. Harsh, G. Cruz, J. C. Tewari, K. Cadoret, and L. J. Maldonado
2001 *The* Prosopis juliflora–Prosopis pallida *Complex: A Monograph*. Henry Doubleday Research Association, Coventry, England.

Pedersen, Asbjorn
1976 El Ajuar Funerario de la Tumba de la Huaca Menor de Batán Grande, Lambayeque, Perú. *Actas del 41er Congreso Internacional de Americanistas* 2:60–73.

Proyecto Algarrobo
1997 *Inventorio Forestal de los Bosques Secos*

de *Batán Grande*. Universidad Nacional Agraria la Molina, Lima.

Ramírez, Susan E.
1996 *The World Turned Upside Down: Cross Cultural Contact and Conflict in Sixteenth-Century Peru*. Stanford University Press, Stanford.

Risco, J. J.
1998 Tecnologías Locales Mejoradas en el Uso de los Rescuros del Bosque Seco en un Contexto Familiar: Cocinas Mejoradas y Algarroberas. In *Bosques Secos y Desertificación*, edited by A. B. Cuba Salerno, A. Silva Peralta, and C. Cornejo Flores, pp. 129–138. Instituto Nacional de Recursos Naturales, Lima.

Rogers, Ken E.
2000 *The Magnificent Mesquite*. University of Texas Press, Austin.

Rostworowski de Diez Canseco, María
1981 *Recursos Naturales Renovables y Pesca Siglos XVI y XVII*. Instituto de Estudios Peruanos, Lima.

Sawyer, Alan R.
1968 *The Mastercraftsmen of Ancient Peru*. S. R. Guggenheim Foundation, New York.

Shimada, Izumi
1978 Economy of Prehistoric Urban Context: Commodity and Labor Flow in Moche V Pampa Grande, Peru. *American Antiquity* 43:569–592.
1981 The Batán Grande–La Leche Archaeological Project: The First Two Seasons. *Journal of Field Archaeology* 8:405–446.
1982 Horizontal Archipelago and Coast–Highland Interaction in North Peru: Archaeological Models. In *El Hombre y su Ambiente en los Andes Centrales*, edited by Luis Millones and Hiroyasu Tomoeda, pp. 185–257. National Museum of Ethnology, Suita.
1985 Perception, Procurement and Management of Resources: Archaeological Perspective. In *Andean Ecology and Civilization*, edited by Shozo Masuda, I. Shimada, and Craig Morris, pp. 357–399. University of Tokyo Press, Tokyo.
1990 Cultural Continuities and Discontinuities on the Northern North Coast, Middle–Late Horizons. In *The Northern Dynasties: Kingship and Statecraft in Chimor*, edited by Michael E. Moseley and A. Cordy-Collins, pp. 297–392. Dumbarton Oaks, Washington, D.C.
1994a *Pampa Grande and the Mochica Culture*. University of Texas Press, Austin.
1994b Prehispanic Metallurgy and Mining in the Andes: Recent Advances and Future Tasks. In *In Quest of Mineral Wealth: Aboriginal and Colonial Mining and Metallurgy in Spanish America*, edited by Alan K. Craig and Robert West, pp. 37–73. Geoscience and Man 53. Louisiana State University, Baton Rouge.
1994c The Role of Metals in Middle Sicán Society. In *The Illustrated Encyclopedia of Humankind, vol. 4: New World and Pacific Civilizations*, edited by Goran Burenhult, pp. 94–95. Weldon Owen Pty Limited, Sydney.
1995 *Cultura Sican: Dios, Riqueza y Poder en la Costa Norte del Perú*. Fundación del Banco Continental para el Fomento del la Educación y la Cultura, Edubanco, Lima.
1996 Sicán Metallurgy and Its Cross-Craft Relationships. *Boletín del Museo del Oro* 41:27–61.
2005 Experimental Archaeology. In *Handbook of Theories and Methods in Archaeology*, edited by Herbert D. G. Maschner and Christopher Chippindale, pp. 603–642. AltaMira Press, Walnut Creek, California.

Shimada, I. (editor)
1994d *La Tecnología y la Organización de Producción de las Ceramicas en los Andes Prehispánicos*. Fondo Editorial, Pontificia Universidad Católica del Perú, Lima.
1998 *Andean Ceramics: Technology, Organization and Approaches*. MASCA Research Papers Vol. 15. MASCA, University of Pennsylvania Museum for Archaeology and Anthropology, Philadelphia.

Shimada, Izumi, Victor Chang, Ursel Wagner, Rupert Gebhart, Hector Neff, Michael Glascock, and David Killick
1998 Formative Ceramic Kilns and Production in Batán Grande, North Coast of Peru. In *Andean Ceramics: Technology, Organization, and Approaches*, edited by Izumi Shimada, pp. 23–62. MASCA Research Report Vol. 15. MASCA, University of Pennsylvania Museum for Archaeology and Anthropology, Philadelphia.

Shimada, Izumi, Stephen M. Epstein, and Alan K. Craig
1982 Batán Grande: A Prehistoric Metallur-

gical Center in Peru. *Science* 216:952–959.

Shimada, Izumi, David J. Goldstein, Werner Häusler, José Sosa, and Ursel Wagner

2003a Early Pottery Making in Northern Coastal Peru: Part II: Field Firing Experiments. In *Mössbauer Spectroscopy in Archaeology*, edited by U. Wagner, pp. 91–105. Hyperfine Interactions 150(1–4). Kluwer Publishing, New York.

Shimada, Izumi, David J. Goldstein, Ursel Wagner, and Anikó Bezúr

In press Pre-Hispanic Sicán Furnaces and Metalworking: Toward a Holistic Understanding. *Boletín del Instituto Frances de Estudios Andinos (Lima).*

Shimada, Izumi, Adon Gordus, and Jo Ann Griffin

2000 Technology, Iconography, and Significance of Metals: A Multi-Dimensional Analysis of Middle Sicán Objects. In *Pre-Columbian Gold: Technology, Iconography, and Style*, edited by Colin McEwan, pp. 28–61. British Museum Press, London.

Shimada, Izumi, Adon Gordus, Jo Ann Griffin, and John Merkel

1999 Sicán Alloying, Working and Use of Precious Metals: An Interdisciplinary Perspective. In *Metal in Antiquity*, edited by S. M. M. Young, A. M. Pollard, P. Budd, and R. A. Ixer, pp. 301–309. BAR International Series 792. Archaeopress, Oxford.

Shimada, Izumi, and Jo Ann Griffin

1994 Precious Metal Objects of the Middle Sicán. *Scientific American* 270(4):60–67.

Shimada, Izumi, Werner Häusler, Th. Hutzelmann, and U. Wagner

2003b Early Pottery Making in Northern Coastal Peru: Part I: Mössbauer Study of Clay. In *Mössbauer Spectroscopy in Archaeology*, edited by U. Wagner, pp. 73–89. Hyperfine Interactions 150(1–4). Kluwer Publishing, New York.

Shimada, Izumi, and John Merkel

1991 Copper Alloy Metallurgy in Ancient Peru. *Scientific American* 265(1):80–86.

Shimada, Izumi, Jorge A. Montenegro, Werner Häusler, M. Jakob, J. Riederer, and U. Wagner

2003c Early Pottery Making in Northern Coastal Peru: Part IV: Mössbauer Study of Pottery from Huaca Sialupe. In *Mössbauer Spectroscopy in Archaeology*, edited by U. Wagner, pp. 125–139. Hyperfine Interactions 150(1–4). Kluwer Publishing, New York.

Shimada, Izumi, Crystal B. Schaaf, Lonnie G. Thompson, and Ellen Mosley-Thompson

1991 Cultural Impacts of Severe Droughts in the Prehistoric Andes: Application of a 1,500-Year Ice Core Precipitation Record. *World Archaeology* 22(3):247–270.

Shimada, Izumi, and Ursel Wagner

2001 Peruvian Black Pottery Production and Metal Working: A Middle Sicán Craft Workshop at Huaca Sialupe. *Materials Research Society Bulletin* 26(1):25–30.

2007 Craft Production on the Pre-Hispanic North Coast of Peru: A Holistic Approach and Its Results. In *Archaeology as Anthropology: Theoretical and Methodological Approaches*, edited by James Skibo, Michael Grave, and Meriam Stark, pp. 163–197. University of Arizona Press, Tucson.

In press Tecnología y Organización de la Producción de Cerámica Negra en la Costa Norte del Perú. *Corriente Arqueológico* 2.

Sillar, Bill

2000 Dung by Preference: The Choice of Fuel as an Example of How Andean Pottery Production Is Embedded within Wider Technical, Social, and Economic Practices. *Archaeometry* 42(1):43–60.

Taylor, Sarah

2000 Artisan Autonomy in the Middle Sicán State: Variability in Mold-Made Ceramic Production. Unpublished M.A. thesis, Department of Anthropology, Southern Illinois University, Carbondale.

Tosi, J. A., Jr.

1960 *Zonas de Vida Natural en el Perú: Memoria Explicativa sobre el Mapa Ecológico del Perú.* Boletín Técnico No. 5. Instituto Interamericano de Ciencias Agricolas de Organización de Estados Americanas, Zona Andina, Lima.

Tschauner, Hartmut

2001 Socioeconomic and Political Organization in the Late Prehispanic Lambayeque Sphere, North Coast of Perú. Unpublished Ph.D. dissertation, Department of Anthropology, Harvard University.

Wagner, Ursel, R. Gebhart, E. Murad, Josef Riederer, Izumi Shimada, C. Ulbert, and Fritz E. Wagner

1998 Production of Formative Ceramics: Assessment by Physical Methods. In *Andean Ceramics: Technology, Organization, and Approaches*, edited by Izumi Shimada, pp. 173–198. MASCA Research Report Vol. 15. MASCA, University of Pennsylvania Museum for Archaeology and Anthropology, Philadelphia.

Wertime, Theodore A.
1964 Man's First Encounters with Metallurgy. *Science* 146(3649):1257–1267.

Wheeler, Elisabeth A., and Pieter Baas
1998 Wood Identification—A Review. *IAWA Journal* 19(3):241–264.

Winterhalder, Bruce, R. Larsen, and R. B. Thomas
1974 Dung as an Essential Resource in a Highland Community. *Human Ecology* 2(2):89–104.

3

A Community of Potters or Multicrafting Wives of Polygynous Lords?

WILLIAM H. ISBELL

Who were the craftspeople of the past, and how did producing crafts change their lives and identities? There can be no question that the development of crafts underlies the enhancement and diversification of material culture, from tools to clothing and symbols. On the production side crafts encourage activity specialization, the development of expertise, and differentiation in products that, in turn, increase economic interdependence and cultural complexity. On the consumption side they promote a more complicated material landscape characterized by new opportunities and choices, sometimes presenting novel cost/benefit considerations that can amplify social difference by increasing wealth, contributing to new identities and legitimizing inequality. The role of crafts comes full circle in evolving cultures as new craft producers become new kinds of people who require new lifestyles and the construction of new identities, in new ways, employing new material culture.

Archaeologists must account for crafts and services as they model the evolution of complexity in archaic cities and states. But they must be careful that their accounts are empirically based and not ideal types derived from evolutionary theories or analogies that project modern social formations into the remote past. In this article I explore a case that was originally conceptualized as a prehistoric community of artisans, earning its livelihood by producing pottery for trade. These craftspeople seemed obvious examples of economically rational agents in the past. However, more extensive excavations have revealed pottery production in a broader context, telling us something different about this ancient community, the motivation for crafts, the multicrafting combination of activities, the people engaged in production, the role of gender, and the way crafts transformed peoples' lives and identities.

THE CASE STUDY OF CONCHOPATA, PERU

Conchopata was a small, Middle Horizon (A.D. 650–1000) city whose ruins are located in the Valley of Ayacucho in the south-central highlands of Peru, on the edge of the modern provincial capital also named Ayacucho (Figure 3.1). Since the 1950s portions of Conchopata have been destroyed by urban expansion, beginning with an airport (Lumbreras 1969, 1974, 1981). By 1998, of a city estimated to have originally encompassed some 20 to 40 ha (Isbell 1987, 2001), about 2.5 ha of the site remained more or less intact. In desperation, that year a group of archaeologists banded together to rescue as much of Conchopata as possible, through emergency excavations.[1] Those excavations provide the basis for the reevaluation of pottery production at Conchopata.

In the 1990s Conchopata was considered one of the better-documented examples of

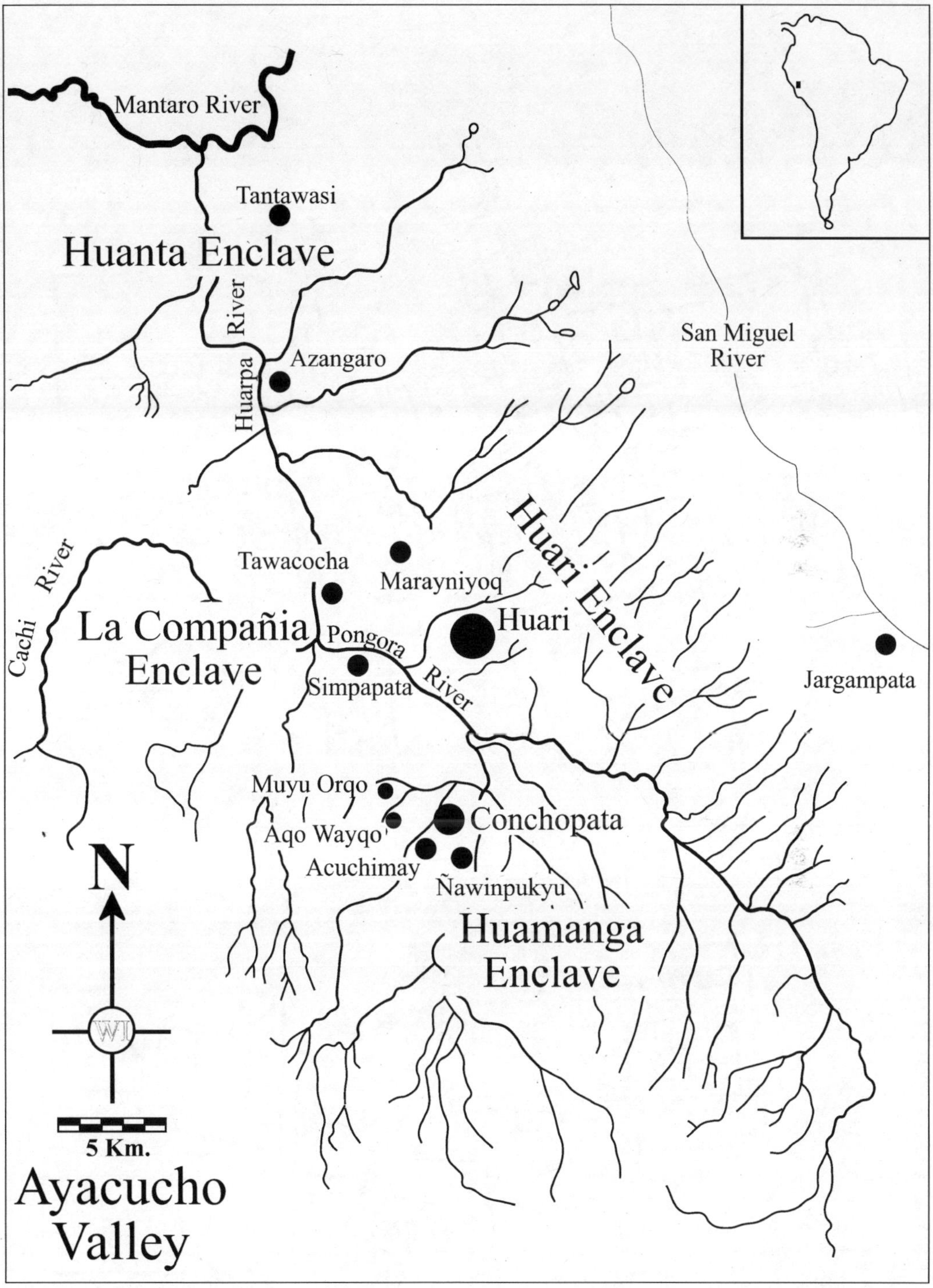

FIGURE 3.1. Map of the Ayacucho Valley.

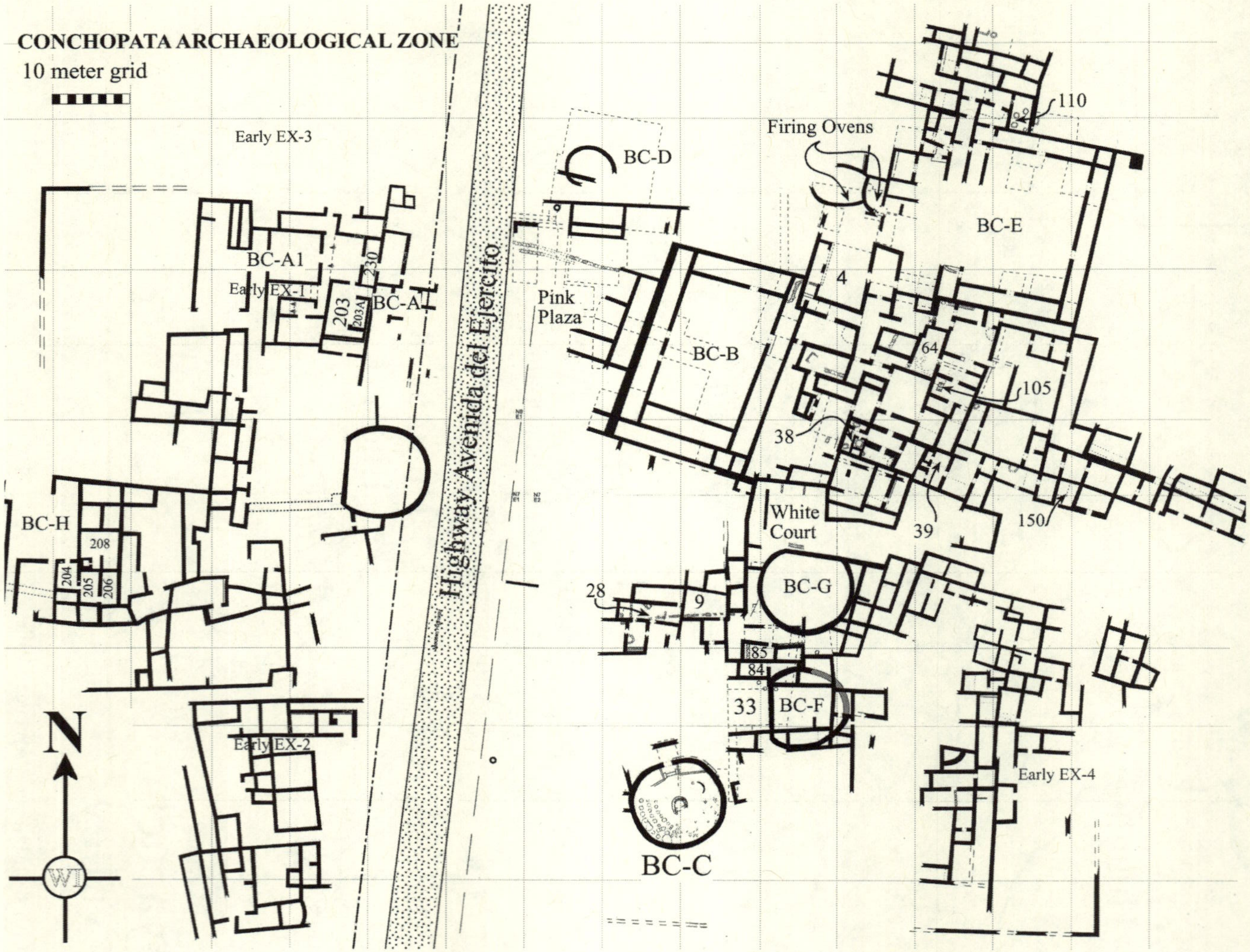

FIGURE 3.2. Map of the Conchopata site.

prehistoric pottery production in the Andes. Publications describe tools, technology, and firing ovens. However, little of the site had been excavated, so interpretations were based on small exposures, on analyses of tools and artifacts, and on ethnographic analogies.

In the 1980s, the first area excavations at Conchopata were conducted by Denise Pozzi-Escot B. (1982, 1985, 1991, 2001; Pozzi-Escot B. et al. 1994, 1998, 1999). She dug in two locations (Figure 3.2, areas labeled Early EX-1 and Early EX-2). By current standards her units were small and shallow, exposing mostly wall tops and only a few buildings to their floors. But she found thousands of pottery-making materials.

Pozzi-Escot employed ethnoarchaeology to show that the tools and raw materials she had excavated were similar to those employed by potters of the modern village of Quinoa, about 15 km from Conchopata. Quinoa potters explained how many of the tools were used and how they fit into a sequence of production. Pozzi-Escot B. (1991) affirms that almost every architectural space she excavated had tools and other pottery-manufacturing materials, including clay and pigments, scraper/polisher tools (Figure 3.3A), potters' plates (Figure 3.3B), paddles and anvils for shaping vessels (Figure 3.3C), positive and negative molds for effigy vessels and figurines (Figure 3.3D–F), and hoes for digging and mixing clay.[2] Astonishing quantities of broken pottery seemed to document production on a grand scale, for trade.

Conchopata was inferred to have been a "community of potters" much like modern Quinoa, except that instead of the modern market economy it participated in the complex prehistoric exchange system of the Wari Empire. Because the frequency of metal artifacts, including silver and gold objects, was high at Conchopata, archaeologists went on to argue that metals must also have been worked, although direct evidence was not found. However, ceramic manufacture and metallurgy are both fire-based crafts, so co-occurrence seemed reasonable (see Goldstein and Shimada, this volume), especially in a "community of potters" making economically motivated decisions and unlikely to be able to afford such expensive goods.

Following Pozzi-Escot's research, Ismael Pérez and José Ochatoma (Pérez Calderón 1998; Pérez Calderón and Ochatoma Paravicino 1998; Ochatoma Paravicino 1998; Ochatoma Paravicino and Cabrera Romero 2000) excavated ceramic firing ovens in other sectors of Conchopata (Figure 3.2, Early EX-3 and Early EX-4). Large firing features were located within architectural compounds that appeared to have housed sizable residential groups, so Conchopata potters' workshops were also homes shared with extended families. Combined with Martha Anders's (1990; Anders et al. 1994; Anders et al. 1998) discovery of pottery ovens at the contemporary site of Maymi in the Pisco Valley on the south coast of Peru that were full of stylistically related ceramics, the past seemed clear. Conchopata and Maymi were archetypical "artisan communities," conceived as similar to villages and barrios that still exist in Peru, where many of the resident families engage in a particular craft.

This image of Conchopata must be questioned. There are reasons to suspect that today's specialized craft communities are not continuous developments from the prehistoric past (see Feinman and Nicholas, this volume; Isbell 1997). But more importantly, since 1998, 2 ha of nearly continuous urban landscape have been revealed, providing much greater contextual understanding of the ancient city. The new discoveries confirm pottery manufacturing and associated pottery making with other crafts in a multicrafting complex, but not at all like the case of modern craft communities.

Contextualizing Conchopata's Multicrafting Environment

Recent excavations at Conchopata have revealed little or no evidence for the diversification of crafts such as metallurgy. The new evidence implies instead a different set of complementary specializations. In addition to pottery making, brewing maize and *molle* (*Schinus molle*) beer was very important (Green and Whitehead 2006; Sayre and Whitehead 2003). The preparation of feasts, hosting and

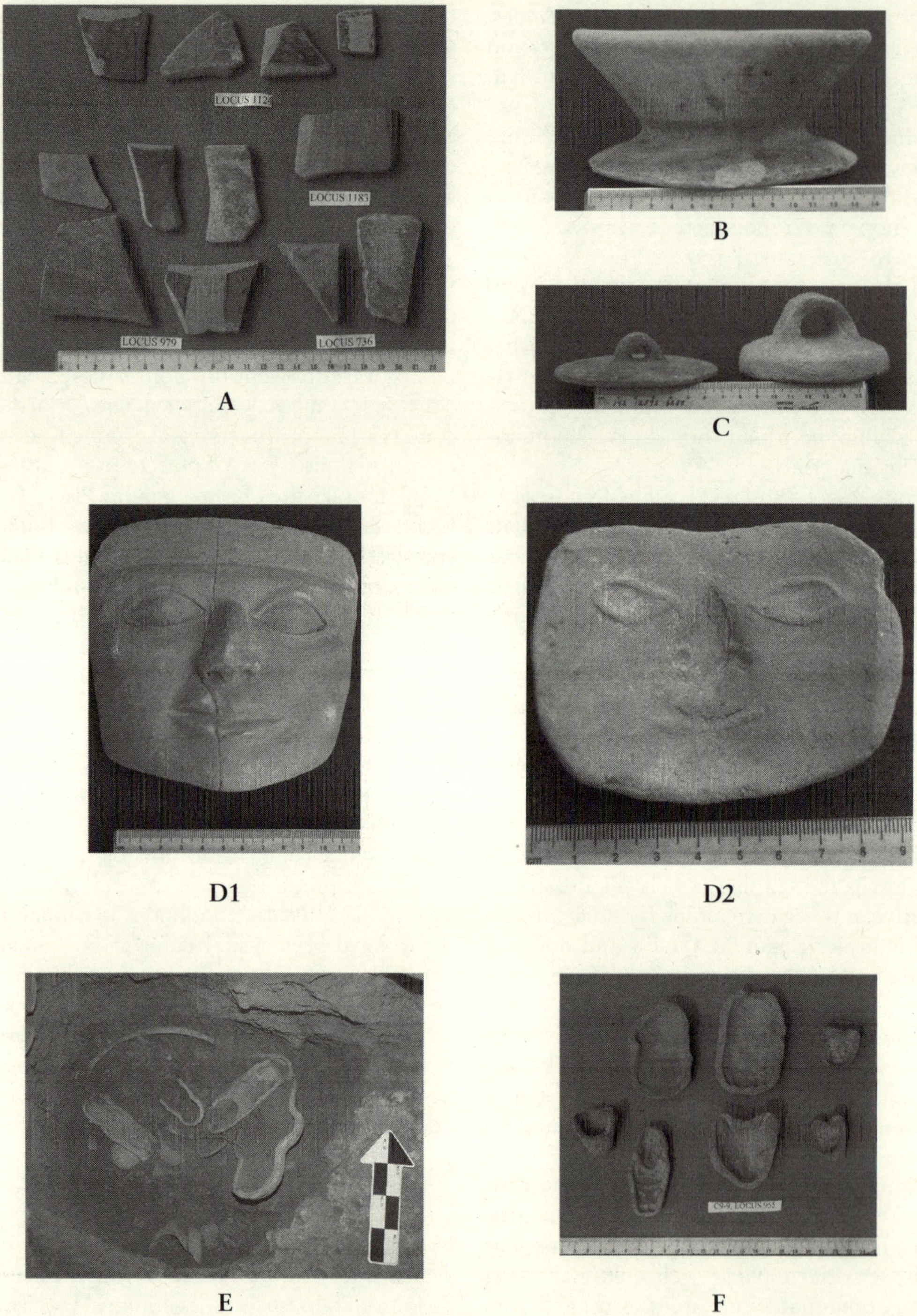

FIGURE 3.3. Ceramic-manufacturing tools from Conchopata. A. Sherds reused as scrapers and polishers. B. Potter's plate. C. Anvils for shaping pottery with a paddle. D1.–D2. Modeled face positive and negative molds. E. Cache of manufacturing molds in situ. F. Manufacturing molds for whistles and small figurines from the late phase at Conchopata.

serving banquets (Cook and Benco 2001; Cook and Glowacki 2003; Isbell 2001; Isbell and Cook 2002), and the creation of numerous offerings that included, among other things, smashing large quantities of pottery (Cook 1994; Groleau 2005; Isbell 2001; Menzel 1964; Nash 2006) were also important.

Conchopata's social context does not seem to have been independent families of middle-class artisans but, rather, elite palace complexes, within which wives and concubines appear to have labored in crafts and services to create social events of aggrandizement for lords sponsoring competitive and status-building feasts, parties, and drinking bouts. This interesting version of multicrafting was surely sequential, not simultaneous, with pottery being produced at one moment, probably to meet needs for specific upcoming events. This was followed by use of the ceramic equipment for brewing and feast preparation. Other ceramics were then employed in banquet serving—vessels that may have been designed specially to communicate meanings, to serve appropriate dignitaries, and to promote symbolic negotiations. Finally, and perhaps as part of the same sequence, pottery and other items were sacrificed in ritual offerings. Vessels were smashed and buried, sometimes in pits but sometimes on building floors, occasionally filling entire rooms. Of course, to the degree that the ceramic equipment for a particular event was smashed, a new set was required for the next event.

There is some evidence for the presence of charcoal, perhaps used in firing pottery. If it is present, it must have been manufactured elsewhere. Additional crafts are implied from a study of Wari artifacts from other Middle Horizon contexts, but apparently none of these specializations was located in the excavated portion of Conchopata.

Artifacts from recent excavations at Conchopata are undergoing analysis, so this discussion is based mostly on observations made during excavation. However, several crafts, including metallurgy, that were anticipated as at least incipient specializations are convincingly absent from the new Conchopata archaeological record. Metal artifacts are being studied by Heather Lechtman. The number of these objects and the variety of forms are considerable. Arsenical bronze that required sophisticated metallurgical skill is well represented (Lechtman and Frame 2005). The frequency of silver and gold in what was formerly considered a community of potters led early investigators to erroneously infer metalworking. Rather, Conchopata's residents were wealthy enough to consume these metal artifacts, although the most common objects are *tupu* pins generally thought to be gender-specific items of women's dress.

Weaving, by proxy of tools such as ceramic spindle whorls and diagnostic bone artifacts, was not a significant activity at Conchopata. In fact, appropriate artifacts are scarce enough to imply that weaving was unusual, even household production for domestic use, which archaeologists take for granted throughout the Andes. By implication, Conchopata's residents must have been supplied with cloth from elsewhere, something that constitutes indirect evidence for its specialized production. And indeed, textile analysts argue that at least some of Wari's impressive cloth was woven in workshops by expert artists (Conklin 1970, 1986, 2004). But none was located in the excavated portions of Conchopata.

An analysis of Conchopata's lithic artifacts is in progress by Catherine Bencic (2001). She has found a mix of expedient and formal tools, manufactured from local as well as more valuable imported raw materials. Virtually all obsidian came from the high-quality Quispisisa source (Catherine Bencic, personal communication regarding instrumental neutron activation analyses conducted on 100 Conchopata obsidian flakes selected by Bencic for analysis by Richard Burger and Michael Glascock, 2006), which supports specialization. However, there is no evidence of lithic manufacturing in the excavated portion of Conchopata, even though there was a significant increase in the relative frequency of diagnostic tools during the occupation. Distinctive T-shaped or "sock-shaped" artifacts usually called "hoes" show the greatest increase. Pozzi-Escot B. (1991) argues that these artifacts were used by potters to paddle vessels into shape and also to dig and mix clay. But they are generalized tools useful for many purposes including agriculture.

It seems likely that potting and the associated multicraft feasting complex began at Conchopata with the appearance of palace compounds, around the beginning of the Middle Horizon. Leaders were constructing large facilities to serve as residences, administrative/ceremonial centers, and assembly places. Specialized ceramic manufacturing was part of a related new feasting complex. Immense brewing jars (125–140 cm tall and 75–95 cm in diameter) were produced, as were special vessels for preparing banquets and elaborately decorated containers of many kinds for serving food and drink to increasingly differentiated and status-conscious guests.

The new social role of ceramics was associated with a new firing technology employing kiln-like pit ovens, perhaps fired with charcoal, as well as a new technology of polychrome slip painting. Curiously, the designs on some fancy ceramics were painted over earlier decorations. Pending further study, this may suggest that vessels were fired more than once, with decorations perhaps being painted and fired on after an initial firing. If this were the case, then producing immense vessels, including urns a meter tall and equally as wide, with elaborate polychrome decorations may have been less difficult than imagined, although perhaps more costly in fuel.

A new religious iconography associated with Tiwanaku was also adopted at Conchopata, soon after palaces and the feasting complex made their appearance. The new images were depicted on great brewing and serving vessels and eventually on smaller ceramics. They also appear on fine textiles, some so similar to Conchopata's ceramic iconography that pot painters and weavers must have seen one another's work (see Conklin 1970, 2004). But Conchopata has yielded few weaving tools.

Based on her 1980s excavations at Early EX-1 (Figure 3.2), Pozzi-Escot B. (1985, 1991) inferred an elite residential function for the area. She has suggested that the smaller and more modest rooms to the south, where she excavated around Early EX-2, were more associated with lower-class artisans. However, the high rank of northern compound BC-A was not fully appreciated until excavations cut though floors to reveal monumental tombs in room 203 (Figures 3.2, 3.4). Burials under floors appear in other domestic compounds at Conchopata, confirming the residential function. However, comparative study of hierarchy in Conchopata/Wari mortuary practices (Isbell 2004) shows that the sepulchers in room 203 rank high enough in the regional status structure that they must represent the graves of petty kings or governors. In fact, it looks as though this compound played a key role in the early development of palaces and feasting at Conchopata.

Early in the Middle Horizon Conchopata was focused around an open space covered with pinkish sand, named the "Pink Plaza" (Figure 3.2). To the west was the large and elite residential compound, BC-A, with monumental tombs that probably contained the remains of lords who had resided in, and ruled from, this great house. To the east was another compound enclosing a sizable court, BC-B, which is best identified as an assembly and feasting hall (Figures 3.2, 3.5). The functional ascription is based on decorated serving vessels from the floors of BC-B and on the highest concentration of obviously butchered camelid bones at Conchopata, found in one of BC-B's small southern-corner rooms (Rosenfeld 2004, 2006). But the most convincing discovery is a great urn that had been permanently seated next to the western doorway of BC-B (Figure 3.5), where guests could be served as they entered the courtyard (Cook and Glowacki 2003). Significantly, the urn is undecorated, perhaps indicating that it predates the polychrome painting and Tiwanaku-related iconography so intimately associated with most of Conchopata's feasting. Be that as it may, I believe that the pair of buildings facing one another across the Pink Plaza—BC-A and BC-B—represents the initial development of a new kind of Wari architecture at Conchopata, the palace complex.

The buildings bordering the Pink Plaza on the east and west were rectangular compounds

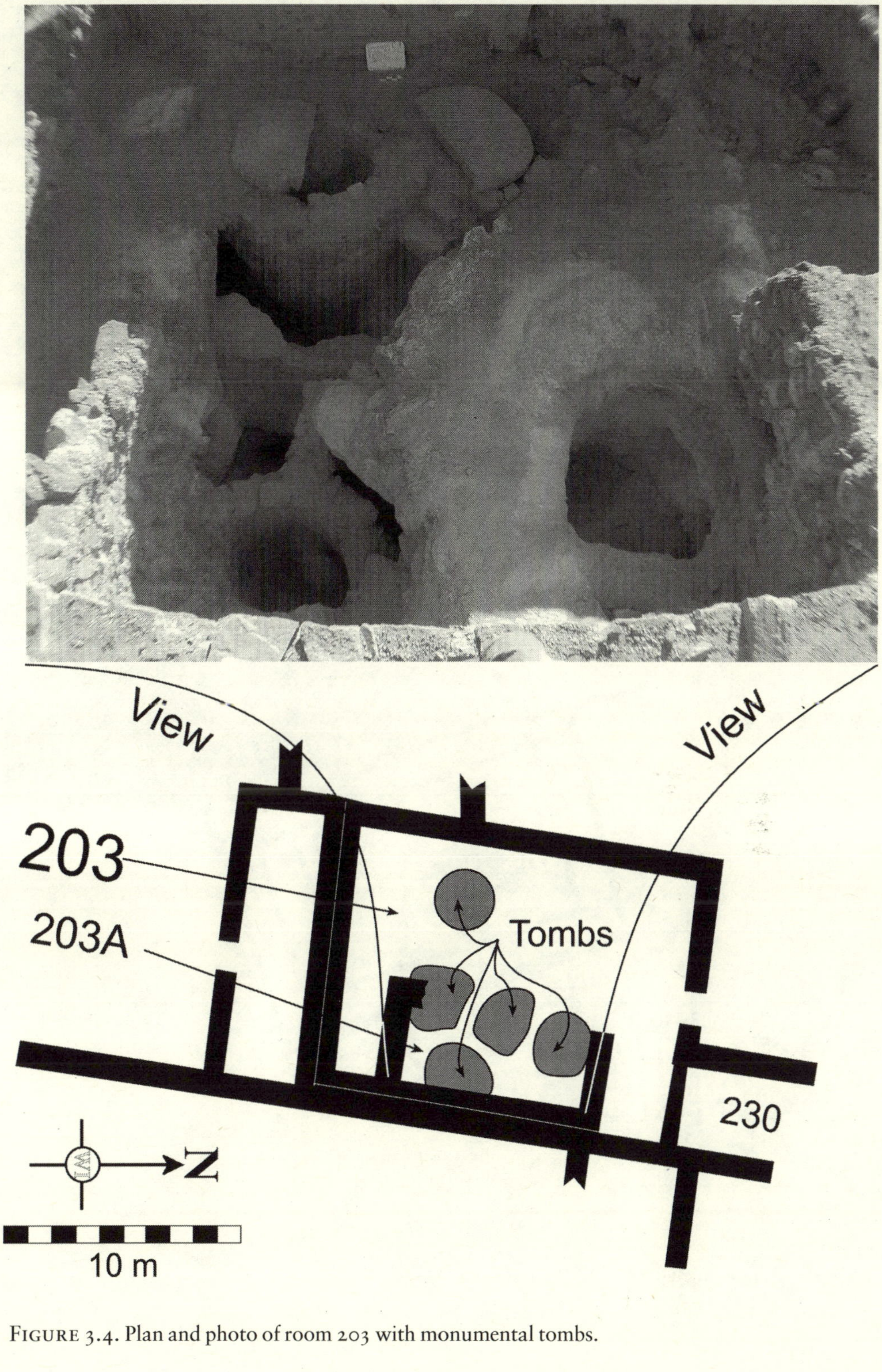

FIGURE 3.4. Plan and photo of room 203 with monumental tombs.

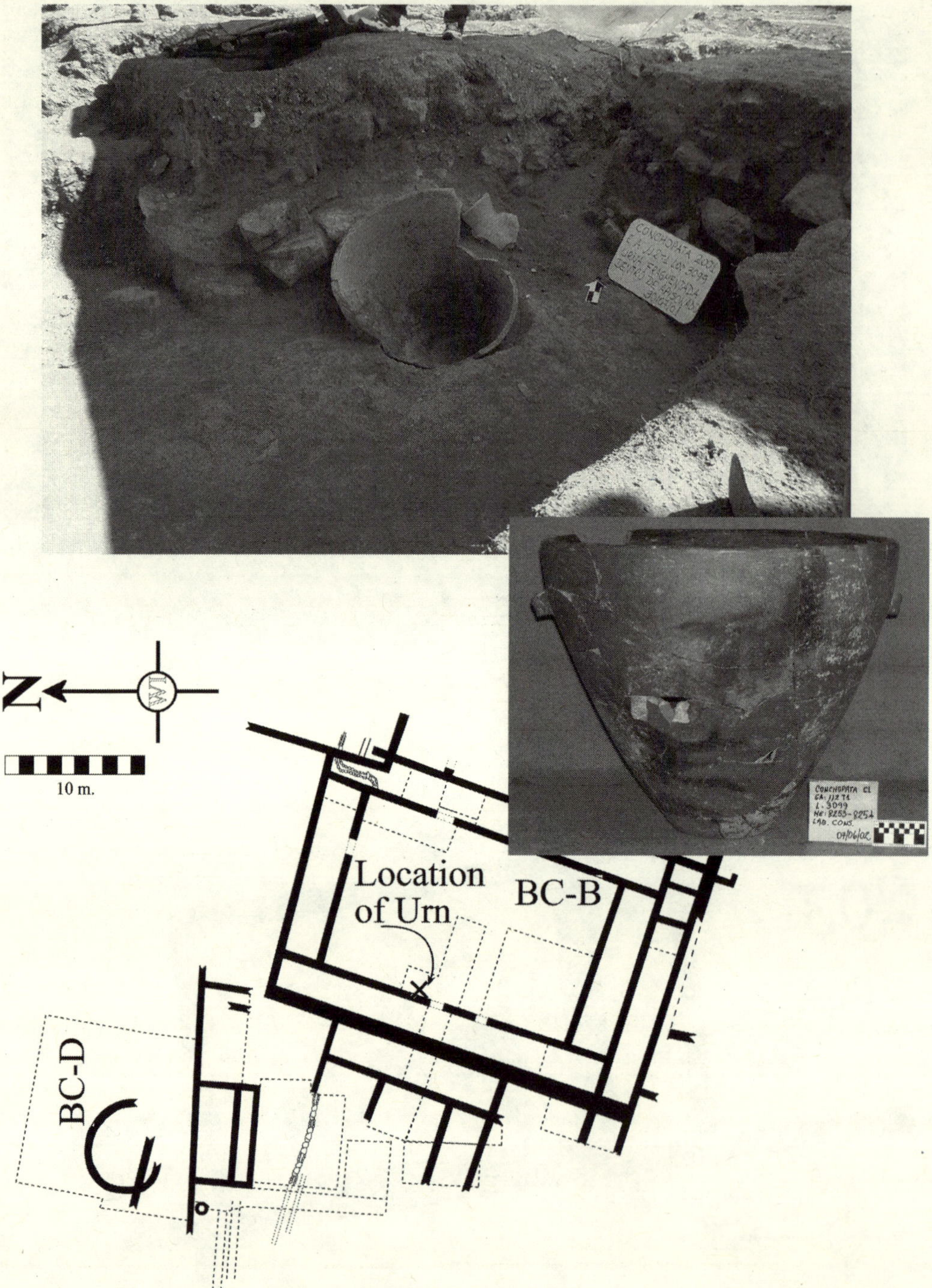

FIGURE 3.5. Feasting hall BC-B with giant urn at main entrance.

belonging to Conchopata's elite lord. They represent political and residential functions, as discussed above. By contrast, the buildings bordering it on the north–south axis were circular or D-shaped buildings that can probably be identified as temples (Cook 2001; Isbell 2001; Isbell and Cook 2002). Temple complex BC-C (Figure 3.6) faced north toward the Pink Plaza, with an entrance descending two steps to a sunken floor. At the center was a walled, circular depression with oddly shaped stones placed on end—surely objects of veneration. Adjacent to a northern section, several articulated leg and toe bones of deer were discovered. Because deer remains are extremely scarce at Conchopata, it seems likely that these bones were attached to skins employed as costumes in ritual performances.

In the southeastern portion of circular temple BC-C, the floor was littered with burned human skulls as well as bones from human hands (Figure 3.6). Cut marks show that these human parts were defleshed and disarticulated (Tung 2003; Tung and Cook 2006). By contrast, the western sector of the building floor has three rows of holes, totaling about 15, that are the right size for the pointed bases of large vessels. Given that Conchopata's great brewing jars can contain more than 25 gal, and if the holes represent a collection of brewing vessels, the beer produced in this "spiritually transformative" temple could have totaled 300–400 gal at a time.

To the north of the Pink Plaza was a smaller circular or D-shaped building, BC-D, much disturbed in both the Middle Horizon and modern times. Unfortunately, the intensity of modern disturbances—including power poles, the highway, a large water pipe, a new building, and foundation trenches for additional construction—makes it difficult to understand the Middle Horizon disturbances, which seem significant as well (Figures 3.2, 3.7). However, several extravagant ceramic offerings were found nearby, supporting inferences about the religious nature of the building.

Spectacular ceramic offerings were made at Conchopata that required the deliberate smashing of ceramic vessels, including giant urns about a meter tall and jars more than 1.25 m in height, followed by their interment in pits 1.5 to 2 m across and at more than a meter deep (Menzel 1964, 1968). An example consisting of 20 to 25 giant jars (Figure 3.8C) painted with images focused around the newly adopted Tiwanaku-related Staff God icon (Cook 1987; Isbell 1987; Isbell and Cook 1987) was accidentally discovered by construction workers in 1977. The jars had been smashed into thousands of fragments and buried in a pit below the Pink Plaza. In 2000 a similar offering, labeled Offering 2000B, was excavated a few meters north of Offering 1977 (Figures 3.2, 3.7), except that these large and deliberately smashed jars had only local religious icons (Figure 3.8A–B). Most likely, the 2000B offering predates the adoption of Tiwanaku-related themes, whereas Offering 1977 represents an early moment in the conversion to the new religious ideology. Surely, both offerings commemorate a feast, for which the jars were manufactured and used to brew vast quantities of beer. Their destruction probably constituted a spectacle of termination while also creating new demand for giant brewing vessels.

A third ceramic find consisting of thousands of jar fragments, labeled Offering 2000A, appeared farther to the northeast, adjacent to the little D-shaped temple BC-D (Figure 3.7). This discovery reveals a problem that is emerging in the study of Wari ceramic offerings. As more Wari ruins are excavated and new ceramic offerings are found, some diverge so much from Menzel's (1964, 1968, 1977) original concept that the nature of the related activities requires rethinking. I (2001:37–43) have outlined a new, four-class typology recognizing much more variation in Conchopata ceramic offerings, but even this is proving inadequate. Discoveries at Conchopata and other sites show that offering practices varied widely and are often difficult to distinguish from the discard of rich refuse (Groleau 2005; Nash 2006).

Is Conchopata's immense pottery concentration labeled Offering 2000A really a ceramic offering, like others? Perhaps it represents one or more disturbed offerings. But perhaps it is refuse, such as large sherds discarded after being used to enclose pots while

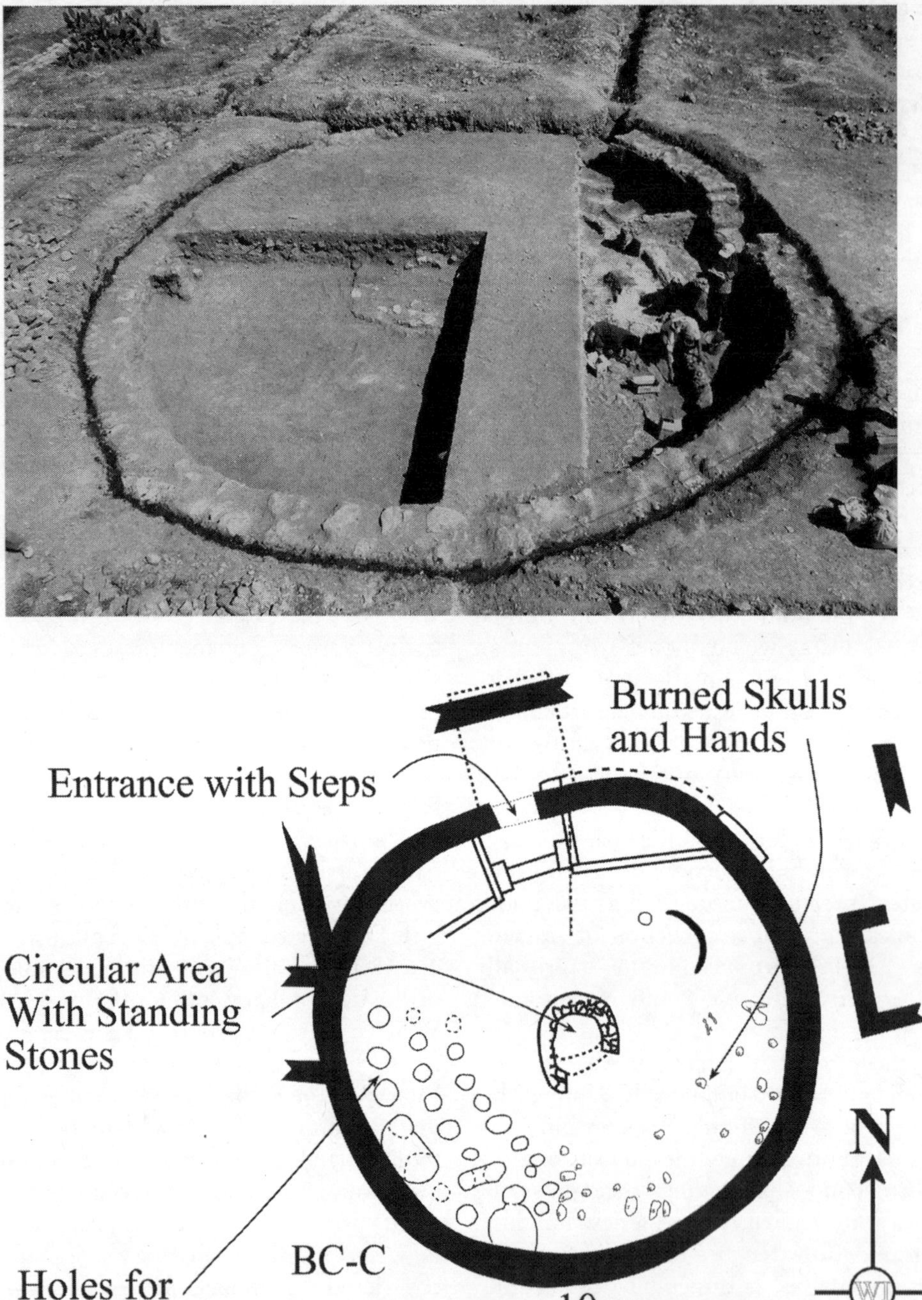

FIGURE 3.6. Plan and photo of BC-C, a D-shaped temple building.

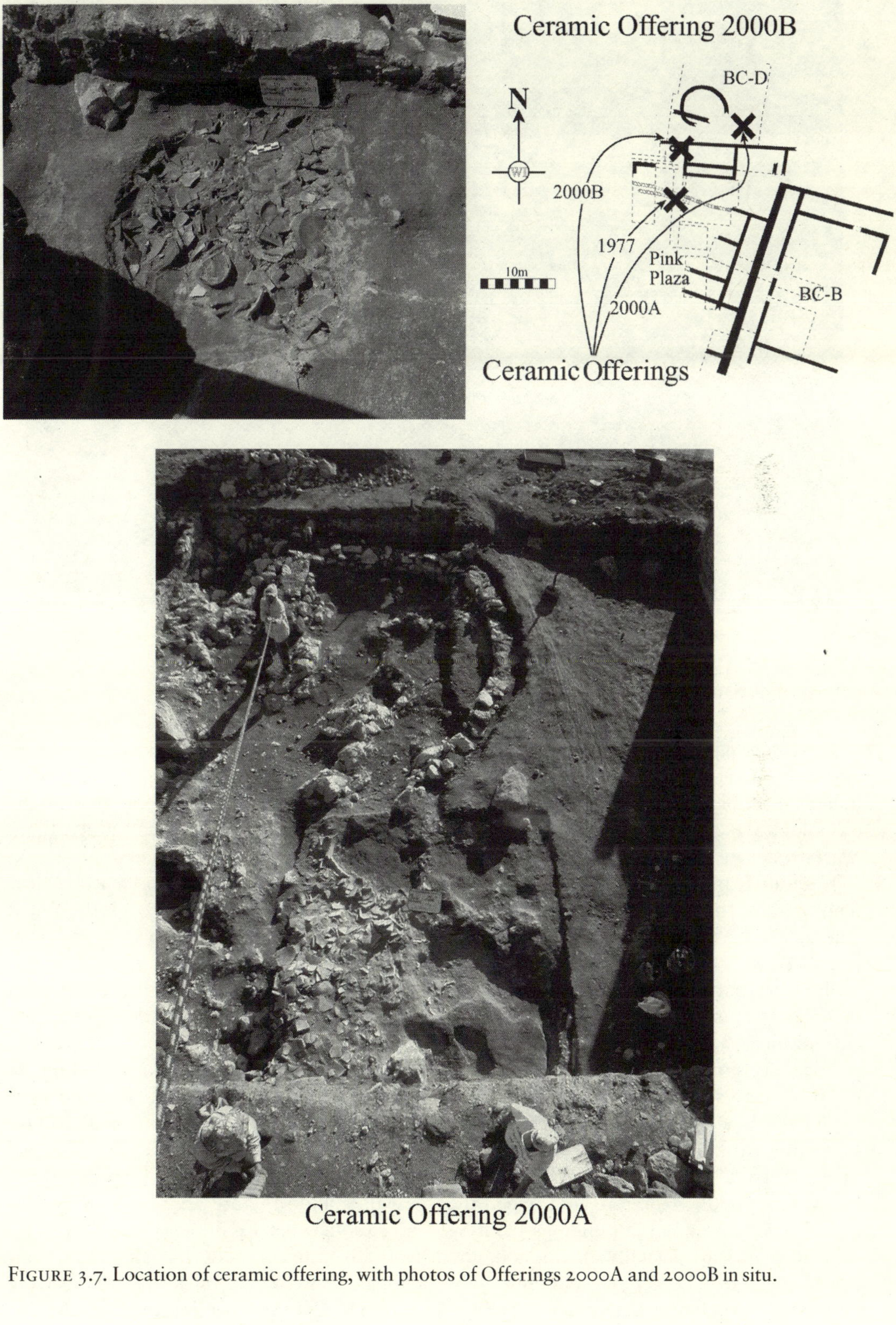

FIGURE 3.7. Location of ceramic offering, with photos of Offerings 2000A and 2000B in situ.

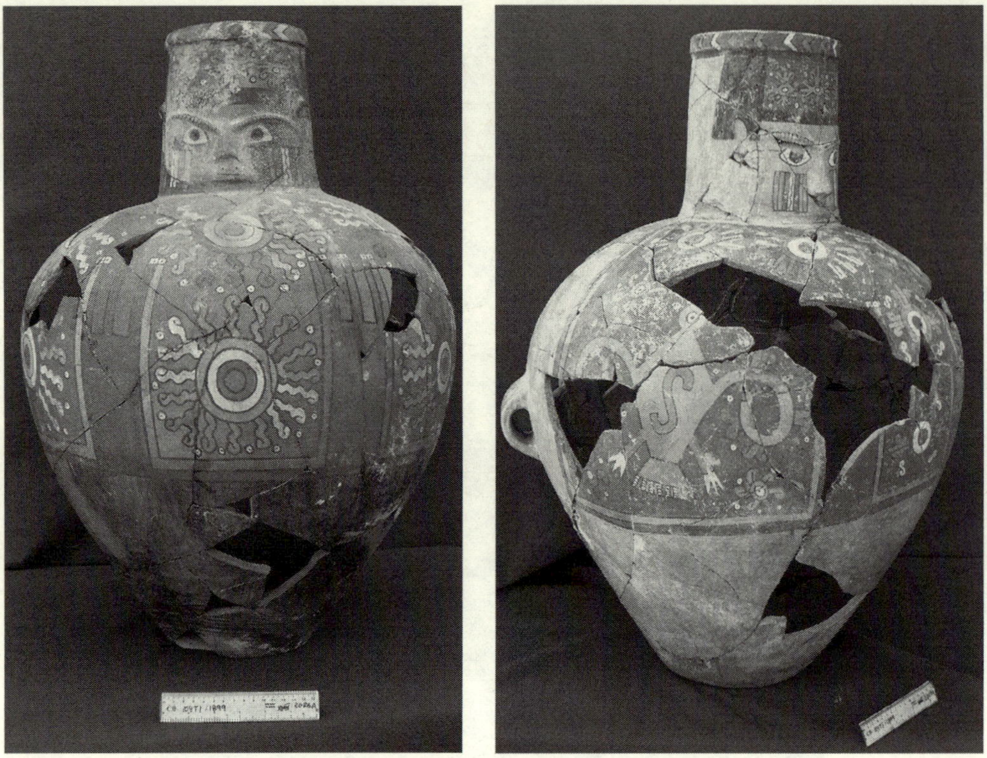

A B

Figure 3.8. A.–B. Giant jars from Offering 2000B. C. Giant jar from Offering 1977, with Tiwanaku-related Staff God image.

they were firing in pit ovens, or some other explanation. This "offering" versus "trash" issue is a problem for understanding the archaeological record at Conchopata and other Wari sites.

Ceramic-manufacturing tools are abundant in all of the spaces discussed above, including the Pink Plaza, elite residential compound BC-A, and feasting hall BC-B. If the tools represent primary refuse, then they document pottery manufacturing in spaces where elite residence, assembly, religious activities, political decisions, and feasting took place. If the tools represent offerings, then they document the ritual and conceptual association of potting with these spaces. Both inferences may be correct.

As the Middle Horizon continued, the Pink Plaza as well as the buildings fronting on it were at least partially abandoned, and a new focus of elite activity, building compound BC-E, was constructed to the east (Figure 3.2).[3] This compound is formally very similar to old feasting hall BC-B, and in fact, fragments of large urns like the one by the doorway into BC-B were found in BC-E, including a fine example smashed in place next to a northern doorway into the central courtyard (Figure 3.9). Significantly, it is decorated with polychrome religious icons derived from Tiwanaku-related themes.

Unlike its predecessor, BC-E was built in a single, planned episode, and it does not appear to have faced a plaza with an elite residential complex and sepulcher opposite it. Rather, buildings best interpreted as elite residential groups were attached to the north and south of the BC-E enclosure. Furthermore, several have rooms with monumental sepulchers below their floors (Figure 3.2, rooms 38, 39, and 110; a somewhat less impressive bedrock cavity tomb occurs in room 105). I infer that Conchopata's late palace integrated the feasting hall and the elite residential compound(s) into

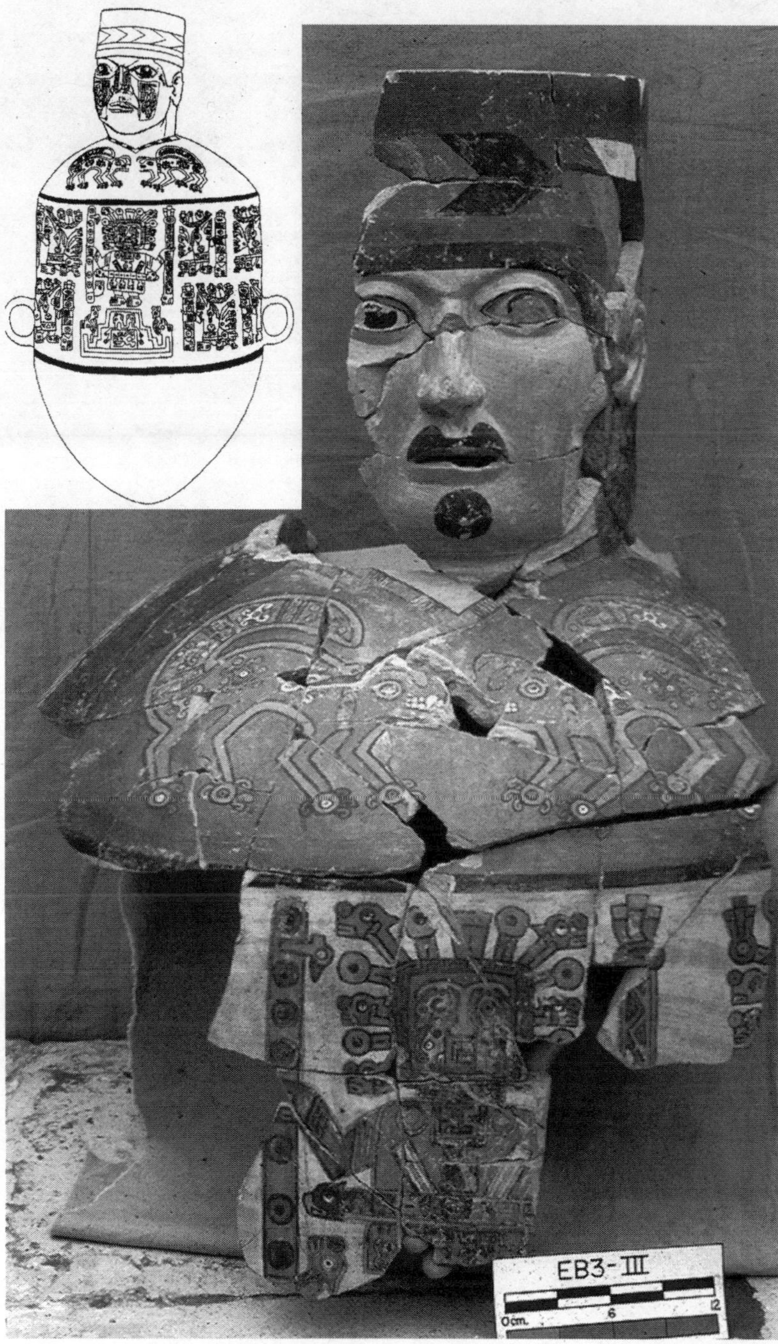

C

a single complex, a new kind of palace. Perhaps a succession of three lords ruled from BC-E who were subsequently buried in the impressive graves before power moved elsewhere within the little city.

A surface of shiny white plaster can be traced to the south of the new palace, where several D-shaped temples were constructed (Figure 3.2, BC-F and BC-G). Before long, BC-F was filled with what appears to have

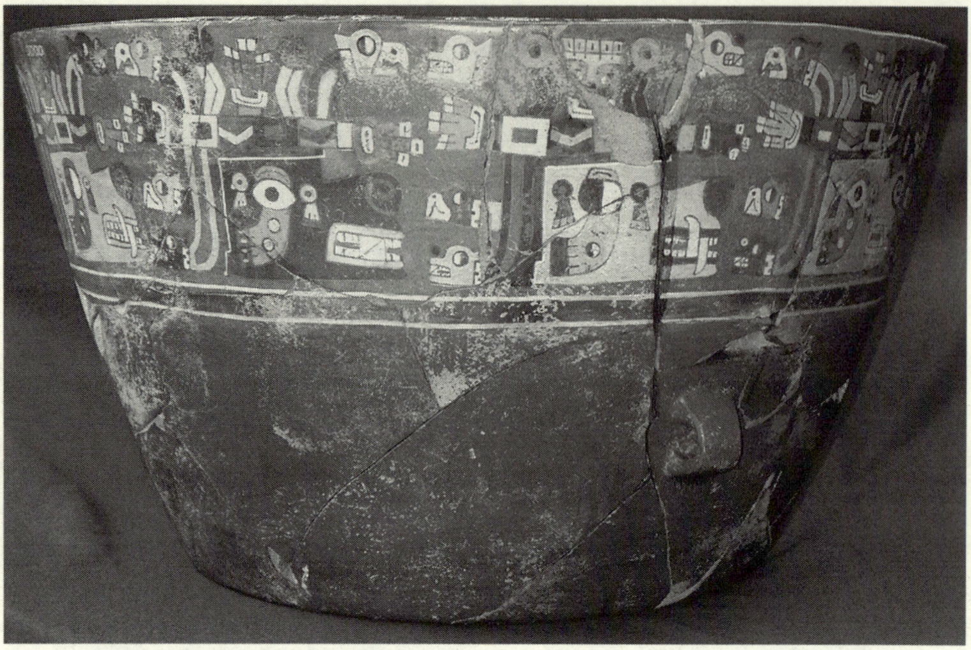

A

B

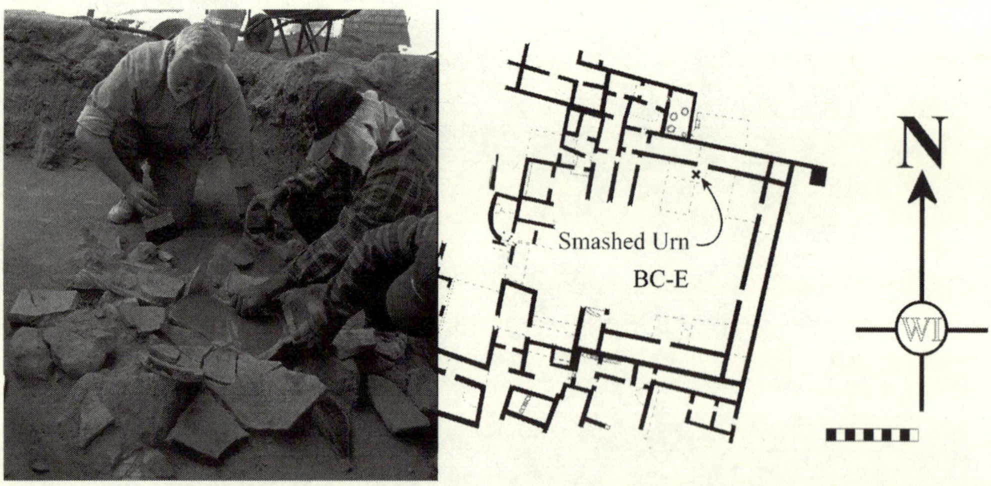

Figure 3.9. A. Giant urn with Tiwanaku-related imagery, from feasting hall BC-E. B. Location and excavation of the giant urn shown above, in feasting hall BC-E.

been trash and was built over. BC-G was ritually closed, for its floor was covered with offerings of elaborately painted oversize pots, completely smashed. Like the earlier temple BC-C, its floor had a circle with an upright stone, human skulls, and the bases of large jars set into holes in the western edge of the floor (Ochatoma Paravicino and Cabrera Romero 2001b, 2001c, 2002). Nearby, in rooms 85 and 84, caches of ceramic molds were found, along with other potting tools. The adjacent space 33 had the bases of jars set into its plaster floor. Perhaps this was a potters' workshop, but the configuration of spaces and doorways does not seem consistent with a dedicated workshop. Rather, the occurrence of potting tools almost everywhere suggests that ceramic manufacturing was periodic throughout the year and was conducted in any space available at the moment.[4] Apparently, potters stored tools in residences but set up operations in patios and important public and ceremonial spaces when they were free.

The final century or so of occupation at Conchopata, A.D. 900–1000, is poorly understood, perhaps because of ceramic seriation errors. There is no evidence that palaces were still functioning, and no pit ovens have been identified as belonging to this time. Giant-sized pots were scarce and may have disappeared. Tiwanaku-related iconography was no longer popular. Multicrafting may have been replaced by simple ceramic production, emphasizing small effigy figures and figurine whistles made in molds. If so, pottery making at Conchopata survived, but probably in a new form.

IDENTIFYING CONCHOPATA'S MULTICRAFTERS

I believe that the large building complexes at Conchopata were palace compounds occupied by lords who governed the city. Where architectural preservation is good in the late palace we can recognize groupings of interconnected rooms centered around an open patio that appears to have served as a central work area. Frequently, several of the adjoining rooms have a raised floor section, probably a sleeping space. Smaller rooms with concentrations of ash were kitchens, although hearths existed in at least some of the open patios as well (see Figure 3.2). I infer that each of these room groupings represents a residence, probably for some kind of extended family that included servants (see also Blacker 2001; Blacker and Cook 2006).

Many of these late palace room groupings have tombs below their floors. They range from simple pits containing one or more skeletons to monumental graves clustered together so that they fill an entire room. The former have no, or very few, offerings, whereas the grandest graves were elaborately furnished, including gold. But of course, fancy graves have been looted, so their contents can only be interpreted with caution. One important observation is that Conchopata's graves were reopened periodically, so that additional bodies could be interred and perhaps some bones of long-dead individuals could be removed. I (2004) have speculated that the addition of the newly dead to already occupied tombs might indicate that family members were added to the tombs of their close relatives when they died, within, or close to, their place of residence.

An *unlooted* bedrock cavity tomb in room 105 provides especially valuable insights about the composition of Conchopata's palace residential groupings, at least the elite members. It is the most elaborate undisturbed burial excavated by archaeologists at the site. The subfloor chamber contains the remains of 15 bodies. Near the bottom of the tomb was a male between 23 and 27 years of age with a healed cranial fracture on the left parietal, suggesting that he was a warrior (Tung 2003). His body was seated on fragments of a wooden stool, or perhaps a shield, and surrounded by organic traces that appear to have come from a feather-covered textile. Other skeletal remains include two fetuses in jars, three infants, part of a child, a juvenile, and six adult females of various ages, as well as a seventh skeleton too incomplete to be sexed but probably also an adult female (Tung 2003; Tung and Cook 2006).

Given the sex and age ratio in the room 105 tomb, I infer that the grave represents a polygynous family—an elite male buried at death and accompanied by his wives and concubines, as well as their children who died very young. At least some bodies were added later by successively reopening the tomb.

If polygyny was a prominent pattern for Conchopata's palace elites, then a small number of noblemen lived with a much larger number of women. Each grouping of interconnected rooms was probably the residence

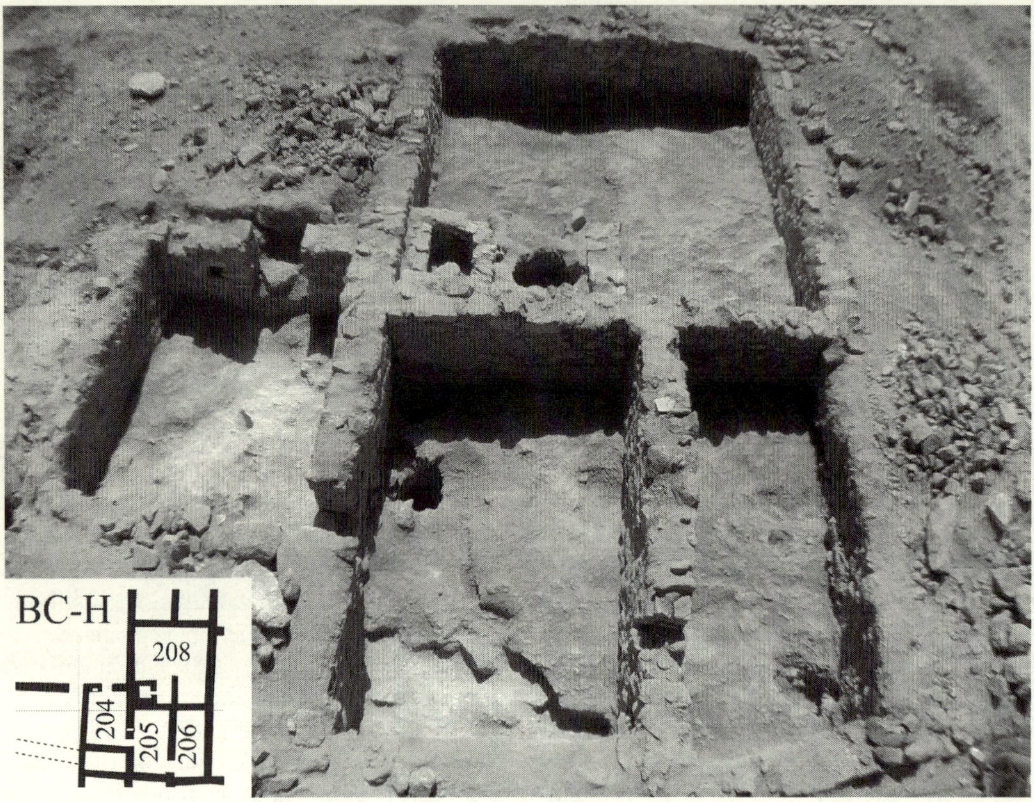

Figure 3.10. Photo and plan of a multiroom residence in the southwestern corner of Conchopata.

of an important courtier, including the lord's principal and secondary wives and brothers and so on with their wives. Palace compounds concentrated women and created a new kind of labor pool that made females available to become the new multicrafters of Conchopata.

The "polygynous wives and concubines" hypothesis would be supported if other elite burials from Conchopata contained a single male with several female bodies. Unfortunately, most of the elite burials have been looted. However, site totals offer some insight. Of 199 adult burials from the Conchopata excavations that have been studied by Tiffiny Tung (2003), 16 sexed adults belong to the transition to the Middle Horizon. A total of 12.5 percent are male, and 87.5 percent are female. Of 53 sexed adults from Middle Horizon graves, 38 percent are male and 62 percent are female.

A Woman Remembered in Death

The high frequency of female burials at Conchopata supports but does not prove the polygynous wives and concubines hypothesis. However, a complementary set of qualitative information comes from a female whose individual grave can, with security, be associated with offerings identifying her as a potter and brewer.

The well-preserved, intermediate-level burial of a woman was discovered in Conchopata's room 205, a house of three interconnected spaces (Figure 3.10) that was abandoned sometime after the interment.[5] The outermost room 204 contained a great deal of pottery, much of it apparently refuse tossed in through the exterior doorway after abandonment. This trash was concentrated along the west side of the room, having apparently been cleared from the east wall, where an opening through the wall led to the woman's grave.

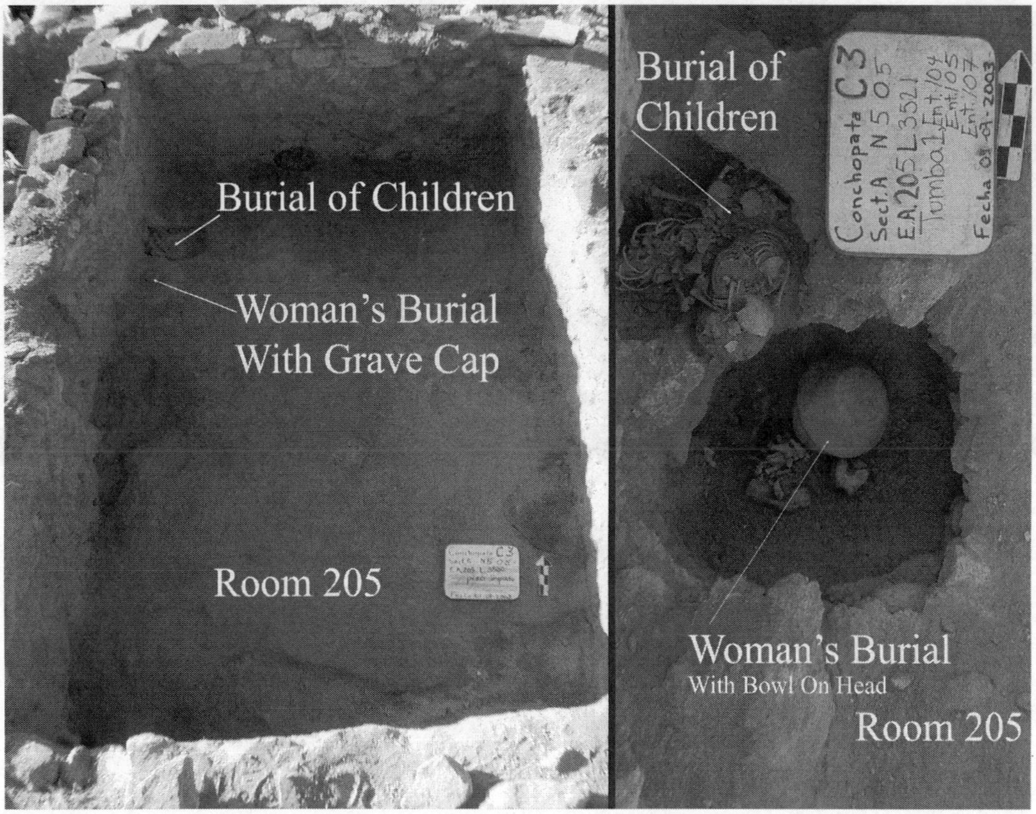

FIGURE 3.11. Late floor of room 205 with burials of an adult woman and two children.

Room 206, the innermost space (Figure 3.10), contained much less material, but most of it is consistent with offerings rather than trash. The single floor had a metal tupu under an elongated stone lying across the middle of the room, several camelid scapulae tools, a bone perforator, and some yellow pigment. Fifteen ceramic smoothing and polishing tools made from reutilized sherds, pigment, and two polishing stones document ceramic manufacturing. The room was probably ritually closed when the fragments of eight jars were placed on the floor. Several were about 50 cm tall with sooted exteriors, whereas others were smaller "face-neck" vessels. All were semicomplete, so it is not clear whether they were smashed in situ or brought from elsewhere already broken. Among these sherds was a copper needle, a mano (a fist-sized cobble used for grinding or pounding) and mortar, seven smoothing sherds, two ceramic molds, two potter's plates, and a stone hoe sometimes interpreted as a paddle for shaping pottery.

By far the most interesting remains are from room 205, located between 204 and 206, although both doorways were sealed by the end of the occupation. An early floor in room 205 was associated with several spindle whorls, numerous sherds used to smooth pottery, a mortar, a ceramic mold, and fragments of ceramic spoons. A later floor covered this surface. A burial of intermediate status (Isbell 2004) was excavated through the late floor, against the west wall (Figure 3.11). A circular cap of clay with a central "offering dimple" covered the grave. The remains of a small animal wrapped in cloth lay next to the dimple.

The stone-lined grave contained the body of a seated and tightly flexed woman in her forties (Finucane 2003). She had a bowl over her head, a blackware face-neck jar between her feet, and a miniature face-neck jar behind

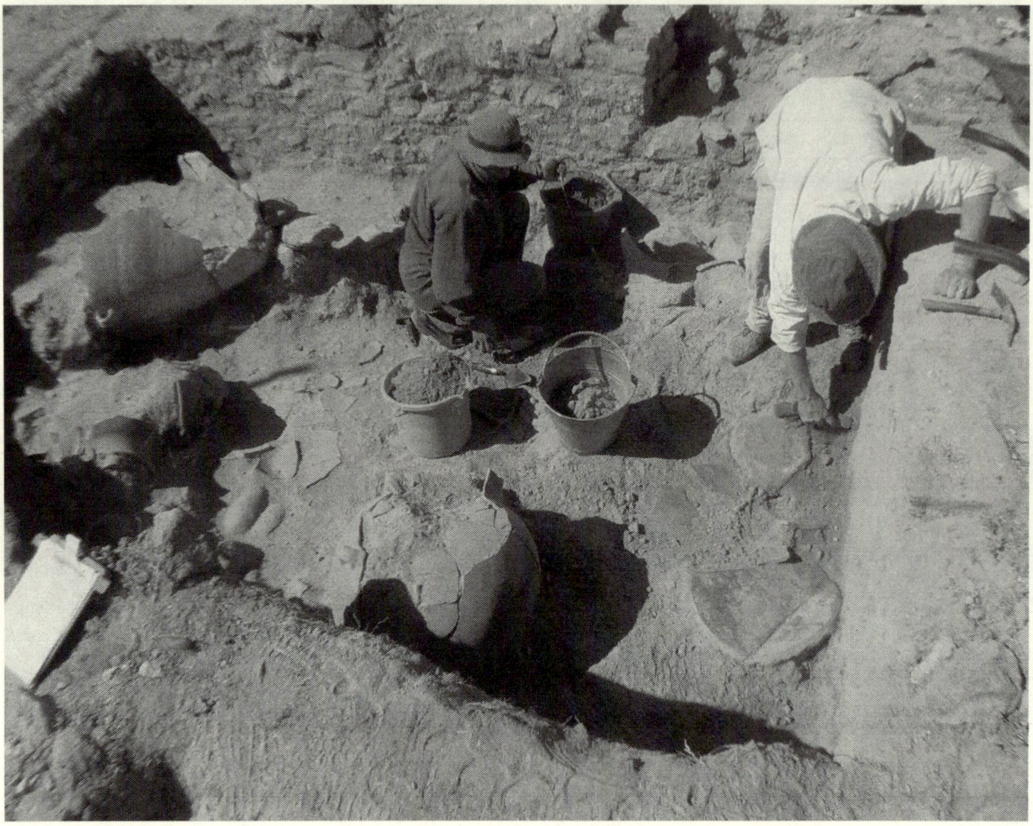

Figure 3.12. Excavation of ceramic offering that filled room 205 and included large vessels broken on the spot.

her back. Below the clay cap of the grave a hole was dug, penetrating through the wall into room 204—apparently so offerings to the deceased could continue after room 205 was closed. Directly to the north and merging with the woman's grave is another pit containing two young children with no grave furnishings (Figure 3.11).

Sometime after the interments, offerings were placed in pits along the edges of room 205. Then the floor was covered with broken pottery, including three very large brewing jars and several medium-sized pots broken on the spot (Figure 3.12). The shapes are appropriate for several stages of chicha beer making, and the three largest vessels were so big (about 125–140 cm tall and 75–95 cm wide) that they did not fit though the doorway. The east wall had to be dismantled to carry out the offering. Among the vessel fragments were numerous ceramic smoothing and polishing tools, ceramic molds, a complete small jar, a grinding stone, a mortar, and a copper tupu pin.

The burial of the woman and children, the offerings, the closure of room 205, and the abandonment of the house are surely related to one another. Perhaps the death of the woman and her children initiated closure of her house. Be that as it may, survivors who organized the events were clear in their material identification of the woman as a potter and a brewer. Furthermore, her memorial was insured by the offering hole into room 204 that was left open.

Expression of Identity by Female Multicrafters

Multicrafting by women supported the political ambitions of male lords, a fact that could not have escaped either. Some activities put women and their crafts at the focus of key social events, even if their position was subordinate. Pottery and pot painting provided an arena for Conchopata's women to represent and negotiate new identities for themselves.

When new techniques of representing realistic images were adopted, the dominant themes were male deities and male warriors (Ochatoma Paravicino and Cabrera Romero 2001a, 2002). Ceremonial serving vessels were often represented as elite male effigies (Figure 3.8A–C). However, Conchopata's women identified with the crafts and products they produced, gradually promoting a metaphorical relationship between their bodies and the bodies of brewing vessels and perhaps all ceramic containers.

We are poorly prepared to identify women in Conchopata art, for we know little about conventions of attire or the representation of gender. But we must begin somewhere, so it seems that from the adoption of Tiwanaku-related iconography (Isbell and Knobloch 2006) females are among the mythical, or perhaps royal, figures.

One early representation from Conchopata shows a pair of mythical beings (Figure 3.13 B–C). The two figures were painted with their bodies in front view, but the faces and the feet are in profile. Each figure holds out its arms to grasp a vertical object in each hand that is probably a variant of the staff better known in later Tiwanaku sculptures. One figure wears a tunic belted at the waist, whereas the garment of the other is looser and fuller. I believe that the first is male and the second is female. Their clothes resemble male and female costumes depicted on better-preserved Robles Moqo pottery from Pacheco (Figure 3.13A), where a similar pair—male and female deities—appears on oversize urns belonging to this same Conchopata-Wari tradition (Menzel 1964, 1977). The female is distinguished by her loose garment and cape-like outer wrap, draped over her shoulders and fastened below her chin, as well as ears of corn in her rayed headdress and on her dress.

Tupu pins are believed to be women's clothing (Gero 1990), an argument supported by our room 105 tomb. Consequently, an early representation of the Tiwanaku-related Staff God, shown standing on a small pyramid (Figure 3.14; see also Figure 3.8C), may be female. The anthropomorphic figure has circular objects just below its shoulders that seem to represent tupu pins. This important figure was painted on large jars with modeled male faces on their necks that were smashed and buried in an offering cache dug through the Pink Plaza, discovered in 1977 (Cook 1987, 1994, 2001; Isbell 1987; Isbell and Cook 1987). The central figure wears a belted garment that might imply male gender, but it is looser than those of most other staff figures. Perhaps it is best to concede that the gender of this figure remains in question but may be female.

Conchopata's female deities, or perhaps female royals, constitute a set of women who, like elite males, wear elaborate headdresses and grasp status-defining objects in their hands. However, another group of women tends to be shown with bare heads, two prominent braids or bunches of hair (by contrast, at least some males have multiple, small braids), and empty hands. An early example of such a woman is probably the most securely identified female at Conchopata, for she was represented partially nude, with prominent breasts.

The nude woman was painted on the inside of an oversize urn of very high quality that has male warriors painted on the outside (Figure 3.15B). She is very realistically represented, although only her upper half is preserved. A small feline dangles upside down from the upper right, to suckle at her breast. Her face is in profile, with painted triangles below her eyes, a nose ring, and ear spools. The body seems to be in three-quarters view and animated, with empty hands, the left one raised and the right down to her side. Her hair trails out behind her in two bunches.

The tupu pin identifies a pair of women painted on a large jar (Figure 3.15A) discovered in one of Conchopata's firing ovens (Figure 3.2). This broken brewing jar was less carefully polished than the urn discussed above, but the figures are as realistically drafted as the nude woman nursing the feline. Perhaps this represents mythical narrative. Although incomplete, this scene shows dog- or fox-like animals devouring flesh from dismembered human limbs and vultures consuming what

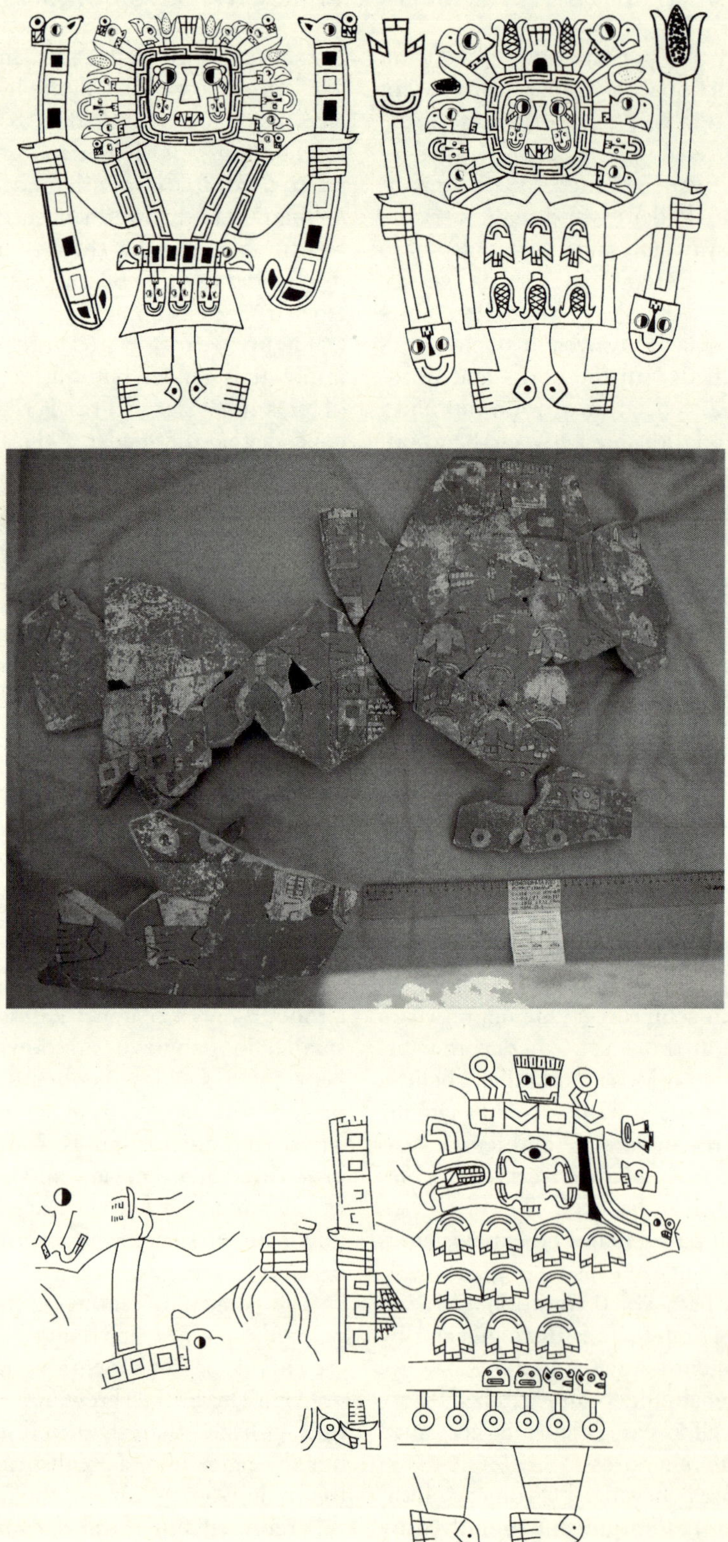

A

B

C

FIGURE 3.13. A. Male and female deities from Pacheco (redrawn from Morris and von Hagen 1993: 112, with modifications based on de Lavalle 1984:136 and Menzel 1977: 12g, figs. 122–123). B.–C. Male and female deities from Conchopata, photo and partially reconstructive drawing.

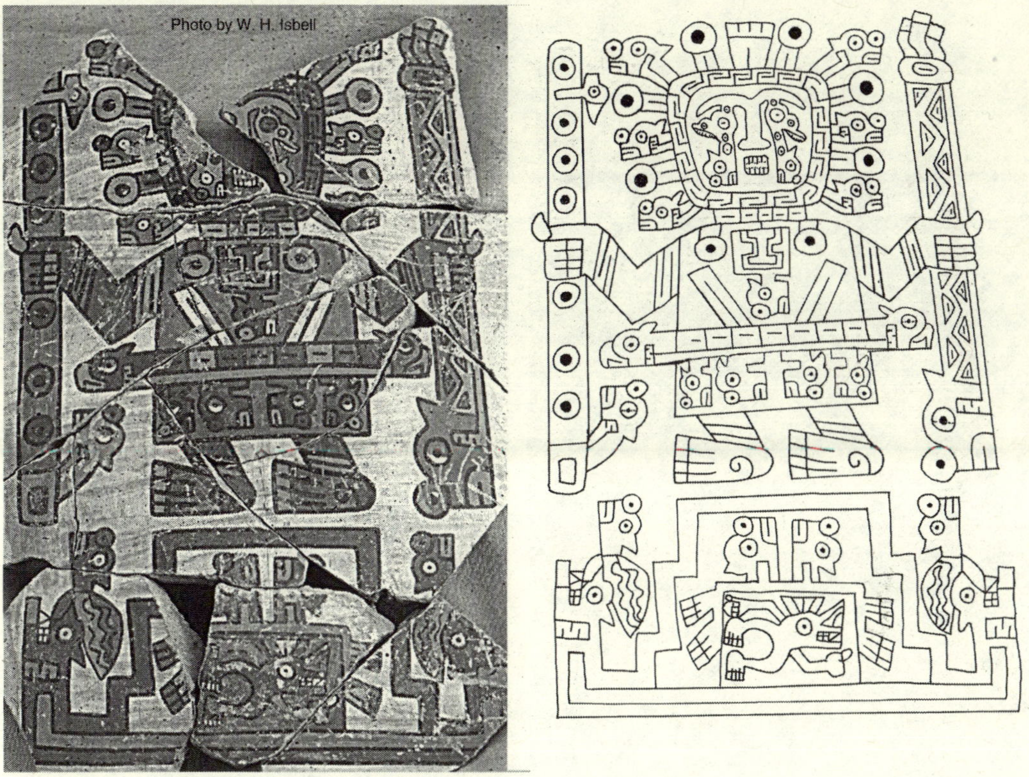

FIGURE 3.14. Photo and drawing of Staff God from giant jars discovered in Offering 1977. Circular objects at each shoulder might represent *tupus*, items usually associated with women's clothing. Drawing by P. Knobloch.

are probably entrails. Two women are untouched by these carnivores. Their garments are long, full, and plain except for a band design at the waist, probably a sash. On a well-preserved shoulder a tupu can be seen pinning the garment together. The two women stand facing one another in profile, each grasping a severed head by one of two dark braids. Another fragment of the same vessel reveals that the face of the severed head was luminescent white. If the pair of braids signals that the severed head or skull is female, it is tempting to associate this iconography with a myth about women and a female moon.

In addition to these fine ceramics, numerous large and coarse pots, probably brewing jars, were hastily painted with figures that share certain attributes, including long full dresses, not belted, of a single color with little or no decoration (Figure 3.16). I infer that they are women. They are usually shown with empty hands and bare headed except for two bunches of hair or braids. One example has a smaller image with the same attributes to her right and to her left, perhaps two daughters or female assistants. In many cases these women are associated with distinctive rayed filler designs that are more asymmetrical than the fillers associated with male figures (Figure 3.16).

If I am correct in these gender identifications, women at Conchopata decorated a great deal of fancy pottery with images of men clad in elegant outfits and grasping staffs, weapons, severed heads, and similar symbols. But they increasingly represented women, including women of modest status. Sometimes females were painted on fine vessels. These images with headdresses and objects in the hands seem to refer to goddesses or royals. Others with less paraphernalia, very realistically depicted, may represent mythical narratives. But coarse vessels, particularly great brewing

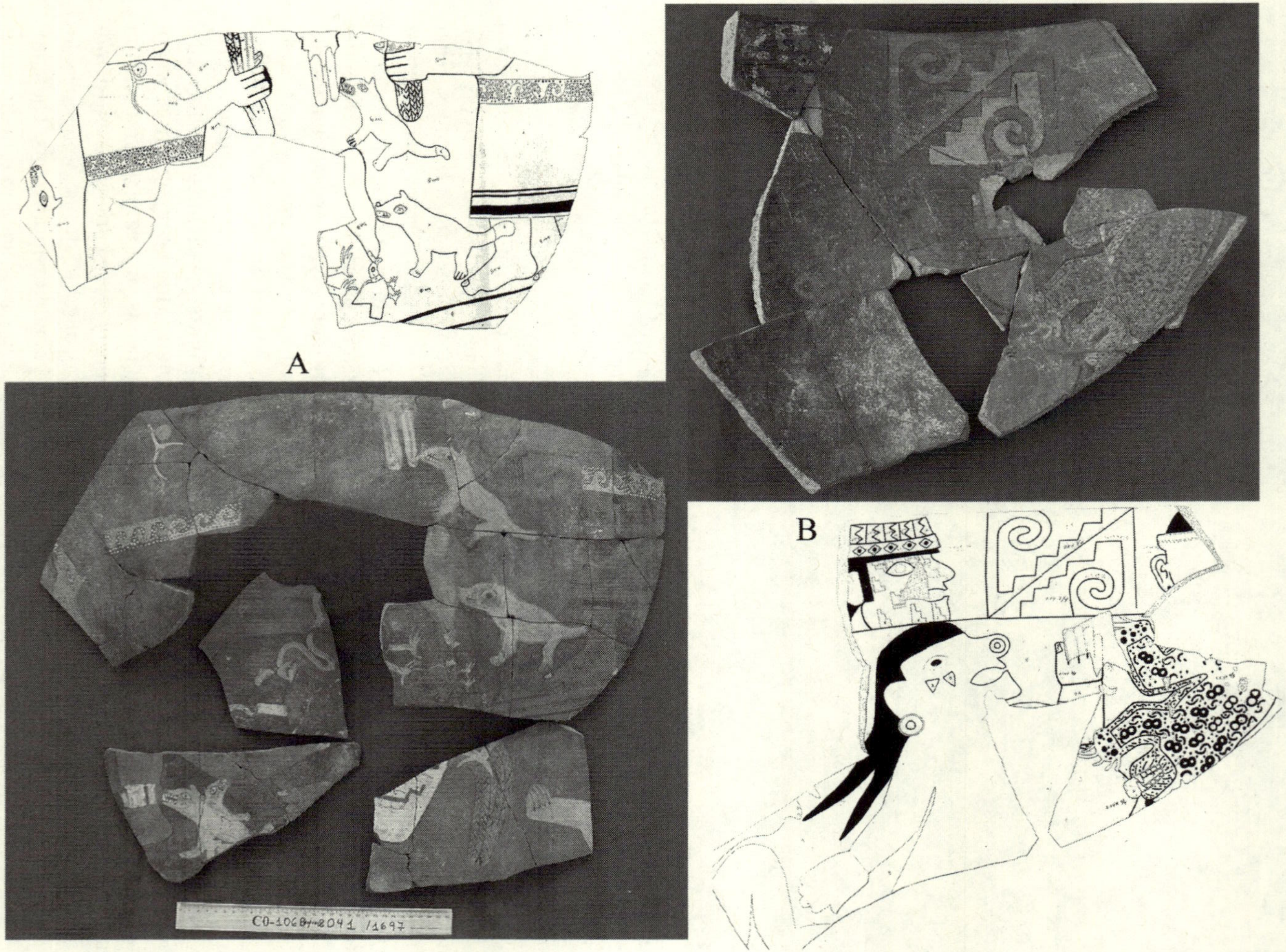

FIGURE 3.15. A. Women grasping a severed head or skull by two braids, with carnivores and carrion-eating birds consuming human parts. B. Nude woman with a feline suckling at her breast.

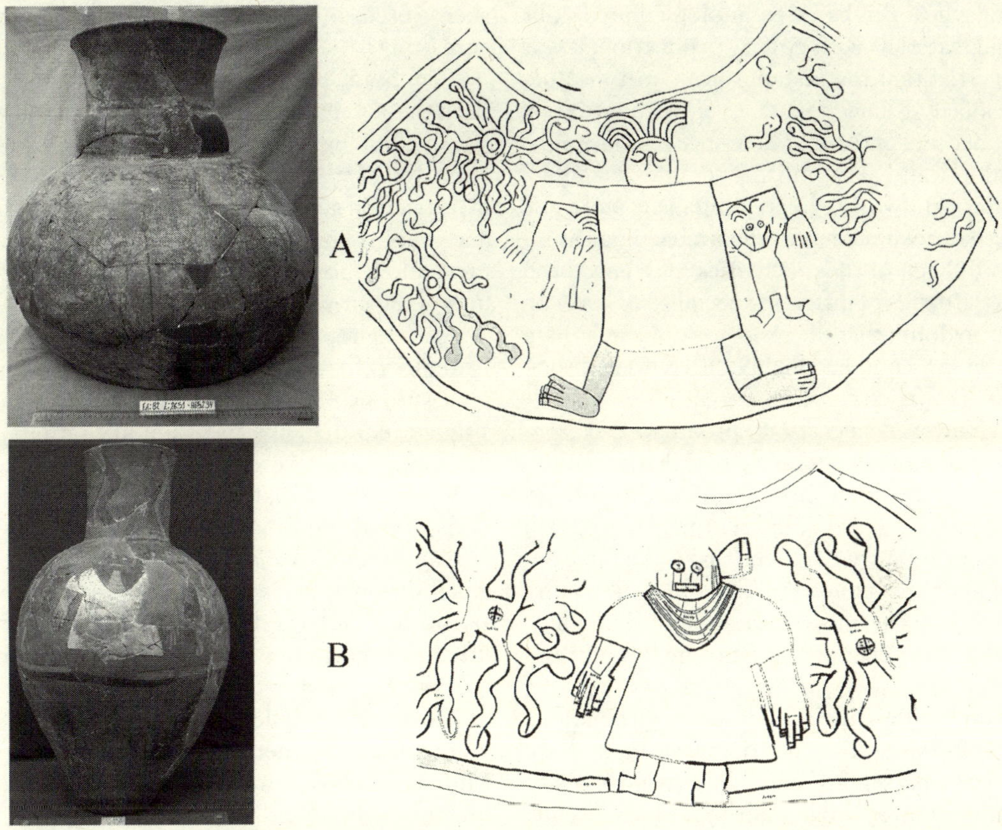

FIGURE 3.16. A.–B. Utilitarian brewing jars form Conchopata coarsely decorated with anthropomorphic images that have loose, unbelted garments; bare heads with two braids or bunches of hair; and empty hands. This combination of features seems to identify females. If so, jar A shows a woman accompanied by two smaller but similar images, perhaps daughters or female assistants.

jars, were increasingly embellished with cursive images that suggest modest women from daily life—probably the potters themselves. Women, it would seem, were claiming the large beer-making jars, and the product, both symbolically and in reality. This ceramic art projected women's agency into important social contexts (Gell 1998).

CONCLUSIONS

Until recently, Conchopata's ancient ceramic making was compared with modern Peruvian pottery production, especially at Ayacucho's Quinoa (Pozzi-Escot B. 2001; Pozzi-Escot B. et al. 1999). Comparisons were also made with Santa Ana, a craft barrio within the modern city of Ayacucho, most famous for weaving. Santa Ana's tradition of craftspeople has diversified in recent decades to include painting, stone carving, *retablo* production and sales, and even some pottery manufacturing.

Many popular approaches to the past model ancient people as rational decision makers. As at Santa Ana, multicrafting is a likely development from an initial craft specialization. One craft may not meet all subsistence needs, and craftspeople develop many talents—from specialized motor skills, to raw material procurement, to knowledge of marketing—that may apply to other crafts. Diversification into multicrafting is an expectation of many evolutionary theories. Consistent with this, when metals appeared at Conchopata that seemed too abundant and costly for middle-class artisans, and whose fire-based technology shared skills and materials with pottery production, it was inferred that the metallurgical craft also developed at Conchopata, even without direct evidence; Pozzi-Escot B. writes: "Besides agriculture and pastoralism, ceramic manufacture, metalworking, ceremonialism, and trade are all documented in the archaeological record. Each of these activities may have been in the hands of different specialists. There can be no doubt that the residents of Conchopata lived in a complex, highly differentiated society" (1991:92).

However, since 1998, broad-area excavations at Conchopata have revealed a community whose built environment, elite tombs, mortuary sex ratios, and ceramic offerings tell a contrary story. Multicrafting is clear in Conchopata's expanded archaeological record, but the crafts and services that accompanied pottery production were not metallurgy and trade but, rather, components of a feasting complex that included pottery production, beer brewing, banquet preparation, and appropriately serving guests at impressive feasts. Furthermore, ritual offering, which usually involved smashing large quantities of pottery, is a shadowy activity still inadequately understood. Offering smashes appear to have followed many events, from public ceremonies in Conchopata's Pink Plaza to closing the house of a humble woman. Surely they increased demand for pottery, but more research is needed to understand offerings as well as their effects. Particularly, offerings must be differentiated from refuse. For example, does the huge ceramic collection labeled Offering 2000A represent a ritual smash or a collection of sherds used to enclose decorated pots for secure firing?

The study of Conchopata emphasizes both the potentials and the dangers of ethnographic analogies, especially when archaeological sites are little explored. Conchopata's investigation shows how broad-area contexts and a holistic approach to the built environment allow more comprehensive understandings of the past.

At Conchopata we have discovered a case of multicrafters whose activities and strategies were probably little influenced by subsistence requirements and economic rationalism. Rather, they were driven by a competitive feasting complex that was linked to the ambitions of male elites in an environment of increasing concentrations of political power and social inequality. These crafts were surely subsidized. Perhaps this special kind of subsidized multicrafting retarded the development of greater economic specialization in Wari cities, creating a fragile urban base. But much more must be learned about Huari and other Middle Horizon centers before global interpretations should be proffered.

Conchopata's craftspeople were women, and these women employed the products of their arts to project feminine agency into dynamic collective events where important social and political negotiations took place (Gell 1998). I suspect that women were identifying themselves with the power of brewing vessels and beer itself, as they participated in a rapidly developing urban landscape where new identities, especially a new "city woman," were being constructed (Smith 2003).

Acknowledgments

The Conchopata Archaeological Project is directed by Dr. Anita G. Cook, Dr. Jose Ochatoma P., Lic. Martha Cabrera R. de Ochatoma, and me and administered by Alberto Carbajal. Thanks are due to the sponsors and particularly the National Geographic Society, which was the primary patron from 1998 through 2002. Initial support was a grant from Wenner-Gren in 1997 to Ochatoma. Additional funding has come from the Curtiss T. and Mary G. Brennan Foundation, Dumbarton Oaks, and the Heinz Foundation. The National Science Foundation sponsored the project in 2003 and made doctoral improvement grants for special research by Catherine Bencic. Fulbright dissertation funds were provided to Bencic, Greg Ketteman, and Barbara Wolfe. Excavations were conducted in 1997, 1998, 1999, 2000, 2001–2002, and 2003, with permission granted to Professor Ochatoma (1997–1998) and me (1999, 2000, 2001–2002,

2003). I wish to thank the codirectors and also the archaeology students (especially those from the State University of New York–Binghamton: Catherine Bencic, Juan Carlos Blacker, Juan Leoni, Greg Ketteman, Mike Calaway, Ariela Zycherman, Kris Mearish, Amy Groleau, Dan Eisenberg, Brian Finucane, Meredith Davis, and Bren Torelli; the Universidad Nacional de San Cristóbal de Huamanga: Lorenzo Huisa, Carlos Mancilla, Ismael Mendosa, Maximo Lopez, Teresa Limalla, Irela Vallejo, Alina Alvarado, and Edgar Alarcón; the Catholic University of America: Barbara Wolff, Nikki Slovak, David Crowley, Teresa Carmona, and Eric Schmidt; the Pontificia Universidad Católica del Perú: Gonzalo Rodriguez, Manuel Lizarraga, and Antonio Gamonal; the University of Pittsburgh: Charlene Milliken; the University of North Carolina: Tiffiny Tung; the University of California–Berkeley: Bill Whitehead and Matt Sayre; the Universidad Nacional de San Marcos: Patricia Mayta; and Mabel Mamani and Silvana Rosenfeld from Argentina) who have contributed to the research by directing excavation crews and participating in analyses. Special recognition is due to Alberto Carbajal A., my friend, our project administrator, and an insightful archaeologist. I want to thank Dr. Luis Lumbreras, former director of Peru's Instituto Nacional de Cultura, for his encouragement and help. Through the years unequaled support has come from Dr. Enrique Gonzalez Carré, in his posts as rector of the Universidad Nacional de San Cristóbal de Huamanga, director of the Museo Nacional de Arqueología, and director of Peru's Instituto Nacional de Cultura. Gonzalez had a storage facility added to the archaeology laboratory of the Universidad Nacional de San Cristóbal de Huamanga expressly for materials excavated at Conchopata. I want to thank the directors of Ayacucho's Instituto Nacional de Cultura—Ulyses Lareya, Teresa Carrasco, Mariano Benites, and Severino Castillo—for their help, friendship, and advice. César Alvarez, director of Ayacucho's archaeological museum Hipolito Unanue, also contributed to the success of the Conchopata Archaeological Project, as did many others. I also wish to thank my university, the State University of New York at Binghamton, and my colleagues in anthropology, for support and encouragement. Dr. Cook's Catholic University of America and the Universidad Nacional de San Cristóbal de Huamanga have also contributed generously. Finally, my wife, Judy Siggins, has been a source of continual help and inspiration. She has managed our home and family life during my long research absences; she read and edited my proposals, reports, and manuscripts; and she is the best consultant anyone could hope for.

Notes

1. In 1998, in response to continuing destruction at Conchopata, several archaeologists agreed to cooperate in the salvage archaeology of as much of the site as possible before its destruction. The Conchopata Archaeological Project was established by J. Ochatoma, M. Cabrera, A. Carbajal, A. Cook, and me. See the acknowledgments for additional information. Five seasons of excavations exposed an almost continuous urban landscape of almost 2 ha, linking earlier excavation areas except along the highway that bisects the site (see Figure 3.2).
2. The artifacts in Figure 3.3 were excavated between 1998 and 2003 by the Conchopata Archaeological Project, but they are virtually identical to the potters' tools found by Pozzi-Escot.
3. Unfortunately, building BC-E was partially destroyed by a modern house constructed in its west center. Important information about this palace compound was lost.
4. Ceramic production was probably a function of the ceremonial calendar, which determined when major celebrations were to take place, requiring quantities of beer and food. Conchopata is in a sufficiently arid environment so that ceramics can be shaped, dried, and fired all year round, although the dry season, from about May through October, is best for drying and firing.
5. The burial in room 205 is of Type 3, named "Cist Interment," and considered to represent average citizens, neither wealthy nor poor (Isbell 2004). The next higher step, Type 4, the "Bedrock Cavity Tomb," is inferred to represent minimal nobility. If the woman buried in room 205 was a sister or half sister of a minimal male noble, Type 3 might have been the appropriate mortuary treatment for her. Of course, more research is required.

References Cited

Anders, Martha B.

1990 Maymi: Un Sitio del Horizonte Medio en el Valle de Pisco. *Gaceta Arqueológia Andina* 5:27–39.

Anders, Martha B., Susana Arce, Izumi Shimada, Victor Chang, Luis Tokuda, and Sonia Quiroz

1998 Early Middle Horizon Pottery Production at Maymi. In *Andean Ceramics: Technology, Organization, and Approaches*, edited by Izumi Shimada, pp. 233–251. MASCA Research Papers. University Museum of Archaeology and Anthropology, University of Pennsylvania, Philadelphia.

Anders, Martha B., Victor Chang, Luis Tokuda, Sonia Quiroz, and Izumi Shimada

1994 Producción Cerámica del Horizonte Medio Temprano en Maymi, Valle de Pisco, Perú. In *Tecnología y Organización de la Producción de Cerámica Prehispánica en los Andes*, edited by Izumi Shimada, pp. 249–267. Pontificia Universidad Católica del Perú, Lima.

Bencic, Catherine

2001 Industrias Líticas de Huari y Tiwanaku. In *Boletín de Arqueología PUCP, No. 4, 2000 Huari y Tiwanaku: Modelos vs. Evidencias, Primera Parte*, edited by Peter Kaulicke and William H. Isbell, pp. 89–118. Pontificia Universidad Católica del Perú, Lima.

Blacker, Juan Carlos

2001 Growing Up Huari: An Analysis of Architectural Style, Technique, and History at the Middle Horizon Site of Conchopata, Ayacucho, Peru. Unpublished M.A. thesis, Department of Anthropology, State University of New York at Binghamton.

Blacker, Juan Carlos, and Anita Cook

2006 Wari Wasi: Defining Conchopata Houses. Paper presented at the 71st Annual Meeting of the Society for American Archaeology, Puerto Rico, April 27–30.

Cook, Anita G.

1987 The Middle Horizon Ceramic Offerings from Conchopata. *Ñawpa Pacha* 22–23: 49–90.

1994 *Wari y Tiwanaku: Entre el Estilo y la Imagen*. Pontificia Universidad Católica del Perú, Lima.

2001 Huari D-Shaped Structures, Sacrificial Offerings, and Divine Rulership. In *Ritual Sacrifice in Ancient America*, edited by Elizabeth Benson and Anita Cook, pp. 137–163. University of Texas Press, Austin.

Cook, Anita G., and N. Benco

2001 Vasijas para la fiesta y la fama: Producción artesanal en un centro urbano Huari. In *Boletín de Arqueología PUCP, vol. 4: Huari y Tiwanaku: Modelos vs. Evidencias*, edited by Peter Kaulicke and William H. Isbell, pp. 489–504. Fondo Editorial, Pontifica Universidad Católica del Perú, Lima.

Cook, Anita G., and Mary Glowacki

2003 Pots, Politics, and Power: Huari Ceramic Assemblages and Imperial Administration. In *The Archaeology and Politics of Food and Feasting in Early States and Empires*, edited by Tamara L. Bray, pp. 173–202. Kluwer Academic/Plenum Publishing, New York.

Conklin, William J.

1970 Peruvian Textile Fragment from the Beginning of the Middle Horizon. *Textile Museum Journal* 1:15–24.

1986 The Mythic Geometry of the Ancient Southern Sierra. In *The Junius B. Bird Conference on Andean Textiles*, edited by Ann P. Rowe, pp. 123–136. Textile Museum, Washington, D.C.

2004 The Fire Textile. *Hali* 133(March–April): 94–100.

Finucane, Brian

2003 Conchopata Archaeological Project field and laboratory notes. Stored at the University of Huamanga, Ayacucho, State University of New York–Binghamton, and Catholic University of America, Washington D.C.

Gell, Alfred

1998 *Art and Agency: An Anthropological Theory*. Clarendon Press, Oxford.

Gero, Joan M.

1990 Pottery, Power, and Parties! at Queyash, Peru. *Archaeology* 43(2):52–55.

Green, Kelsey, and William Whitehead

2006 Domesticated Plant Resources at Conchopata, a Wari Site in the Ayacucho Valley, Peru. Poster presented at the 71st Annual Meeting of the Society for American Archaeology, Puerto Rico, April 27–30.

Groleau, Amy

2005 House-Keeping and House-Leaving: A Case Study of Oversize Ceramic Offering and Modes of Abandonment from Middle Horizon, Peru. Unpublished M.A. thesis, Department of Anthropology, State University of New York at Binghamton.

Isbell, William H.

1987 Conchopata: Ideological Innovator in

Middle Horizon IA. *Ñawpa Pacha* 22–23:91–126.

1997 *Mummies and Mortuary Monuments: A Postprocessual Prehistory of Andean Social Organization.* University of Texas Press, Austin.

2001 Repensando el Horizonte Medio: El caso de Conchopata, Ayacucho, Perú. In *Boletín de Arqueología PUCP, No. 4, 2000 Huari y Tiwanaku: Modelos vs. Evidencias, Primera Parte*, edited by Peter Kaulicke and William H. Isbell, pp. 9–68. Departamento de Humanidades, Especialidad de Arqueología, Pontificia Universidad Católica del Perú, Lima.

2004 Mortuary Preferences: A Wari Culture Case Study from Middle Horizon Peru. *Latin American Antiquity* 15(1) 3–32.

Isbell, William H., and Anita G. Cook

1987 Ideological Origins of an Andean Conquest State. *Archaeology* 40(4):27–33.

2002 A New Perspective on Conchopata and the Andean Middle Horizon. In *Andean Archaeology, vol. 2: Art, Landscape and Society*, edited by Helaine Silverman and William H. Isbell, pp. 249–305. Kluwer Academic/Plenum Publishers, New York.

Isbell, William H., and Patricia J. Knobloch

2006 Missing Links, Imaginary Links: Staff God Imagery in the South Andean Past. In *Andean Archaeology, vol. 3: North and South*, edited by William H. Isbell and Helaine Silverman, pp. 307–351. Springer, New York.

Lavalle, José Antonio de

1984 *Culturas Precolombinas: Huari.* Banco de Credito del Perú, Lima.

Lechtman, Heather, and Lisa Frame

2005 Bronze Artifacts from Conchopata, Peru. Paper presented at the 70th Annual Meeting of the Society for American Archaeology, Salt Lake City, March 30–April 3.

Lumbreras, Luis G.

1969 *De los Pueblos, las Culturas y las Artes del Antiguo Perú.* Francisco Moncloa Editores, Lima.

1974 *Las Fundaciones de Huamanga.* Editorial Nueva Educación, Lima.

1981 The Stratigraphy of the Open Sites. In *Prehistory of the Ayacucho Basin, Peru*, edited by R. S. MacNeish, A. Garcia Cook, L. G. Lumbreras, R. K. Vierra, and A. Nelken-Terner, pp. 167–198. University of Michigan Press, Ann Arbor.

Menzel, Dorothy

1964 Style and Time in the Middle Horizon. *Ñawpa Pacha* 2:1–106.

1968 New Data on Middle Horizon Epoch 2A. *Ñawpa Pacha* 6:47–114.

1977 *The Archaeology of Ancient Peru and the Work of Max Uhle.* R. H. Lowie Museum of Anthropology, University of California, Berkeley.

Morris, Craig, and Adriana von Hagen

1993 *The Inka Empire and Its Andean Origins.* Abbeville Press, New York.

Nash, Donna J.

2006 Garbage or Goodies? The Archaeological Study of Wari Residential Remains in Moquegua. Paper presented at the 71st Annual Meeting of the Society for American Archaeology, Puerto Rico, April 27–30.

Ochatoma Paravicino, José, and Martha Cabrera Romero

2000 Excavaciones en un Poblado Alfarero de la Época Huari. Project report. Facultad de Ciencias Sociales, Universidad Nacional de San Cristóbal de Huamanga, Ayacucho.

2001a Arquitectura y Áreas de Actividad en Conchopata. In *Boletín de Arqueología PUCP, No. 4, 2000 Huari y Tiwanaku: Modelos vs. Evidencias, Primera Parte*, edited by Peter Kaulicke and William H. Isbell, pp. 449–488. Departamento de Humanidades, Especialidad de Arqueología, Pontificia Universidad Católica del Perú, Lima.

2001b Descubrimiento del área ceremonial en Conchopata, Huari. In *XII Congreso Peruano del Hombre y la Cultura Andina, Tomo II*, edited by I. Pérez, W. Aguilar, and M. Purizaga, pp. 212–244. Universidad Nacional de San Cristóbal de Huamanga, Ayacucho.

2001c Ideología Religiosa y Organización Militar en la Iconografía de Área Ceremonial de Conchopata. In *Wari: Arte Precolumbino Peruano*, edited by L. Millones, M. Cabrera Romero, E. González Carré, W. H. Isbell, F. Meddens, C. Mesía Montenegro, J. Ochatoma Paravicno, D. Pozzi-Escot, and C. Williams León, pp. 173–211. Fundación El Monte, Seville.

2002 Religious Ideology and Military Organization in the Iconography of a D-Shaped Ceremonial Precinct at Conchopata. In *Andean Archaeology, vol. 2: Art, Landscape and Society*, edited by Helaine Silverman and William H. Isbell, pp. 225–247. Kluwer Academic/Plenum Publishing, New York.

Pérez Calderón, Ismael
1998 Excavación y Definición de un Taller de Alfareros Huari en Conchopata. *Conchopata: Revista de Arqueología (Universidad Nacional de San Cristobal de Huamanga, Oficina de Investigación)* 1: 93–137.

Pérez Calderón, Ismael, and Jose A. Ochatoma Paravicino
1998 Viviendas, Talleres, y Hornos de Producción Alfarera Huari en Conchopata. *Conchopata: Revista de Arqueología (Universidad Nacional de San Cristobal de Huamanga, Oficina de Investigacion)* 1:72–92.

Pozzi-Escot B., Denise
1982 Proyecto Qonchopata: Campaña 1982. Unpublished excavation report.
1985 Conchopata: Un Poblado de Especialistas Durante el Horizonte Medio. *Boletín del Instituto Francés de Estudios Andinos* 14(3–4):115–129.
1991 Conchopata: A Community of Potters. In *Huari Administrative Structure: Prehistoric Monumental Architecture and State Government*, edited by William H. Isbell and Gordon F. McEwan, pp. 81–92. Dumbarton Oaks Research Library and Collection, Washington, D.C.
2001 Viejas Formas, Nuevos Estilos: La Tradición del Barro. In *Wari: Arte Precolumbino Peruano*, edited by L. Millones, M. Cabrera Romero, E. González Carré, W. H. Isbell, F. Meddens, C. Mesía Montenegro, J. Ochatoma Paravicno, D. Pozzi-Escot, and C. Williams León, pp. 273–303. Fundación El Monte, Seville.

Pozzi-Escot B., Denise, Marleni Alarcón G., and Cirilo Vivanco Pomacanchari
1994 Cerámica Wari y su Tecnología de Producción: Una Visión Desde Ayacucho. In *Tecnología y Organización de la Producción de Cerámica Prehispánica en los Andes*, edited by Izumi Shimada, pp. 269–294. Pontificia Universidad Católica del Perú, Lima.
1998 Wari Ceramics and Production Technology: The View from Ayacucho. In *Andean Ceramics: Technology, Organization, and Approaches*, edited by Izumi Shimada, pp. 253–281. MASCA Research Papers. University Museum of Archaeology and Anthropology, University of Pennsylvania, Philadelphia.
1999 *Etnografía Alfarera Wari: Los Artesanos de Conchopata*. Facultad de Ciencias Sociales, Universidad Nacional de San Cristóbal de Huamanga, Ayacucho.

Rosenfeld, Silvana
2004 Zooarchaeology at Conchopata, a Huari Urban Center. Paper presented at the 69th Annual Meeting of the Society for American Archaeology, Montreal, March 31–April 4.
2006 Feasting and Animal Offerings in the Andes: The Case of Conchopata (Ayacucho, Peru). Paper presented at the 71st Annual Meeting of the Society for American Archaeology, Puerto Rico, April 27–30.

Sayre, Matthew P., and William Whitehead
2003 New Paleoethnobotanical Evidence from Conchopata: A Huari Site. Paper presented at the 68th Annual Meeting of the Society for American Archaeology, Milwaukee, April 9–13.

Smith, Monica L.
2003 *The Social Construction of Ancient Cities*. Smithsonian Institution, Washington, D.C.

Tung, Tiffiny A.
2003 A Bioarchaeological Perspective on Wari Imperialism in the Andes of Peru: A View from Heartland and Hinterland Skeletal Populations. Unpublished Ph.D. dissertation, Department of Anthropology, University of North Carolina at Chapel Hill.

Tung, Tiffiny A., and Anita Cook
2006 Intermediate Elites and Their Role in Wari Imperialism as Identified through Bioarchaeological and Mortuary Analysis. In *Intermediate Elites in Pre-Columbian States and Empires*, edited by Christina M. Elson and Alan Covey, pp. 68–93. University of Arizona Press, Tucson.

4

Craft Production in Classic Period Oaxaca: Implications for Monte Albán's Political Economy

GARY M. FEINMAN AND LINDA M. NICHOLAS

For more than a century (e.g., Starr 1900), tourists, explorers, and researchers have traveled to the Valley of Oaxaca, Mexico, to consume, appreciate, and study the region's renowned craft industries, which include pottery making (e.g., Hendry 1992; Shepard 1963), groundstone metate manufacture (Cook 1982), and different kinds of cloth and textile production (e.g., Cohen 1999; Parsons 1936; Stephen 1991), as well as the more recent carving and elaborate painting of wooden animals (Chibnik 2000). Over the last decades, archaeologists have documented that specialized craft manufacture (craft goods made for exchange) has a lengthy history in the valley going back at least two millennia (e.g., Feinman 1986; Feinman and Nicholas 2000; Winter 1984). Clearly, the means, methods, and systems of contemporary Mexican craft producers can help inform us about how such goods were made in the past (e.g., Arnold 2000; Stark 2003). But to what degree, and in which respects? To put it more explicitly, how much can contemporary modes of craft production in the Valley of Oaxaca clue us in about pre-Hispanic times, and, more importantly, what are the key differences?

In this essay, we begin an examination of the aforementioned questions through a discussion of craft production in the Classic period (ca. A.D. 200–800) Valley of Oaxaca. In a chapter of this length, it would be impossible to examine, compare, and contrast all facets of different Oaxacan craft traditions from the past to the present. Rather, our emphasis here is focused on the organization of craft production in the Valley of Oaxaca; much less attention is given to either specific technologies or the step-by-step process of production because both vary greatly depending on the specific craft. Our focus on the scale and organization of production is intended to provide a perspective on the broader regional economy during the pre-Hispanic Classic period when the urban center of Monte Albán (Blanton 1978) is generally thought to have been the political capital of the Valley of Oaxaca. Through an understanding of the organization of Classic period craft production and how it differs from and is similar to that of the present, we aim to enhance our perspective on and understanding of the broader political economy of this key pre-Hispanic region (Palerm and Wolf 1957).

Following a brief review of craft production in contemporary Oaxaca, we discuss the available evidence on Classic period craft production. Our presentation draws from archaeological studies undertaken at various scales from the region, to the site, to the house.

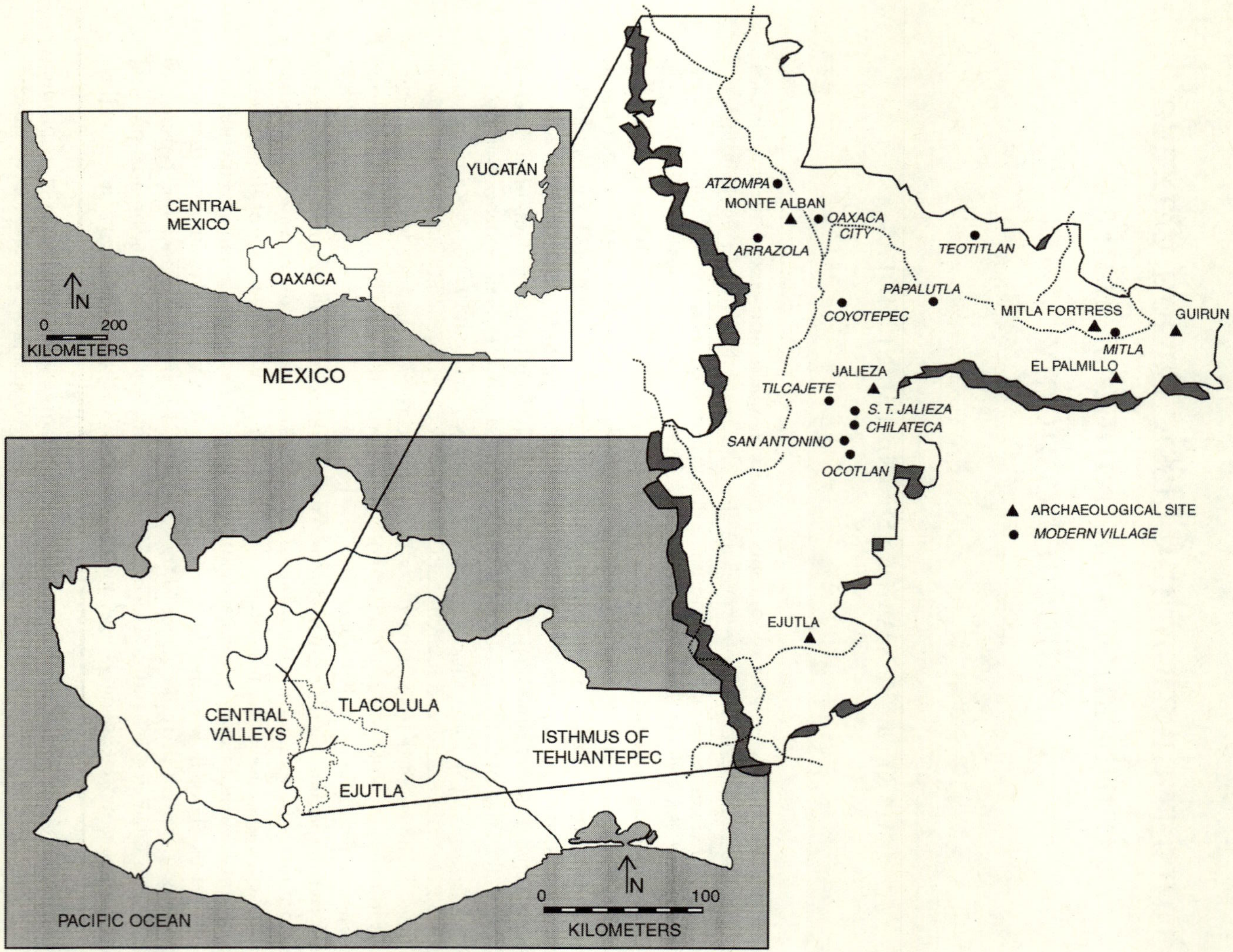

FIGURE 4.1. The Valley of Oaxaca and places and sites mentioned in the text.

In recounting what we know about the past, comparisons are made with more recent patterns of craft manufacture and exchange.

CONTEMPORARY CRAFT PRODUCTION IN THE VALLEY OF OAXACA

From the earliest anthropological accounts of the Valley of Oaxaca (e.g., Beals 1979; Parsons 1936; Starr 1900), the production of craft goods has been described as a noteworthy economic activity engaged in by the region's inhabitants. Earlier colonial-era descriptions also document that nonfood goods produced for exchange were a key aspect of Oaxaca's economy from the time of the Spanish conquest (e.g., Baskes 2000; Spores 1965), with different parts of the valley associated with diverse products. These sources make it evident that the manufacture of nonagricultural goods has a long history in Oaxaca. Clearly, as early as the sixteenth century if not before, craft items and inedible raw materials were exchanged by their producers in markets (Appel 1982; Spores 1965; Taylor 1972), an institution of exchange that has remained a critical building block of the valley's economy into the twentieth and twenty-first centuries (Beals 1975; Cook and Diskin 1976; Drucker-Brown 1982; Kowalewski 2003; Murphy and Stepick 1991).

Today, many villages in Oaxaca are associated closely with specific craft goods. Some of the most prominent include pottery and ceramic figures in Santa María Atzompa, San Bartolo Coyotepec, and Ocotlán de Morelos; rugs in Teotitlán del Valle; woven goods in Santo Tomás Jalieza and San Pablo Mitla; embroidery in San Antonino Castillo Velasco and San Juan Chilateca; carved wooden animals in San Antonio Arrazola and San Martín Tilcajete; and woven baskets in Santa Cruz Papalutla (Figure 4.1). Many Oaxacan villages today specialize in one or two such crafts, with specific families in each community recognized as master artists and artisans who produce for an international market (e.g., Rothstein and Rothstein 2002; Wasserspring 2000).

Although craft production has been economically significant in the Valley of Oaxaca for centuries (if not much longer), the nature of the specialized manufacture of nonfood items for exchange has changed in important ways during the ethnographic present. Current artisan production in the Valley of Oaxaca clearly reflects transport and technological changes that have occurred since the middle of the twentieth century. The construction of the railroad followed by paved highways (particularly the Pan-American Highway in the 1940s), airline connections, and most recently an expressway have changed the relationship between valley communities and the rest of the world (Murphy et al. 1997). The last 75 years have brought some of the greatest economic and demographic transitions in the entire history of this region (Kowalewski 2003; Murphy and Stepick 1991); linking Oaxaca craft workers to a tourist market has had a dramatic impact on craft production. Art and artifacts from Oaxaca can now be found at stores and galleries worldwide.

At the same time, many small-scale craft industries that were situated in Oaxaca during the 1920s and 1930s (Beals 1979; Parsons 1936), including the manufacture of items of clothing, footwear, household implements, and more, have been lost and replaced by cheaper imported goods from Mexico City and beyond (Murphy et al. 1997:239). For example, in the village of San Pablo Mitla, almost half the occupations listed by Elsie Clews Parsons (1936) as being present in 1933 were no longer found 35 years later in 1968 (Beals 1979:168–169). Instead, by 1968, 45 percent of the population (much greater than earlier) was engaged in weaving activities (Beals 1979:169), a percentage that likely is even higher today.[1] A similar pattern has been noted in the reed basket–making village of Santa Cruz Papalutla (Druijven and Kruithof 1992), where a craft once carried out mainly for immediate and local use now is engaged in by 90 percent of the households, accounting for more than 50 percent of the community's income. Some baskets made in the village today are exported to the United States. Given these dramatic changes in Oaxacan craft industries in recent times, a simple comparison (or the presumption of analogical identity) between "contemporary" craft manufacture and that of the Classic period is not possible.

Nevertheless, despite these economic changes in consumption and supply, craft production in the valley still is enacted largely in domestic contexts. Although extended family members frequently are involved today in these domestic production units, the house remains the basic context of artisan manufacture (Rothstein and Rothstein 2002). Thus a basic pattern evidenced in the early decades of the twentieth century (Parsons 1936; Van de Velde and Van de Velde 1939) still is most common today. In looking at past Oaxacan craft manufacture with a model from the present in mind, the context or scale (sensu Costin 1991) of specialized production is an important axis of comparison. That is, was specialized production also carried out in domestic contexts during the Classic period?

Today, many communities in the Valley of Oaxaca are known for specific crafts. Many, if not most, of the households in that community tend to be involved in the manufacture of goods made of one material (either pottery, cloth, wood, or something else). Specific artisans and the craft-producing households in the community focus on that economic activity, perhaps supplemented by agricultural work. In regard to Classic period craft manufacture, the question is whether one sees the pattern that is most evident today, with communities and their component households focused largely on one craft, or whether the pattern was more like that found a half century or more ago, with multiple crafts evident within households and settlements. Frequently, the assumption is made that the twenty-first-century pattern of villages specializing heavily in one craft extends deeply into the past (e.g., Wasserspring 2000:15), but to what degree is that really the case?

CLASSIC PERIOD CRAFT PRODUCTION IN THE VALLEY OF OAXACA: SURVEY

The archaeological perspective on the nature and organization of Classic period craft production in the Valley of Oaxaca is based on information collected at three distinct spatial scales: the region, the site or settlement, and the domestic unit or house. This discussion, which largely synthesizes findings that have been reported in greater depth elsewhere, begins at the most inclusive scale—the region—and then focuses down on sites and houses. As with any scientific observations, the broader-scale view is more representative, whereas the narrower vantage provides additional detail and greater specificity.

Based on the surface findings from the Valley of Oaxaca and its immediate hinterland (Figure 4.1; Blanton 1978; Blanton et al. 1982; Feinman and Nicholas 1990; Kowalewski et al. 1989), we have surmised that craft production activities were distributed differentially at settlements in the region. During the Classic period, production for exchange increased in importance, perhaps largely related to increases in the region's population and an expansion of the area that was under the hegemony of Monte Albán, thereby expanding the potential consumer base for goods produced by specialists (Kowalewski et al. 1989:213, 225, 245). At the same time, the expanded size and power of Monte Albán likely led to greater demands on domestic labor and agricultural production, opening consumption niches for goods produced outside the individual house and community. Many of the largest Classic period sites in the Valley of Oaxaca were upland terraced settlements (Feinman 1997:30; Kowalewski et al. 1989:240–245), situated in locales where corn farming (dependent on rainfall) would have been adequate at best, and a good number of these piedmont and mountain communities had surface indications that one or more craft production activity was important at that location (see Feinman 1997:30; Feinman and Nicholas 1991:253).

Of the largest Early Classic period (ca. A.D. 200–350) settlements in the Valley of Oaxaca and the adjacent Ejutla Valley where survey crews observed evidence of craft manufacture, 17 out of 39 had surface indications of at least two different craft activities (obsidian, local chipped stone, ground stone, marine shell, spindle whorls [fiber], or pottery; Table 4.1). Although chronological and interpretive problems could be an issue for the validity of this multicraft pattern at these sites, one also must consider that many crafts important in pre-Hispanic Oaxaca (such as basketry, feather working, and the making of bone implements) normally would not leave clear archaeological traces on the surface. At the same time, at an eighteenth site (Ejutla), only a single craft activity (shell working) was observed during the regional survey (Feinman and Nicholas 1990). Nevertheless, as amplified below, in the course of subsequent archaeological study at Ejutla several other craft activities were identified in the single house that was excavated (Feinman and Nicholas 2000). Based on these regional-scale findings, the observed presence of more than one craft industry at many larger settlements in Classic period Oaxaca appears to indicate that this likely was a common practice.

The regional-scale findings are supported by intensive site surveys that have been conducted at five large hilltop terrace sites with significant Classic period occupations in the Valley of Oaxaca (Figure 4.1): Monte Albán (Blanton 1978), Jalieza (Finsten 1995), El Palmillo, Guirún, and the Mitla Fortress (Feinman and Nicholas 2004). With the additional time to make many more surface collections (and more detailed observations of artifacts found on the ground) and to map terraces in greater detail than was possible during the regional surveys, a range of craft activities has been well documented at each site. The working of local stone (especially chert) into chipped-stone tools was an important economic activity at Guirún, El Palmillo, and the Mitla Fortress. Other crafts varied more in the intensity at which they were practiced at each site. For example, we recorded the most indications of obsidian working and the processing of xerophytic (drought-resistant) plants at the fortress in Mitla, similar plant processing as well as pottery manufacture at El Palmillo, and the working of soft bedrock into sculptural and architectural materials at Guirún (Feinman and Nicholas 2004). At Monte Albán, Blanton (1978) found surface evidence for a range of different craft activities including ground- and chipped-stone tool production, the working of obsidian, ceramic manufacture, and the crafting of marine shell ornaments. With the exception of shell working, the same craft activities have been noted at Jalieza (Finsten 1995); the production of chipped-stone artifacts from local rock sources was particularly prominent. At each of these five sites, the surface residues of apparent craft work were especially prevalent on terraces, making it likely that such activities were undertaken in domestic contexts.

Classic Period Craft Production in the Valley of Oaxaca: Houses

Based on surface studies, we can surmise that a range of craft production activities was practiced at many larger Classic period sites and that these crafts likely were enacted in domestic contexts. These patterns seem to be relatively widespread in the region. Yet, to affirm that more than one craft was carried out simultaneously in the same domestic unit, a more detailed and focused perspective is needed.

House excavations carried out at the Ejutla site (Feinman 1999; Feinman and Nicholas 2000) and El Palmillo (Feinman et al. 2002, 2006; Haines et al. 2004; Haller et al. 2006) provide just such a vantage. At the former site, one house complex and the adjacent area were uncovered; six domestic residential units, each placed on terraces, were excavated at El Palmillo. The findings from these excavations clearly indicate that Classic period households engaged in specialized crafts, and these activities were markedly different at the two sites. At El Palmillo, different houses had somewhat distinct economic foci, although certain crafts were broadly present across the site. At both sites, at least some domestic units clearly were multicrafting, practicing more than one craft activity contemporaneously.

Craft Activities at the Ejutla Site

The Ejutla site is a large, 1 km^2 settlement located on the floor of the Ejutla Valley at the southern end of the Valley of Oaxaca. Following the discovery of anomalous quantities of marine shell debris on the surface of one sector of the site during regional survey (Feinman and Nicholas 1990), we began a series of

Table 4.1. Classic Period (IIIa) Settlements with Surface Evidence of Craft Production

Subvalley	Chipped Stone	Ceramic	Ground/ Other Stone	Obsidian	Lapidary	Weaving	Shell	Site (Pop ≥100)
Central	X	X	X	X			X	Monte Albán
Ejutla		X		X		X		EJ-LP-LG-1
Ejutla	x	x			x	x	X	Ejutla
Ejutla		X				X		El Vergel
Etla	X							Loma del Trapiche
Etla	X							S. D. Tlaltinango
Etla	X			X			X	S. J. Mogote
Etla	X							Matadamas
Etla				X				Reyes Etla
Ocotlán	X							OC-STJ-SCJ-1
Ocotlán		X				X		S. I. Yatzeche
Tlacolula	X	X		X				Tlacochahuaya
Tlacolula	X	X	x	X		x		El Palmillo
Tlacolula	X							TL-TCH-MAC-1
Tlacolula	X							S. B. Qialana
Tlacolula	X							TL-TL-TL-73
Tlacolula	X			X				Cerro Quiaviní
Tlacolula	X							S. J. Teitipac
Tlacolula	X	X	x	x		x		Mitla Fortress
Tlacolula	X			X				TL-SST-SST-23
Tlacolula	X	X		X				TL-TL-TL-70
Tlacolula	X			X				S. S. Teitipac

TABLE 4.1. (continued)

Tlacolula	X				TL-SJG-SJG-65
Tlacolula	X	X	X		Guirún
Tlacolula	X				S. M. Tlapazola
Tlacolula		X			S. J. Guelavía
Tlacolula	X			X	TL-SJG-SJG-22
Tlacolula	X				TL-MIT-MIT-2
Valle Grande	X	X	X	x	Jalieza
Valle Grande	X	X		X	S. C. Mixtepec
Valle Grande	X				Tejas de Morelos
Valle Grande	X				El Trapiche
Valle Grande	X				S. A. Tlapacoyan
Valle Grande		X			Cuilapan
Valle Grande	X		X		ZA-SRJ-SRJ-3
Valle Grande			X		CE-SMC-SMC-19
Valle Grande	X				ZI-ZI-ZI-2-5, 8
Valle Grande	X				ZI-SG-SG-2-4
Valle Grande	X				ZI-VDF-VDF-3

Note: Small x refers to additional crafts observed during intensive survey or excavations.

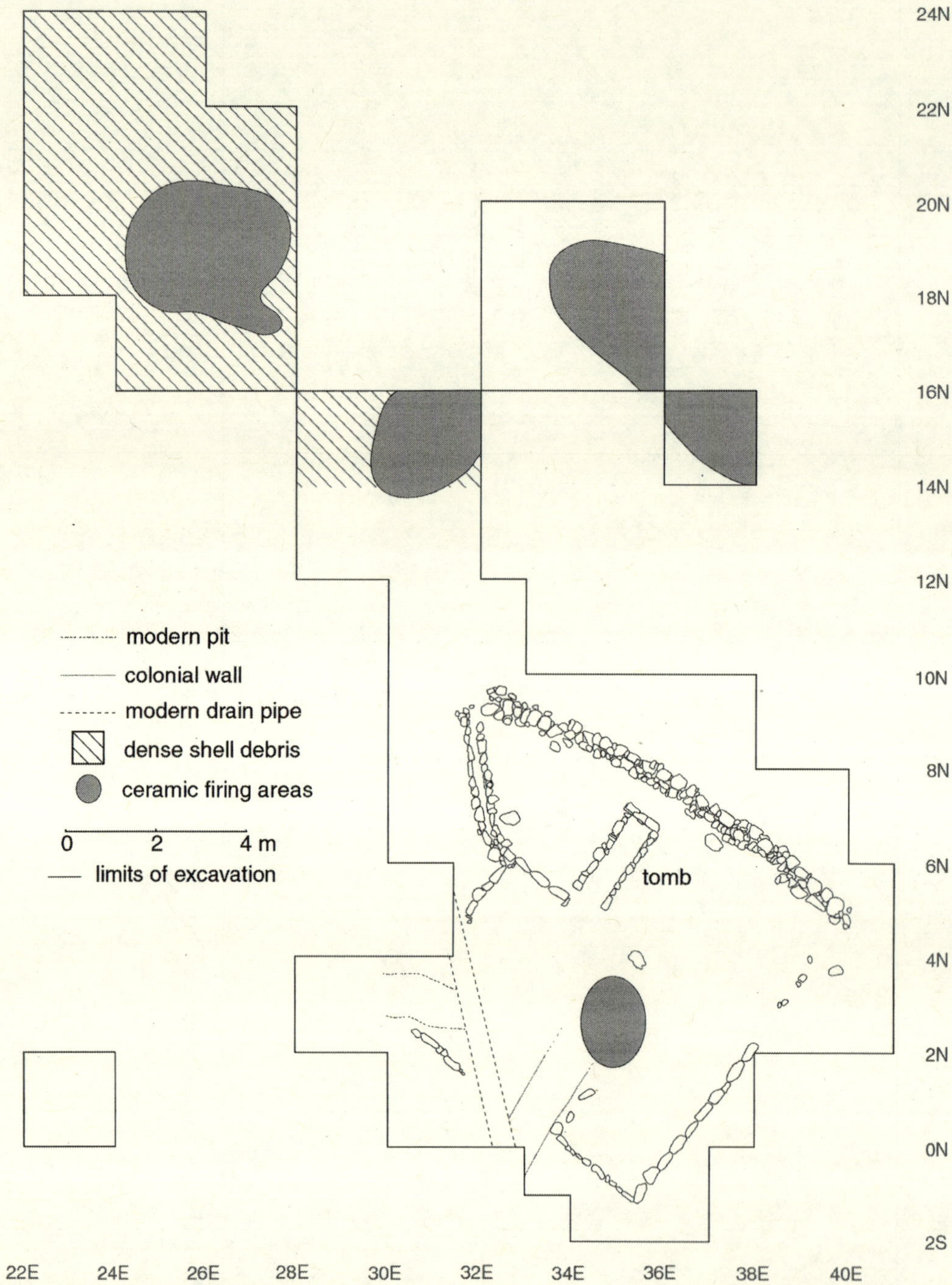

FIGURE 4.2. Excavated Classic period house and outside work areas at the Ejutla site.

excavations to determine the timing and nature of the shell-working activities at the site (Feinman and Nicholas 1993, 1995; Feinman et al. 1991; Feinman et al. 1993). Over four field seasons (1990–1993) of broad horizontal excavations, we exposed a Classic period residential area (Figure 4.2). A small structure, roughly 6 × 4 m with an attached work area to the north, was defined by a stone foundation and a floor of mixed earth and crushed

Table 4.2. Shell and Lapidary Materials at Ejutla and on El Pamillo Terraces

Material	Ejutla	El Palmillo Terrace					
		1162	1163	1147/48	925	507	335
Total shell	23,000+	50	17	17	10	23	45
Shell ornaments	976	15	11	10	10	18	28
Stone beads/ ornaments	35	1	0	2	8	7	12
Lapidary raw material and debris	168	2	6	3	5	34	12
Stone abraders/ polishers	63	20	27	14	36	130	115

bedrock. Several features point to the domestic nature of this complex, including a small subfloor tomb with at least four individuals (Middleton et al. 1998), a shallow fire pit and probable cooking area located immediately outside the structure, and the presence of food remains and other residential trash in adjacent middens. In Classic period Oaxaca, residential complexes often were associated with domestic tombs on the order of the one found in Ejutla (e.g., Feinman et al. 2006; González Licón 2003; Winter 1974, 1995).

The results of these excavations allowed us not only to confirm the survey findings of shell working at Ejutla but also to document the presence of several other craft activities. Of the more than 23,000 pieces of marine shell that were recovered, most were cut, abraded, or in other ways modified. Unfinished and broken ornaments—especially disks, mosaic pieces, and beads—were relatively common, whereas finished ones were rare (less than 1 percent of the assemblage; Table 4.2). Most of the identified shell varieties are native to the Pacific and include taxa (especially *Spondylus*, *Strombus*, and *Pinctada*) that were commonly used to make ornaments across pre-Hispanic Mesoamerica. No known site in the central valleys of Oaxaca, not even the region's capital, Monte Albán, equals this sector of the Ejutla site in terms of the concentration of marine shell, especially debris. Given the abundance of shell debris from ornament manufacture, shell ornaments made at Ejutla clearly were not just for immediate local consumption.

The ancient Ejutleños also engaged in ceramic production (Table 4.3; Balkansky et al. 1997; Feinman and Balkansky 1997). Over 2,000 figurines, many of them broken or malformed, were recovered in association with the house and exterior middens. Almost a quarter of the 77 ceramic molds recovered in the excavations were definitely for figurines, and several molds match commonly noted figurine varieties. In addition, we noted more than 900 pottery wasters (see Feinman 1980; Redmond 1979) from a range of ceramic forms, including figurines, *comales* (tortilla griddles), and *sahumadores* (incense burners), in addition to utilitarian bowls and jars. Petrographic and elemental (inductively coupled plasma [ICP]) analyses indicate that the figurines and other vessels were made from raw clays that can be found on-site or in nearby streambeds (Carpenter and Feinman 1999). Another indicator of ceramic production was the presence of at least four shallow pits, or pit kilns, in external areas adjacent to the structure that were filled with burned rock, charcoal, thick ash lenses, kiln wasters, kiln furniture, and heavy densities of fired clay concretions (Balkansky et al. 1997; see also Stark 1985:176).[2] The simultaneity of the practice of these two crafts is indicated by the nature of the debris found in several pit kilns that were filled with trash after their original use. Generally, these pits (formerly used as firing installations) contained

TABLE 4.3. Ceramic Production at Ejutla and on El Palmillo Terraces

		El Palmillo Terrace					
Material	Ejutla	1162	1163	1147/48	925	507	335
Figurines	2,001	132	79	99	102	162	128
Figurines/urn wasters	185	4	4	9	0	0	0
Bowl wasters	212	56	31	26	17	19	27
Jar wasters	238	39	43	24	21	12	19
Sahumador wasters	34	14	0	3	3	0	1
Comal wasters	6	1	1	1	0	0	0
Other wasters	265	19	27	14	3	7	7
Concretions and coils	1,494	27	12	10	9	10	9
Total molds	77	12	13	13	22	7	1
Figurine/urn molds	31	5	6	10	2	2	1
Molds	74	1	3	4	0	1	0

ceramic firing mistakes and the by-products of pottery production intermixed with shell debris from ornament manufacture.

Based on the quantity of artifactual remains, shell working and ceramic production were practiced at fairly high levels of intensity at Ejutla. Lapidary crafting also was carried out but at a lower volume (Table 4.2). Rectangular onyx plaques and cylindrical drill plugs and chunks and flakes of greenstone and other nonlocal stones were recovered in exterior middens. The diameter of most of the onyx drill plugs matches that of many shell disks, one of the most common shell ornament types recovered in the excavations. Hollow tubular drills (made of cane) seemingly were employed to work both shell and onyx (Figure 4.3), conforming to what has been termed a cross-craft technology (McGovern 1989; Shimada 1996). With high quantities of small chert reduction debris mixed in with the shell debris, it seems likely that the ancient Ejutleños also made the chert microdrills that they used to perforate the shell ornaments.

Several complementary analyses help tie the craft activities to the excavated structure. Microartifactual analyses of the earthen floor have yielded microscopic flecks of shell, chert flakes, and tiny pieces of greenstone that were too small to be swept away during cleaning (Middleton 1998). The recovery of these microartifacts in floor deposits supports the argument that these materials were worked inside the excavated house (Feinman et al. 1993). Chemical analysis (ICP) of soil samples taken from the house floor has revealed traces of repeated firings (possibly from the adjacent fire pits) and signatures from shell working (Middleton and Price 1996).

Most of the excavated artifacts were recovered from the exterior middens, yet one chert microdrill was found on the floor near a flagstone pavement in a work area at the northern edge of the residential complex. Nearby was a small pit filled with sheets of mica, a material often used as an inclusion in the Ejutla ceramics (particularly comales). In the excavated extramural middens, craft debris often was found interspersed with domestic trash, such as food remains and sherds from charred cooking jars. Such remains help tie the craft debris to a domestic context.

As noted above, the high volume of shell debris in the middens, in conjunction with such a small number of finished pieces, indicates that much of the production was for exchange. Surprisingly few finished shell ornaments were found inside the house, and only one small shell bead was recovered from the tomb. There is no empirical evidence to sug-

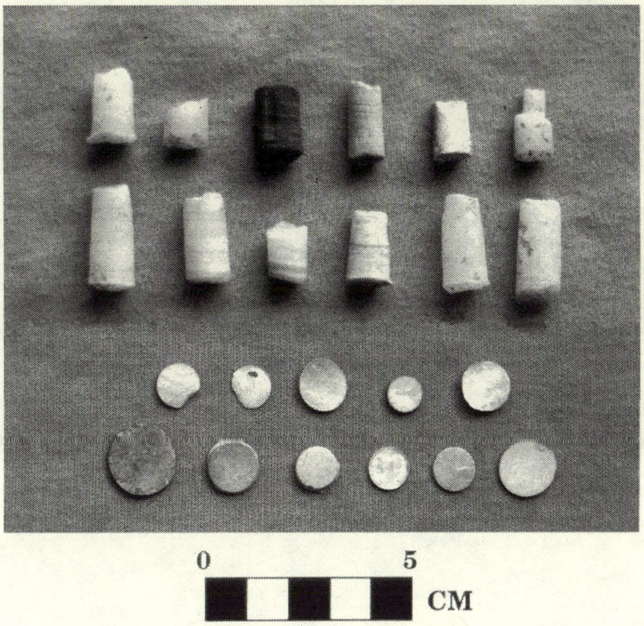

Figure 4.3. Onyx drill plugs and shell disks cut with a hollow tubular drill at the Ejutla site.

gest that this volume of shell ornament production was entirely for immediate household consumption. In addition to the high amounts of ceramic-manufacturing debris, the overall volume of pottery also was great (more than 1.75 metric tons). We were not able to excavate the entire area encircling the stone foundation (which almost certainly included more midden deposits); nor do we know how much production waste was transported away from the house via natural or cultural processes. Nevertheless, from the perspective of what was recovered, it seems highly unlikely that this single household would have used (even over several decades) anything approaching the volume of figurines (or comales and sahumadores) that was produced in this domestic context. Near the base of one excavated firing feature, sherds from at least nine different comales were found in the same 2 × 2-m excavation level. As with the thousands of figurines, the output from this particular production context is unlikely to have been for immediate household needs alone. If the Ejutla potters were supplying only their immediate needs, it seems doubtful that they would have concentrated so heavily on figurines, comales, and sahumadores, as those are not the forms that typically dominate a household's ceramic inventory.

Craft Activities at El Palmillo

El Palmillo is a large, hilltop, terraced site situated on a steep, rocky piedmont ridge at the eastern edge of the dry Tlacolula arm of the Valley of Oaxaca (Figure 4.1). Given their location in the same valley system, El Palmillo's physical setting could not be more different from alluvial Ejutla. The ancient inhabitants constructed over 1,400 terraces at the site, the majority of which were residential. El Palmillo, like Ejutla, was most densely occupied during the Classic period, although both settlements have longer occupational histories.

Over six field seasons (1999–2004), we excavated seven large terraces (six distinct residential compounds), with the goal of examining and defining the nature of spatial and status-related variation in domestic architecture and economic activities, including production and consumption (Feinman et al. 2002, 2006; Haines et al. 2004; Haller et al.

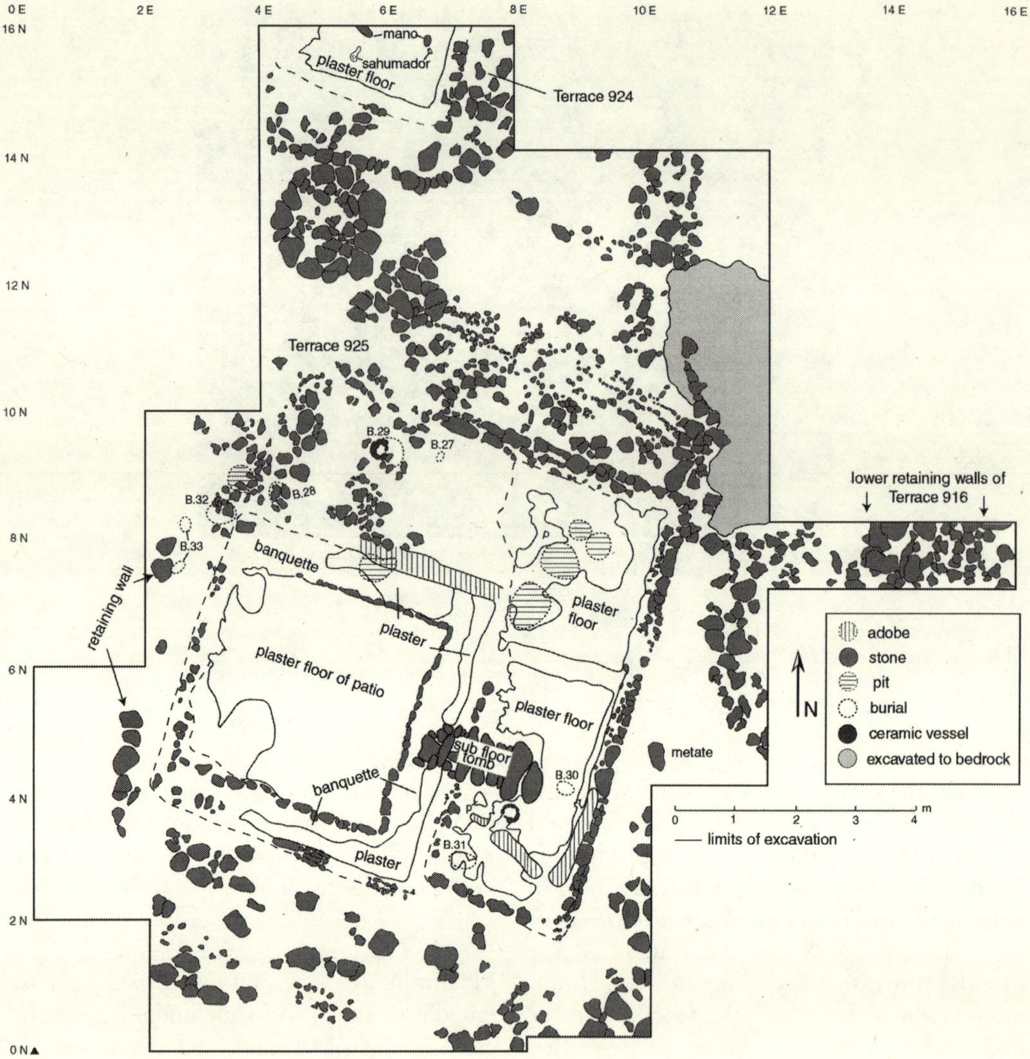

Figure 4.4. Residential complex on Terrace 925 at El Palmillo.

2006). We began by uncovering four contiguous terraces near the base of the site (Terraces 1147, 1148, 1162, and 1163); two separate terraces in intermediate positions on the hill were then excavated (Terraces 925 and 507), followed by a large terrace located adjacent to the public precinct at the apex of the site (Terrace 335). The terraces were excavated to expose as much as possible of the uppermost residential surface on each terrace, including domestic architecture and outdoor work spaces and refuse areas. Except for Terrace 335, each residential complex was composed of a patio group that included two or more rectangular rooms bordering a lime-plastered patio (Figure 4.4; Feinman and Nicholas 2001; Feinman et al. 2001; Feinman et al. 2002). The uppermost terrace (335) had the most elaborate domestic architecture, a residential complex composed of 11 rooms that enclosed two sunken patios (Feinman et al. 2006), similar in form to elaborate residences and palace-like compounds at Monte Albán (Caso 1938; Flannery 1983; Marcus and Flannery 1996: 208–212) and Lambityeco (Lind and Urcid 1983; Paddock et al. 1968). These basic architectural arrangements were retained throughout several minor remodeling and more major rebuilding episodes on each terrace. With the exception of two small adjacent terraces in the lower terrace group that we suspect sustained one domestic unit (Terraces 1147 and 1148), each excavated terrace housed a patio group or domestic unit that was closed to or separated by walls from adjacent terraces/domestic units.

In addition to residential architecture, we found domestic refuse, human burials, and fire installations on all the terraces, including two large ovens on the lowest group of terraces that appear to have been used to roast maguey (Feinman et al. 2002). The 74 burials (with over 92 individuals) uncovered on the terraces were placed in the central patios, under house floors, and in the terrace retaining walls. Subfloor tombs were present on the three uppermost terraces, with the most elaborate and finely made masonry tomb part of the residential complex on the highest terrace (335 [Feinman and Nicholas 2003; Feinman et al. 2006]). Mixed in with the domestic debris on each terrace were artifactual (and occasionally other) indicators of a range of craft activities. Not only were multiple crafts practiced at El Palmillo, but households on each excavated terrace appear to have participated in two or more economic activities likely geared for more than immediate use in that domestic unit.

Households at El Palmillo engaged in a different range of craft activities than their contemporaries at Ejutla. We found much less evidence for ceramic production at El Palmillo. The few firing features at El Palmillo are smaller than those at Ejutla, the overall volume of ceramics is lower, and there were many fewer kiln wasters and other indicators of ceramic production (Table 4.3). For example, the quantity of figurines at Ejutla, many of which were wasters, is higher by more than an order of magnitude than the number recovered on any single terrace at El Palmillo—2,000 figurines in association with one house in Ejutla compared with between 60 and 160 on each residential terrace at El Palmillo. Ceramic production at El Palmillo, in contrast, was largely of utilitarian wares for local consumption only. Shell working was nonexistent at El Palmillo; we have recovered only 162 pieces of shell from the six residential compounds at El Palmillo (more than half of which were finished or broken ornaments), compared with over 23,000 from one domestic complex at Ejutla. The household on one terrace at El Palmillo appears to have made lapidary objects, but the volume of material is much less than that at Ejutla, and the suite of items and materials is different. For example, we have found much less onyx at El Palmillo and no drill plugs.

At El Palmillo, the production of chipped-stone tools from local stone sources was a major economic pursuit. The stone assemblage at the site is dominated by chert and other silicified materials (approximately 92 percent of the stone collected), specifically production waste (about three-quarters of the assemblage). The chipped-stone tools manufactured at El Palmillo include abraders, bifaces, perforators, points, *raspadores* (a large specialized dome-shaped scraper), generalized scrapers, and unifaces. The chert microdrills that were common at Ejutla are extremely rare.

The differences between the stone assemblages at El Palmillo and Ejutla are striking (Table 4.4). The overall quantity of nonobsidian chipped stone at Ejutla is similar to the totals for several terraces at El Palmillo, yet the total weight of this stone assemblage from any one terrace alone at El Palmillo is much greater than that at Ejutla, by at least an order of magnitude on most terraces. At Ejutla we recovered great quantities of tiny chert debris. Tiny chert flakes were created repeatedly by sharpening and using small tools to work shell. The most common chert tools in the Ejutla stone assemblage (microdrills) are very small. In contrast, large cores, tools, and flakes made of silicified materials were prepared at El Palmillo, and large scrapers and raspadores are among the most prevalent tools at that site.

The percentage of formal tools at El Palmillo (approximately 2.5 percent of the entire nonobsidian chipped-stone assemblage) is lower than that at Ejutla (10 percent), which is what we would expect if chipped-stone nodules and chunks were being reduced into cores, usable flakes, and tools at El Palmillo.

Table 4.4. Nonobsidian Chipped Stone at Ejutla and on El Pamillo Terraces

Chipped-Stone Material	Ejutla	El Palmillo Terrace 1162	1163	1147/48	925	507	335
Total weight (kg)	13.5	134.2	126.7	121.5	86.6	80.1	112.2
Weight (g)/m^3	58	1,329	1,474	1,472	677	561	413
Raspadores	8	40	27	22	43	47	18
Thumb scrapers	20	0	5	0	2	16	2
Other scrapers	113	68	49	50	31	70	14
Perforators	108	16	12	6	21	1	4
Microdrills	166	0	0	0	1	0	0
Decortification flakes	—	1,045	247	224	92	109	173
Cores	—	124	63	127	59	57	28
Preforms	—	3	7	1	0	4	3

This mirrors the pattern for the shell debris at Ejutla, where finished artifacts constitute a numerically minute part of the overall assemblage. Thus it appears that the quantity of chipped-stone tools and production waste on some terraces at El Palmillo represents a higher intensity or volume of lithic production, likely for exchange to other households on-site or at other settlements in the region.

Another important activity at El Palmillo was xerophytic plant processing. Today (centuries after the abandonment of pre-Hispanic El Palmillo) the site has a wealth and diversity of xerophytic plants that are of economic importance (or were in the past), particularly maguey, which is the present economic backbone (through the manufacture of the alcoholic beverage mescal) of the contemporary village where the site is located. The remnant plant community at the site appears to be a relict of the pre-Hispanic habitation of the hill (Middleton et al. 2001). For example, we have recorded eight different varieties of maguey on-site, even though only one of these varieties figures prominently in mescal production and only two or three varieties are found on neighboring hills. Many tools at El Palmillo likely were used to process maguey. Scrapers are one of the most abundant chipped-stone tool varieties, including the specialized raspadores or scraper planes that are thought to have served to extract maguey fiber from that plant's pulpy leaves (Hester and Heizer 1972; Robles G. 1994; see also Fish et al. 1992:84; Parsons and Parsons 1990). The production of maguey fiber and textiles was a significant activity at El Palmillo for which we have no evidence at Ejutla. Whereas stoneworking at Ejutla was related to shell ornament manufacture, much of the stoneworking at El Palmillo appears to have been related to processing maguey and other domestic tasks.

Spindle whorls are abundant at both sites; however, the whorls are more variable at El Palmillo. Large ceramic spindle whorls, in the size range typically used for maguey fiber (Parsons 1972; Parsons and Parsons 1990), are more common at El Palmillo and are rare at Ejutla (Table 4.5). Most of the whorls at Ejutla are of the smaller size likely used to spin cotton fiber. Finally, many of the bone tools recovered in the El Palmillo excavations, including battens, needles, perforators, and spindle whorls, are related to textile manufacture (Haines et al. 2004; Middleton et al. 2002). We found more of these bone tools, especially battens and needles, at El Palmillo than at Ejutla (Table 4.5).

Although household members on each terrace at El Palmillo worked with a similar set of raw materials, they did not carry out the exact same suite of activities. Even where different households carried out similar productive tasks, those activities did not have equal

Table 4.5. Bone Tools and Ceramic Spindle Whorls at Ejutla and on El Pamillo Terraces

		El Palmillo Terrace					
Material	Ejutla	1162	1163	1147/48	925	507	335
Bone Tool							
Awls	6	19	7	5	15	26	10
Battens	2	13	5	8	10	0	4
Billets	1	2	2	3	0	0	1
Disks/whorls	0	5	2	1	0	3	0
Needles/perforators	7	10	6	5	9	13	12
Shuttles	0	0	0	0	2	0	0
Total whorls (ceramic)	104	13	15	14	7	22	36
Measurable spindle whorls (small)	80 (80)	9 (3)	12 (2)	9 (4)	6 (4)	20 (12)	32 (22)

importance in each residential unit. For example, given the greater abundance of chert decortification flakes, cores, and preforms on the lower terraces (1162, 1163, 1147, and 1148), households in that sector of the settlement were more actively engaged in stone tool production (Table 4.4), an observation also supported through our intensive survey of this portion of the site, where high volumes of chipped stone but relatively few finished tools were noted (Feinman et al. 2006; Haines et al. 2004). In our excavations, antler billets, key implements in the manufacture of chipped-stone tools (Aldenderfer 1991; Parry 1987), were recovered only from households in the lower terrace group (Middleton et al. 2002). In a preliminary analysis of microdebitage in the 0.5–2 mm heavy fraction we observed a similar pattern. On the lower terraces there were three to four times as many microflakes and other minute pieces of chert debris per kilogram of analyzed sediment.

On the three higher terraces (925, 507, and 335), stoneworking may have largely involved the retouching or sharpening of existing pieces (e.g., Hester and Heizer 1972) or some kind of "nodule smashing" (perhaps with somewhat more design than described by Boksenbaum [1980]) to create stone tools. In contrast, the residents of the lower terraces appear to have been more involved in the manufacture of tools (and possibly preforms and cores) that were intended for house-to-house exchange. Large granitic flakes (which appear to be the by-product of ground-stone tool manufacture) and a greater abundance of metates also were found on Terrace 1162, and it is likely that those residents also engaged in the low-volume manufacture of ground-stone tools (Haines et al. 2004).

House-to-house variation also was noted for other craft industries. Although raspadores, for processing maguey or other xerophytic plants, were recovered in the tool assemblages of all households, they were more abundant on Terraces 507, 925, and 1162 (Table 4.4). Yet the residents of Terrace 925 had the fewest spindle whorls of any household ($n = 7$), and only one small stone spindle bowl was noted there. In contrast, at least 25 ceramic and bone whorls were found on Terraces 507 and 335; and 15–17 whorls were associated with each of the three other residential units. Although the inhabitants of Terrace 925 likely processed maguey on par with the residents of Terrace 507, they may have been less actively engaged in spinning than the residents of other households. On the basis of a high number of battens and the only two possible shuttles identified at the site, it would seem that weaving may have been the more important activity on Terrace 925.

Yet the amount of material spun or woven is not the only axis of variation for spinning activities at El Palmillo. The whorls employed by the residents of the upper terraces (925, 507, and 335) are generally lighter with smaller hole diameters (Table 4.5; Feinman et al. 2002, 2006). Ethnographic studies of textile production in Mexico have documented that variation in the hole diameter of spindle whorls affects the thickness (and therefore quality) of the spun maguey thread (Parsons 1972; Parsons and Parsons 1990). Smaller whorls also are more suitable for spinning finer materials such as cotton. Fine threads and textiles produced from either of these plants likely were valued more highly than cloth made from coarser maguey threads (Berdan 1987:244–245) spun with the larger whorls, which were found in greater proportions on the lower terraces.

Lapidary craft production appears to have been an activity practiced only in specific households (Haines et al. 2004). Evidence for this industry is manifest through both unfinished or broken craft items (beads and ornaments), raw materials and production debitage, and the tools that could be used to manufacture these items (abraders and polishers [Adams 2002:213; Charlton 1993; Charlton et al. 1993; Hayden 1987: tab. 5.2]). On Terrace 507, and to a lesser degree on Terrace 335, we found a larger number of stone artifacts that show abrasion or polishing wear (Table 4.2). On Terrace 507, we also recovered the only two unfinished beads found during the excavations at El Palmillo. These beads were part of a collection of unfinished lapidary items, cut stones, and lapidary raw material recovered from this domestic context, which contrasts sharply with the much lower amounts found on any other terrace (Table 4.2).

We found at least a few kiln wasters and other evidence of ceramic production on all the terraces, although the intensity of production on any one terrace appears to have been much lower than that at Ejulta (Table 4.3). Most of the wasters are from utilitarian bowls and jars, although we did recover a few molds for figurines/urns, especially on the lower terraces. The residents of the lower terraces appear to have been more involved in ceramic production than people living higher up the hill. The residents of the highest terrace do not appear to have manufactured their own pottery. Potters on the lower terraces may have supplied some ceramic forms to their neighbors living higher on the hill. For example, 80 percent of the molds for making figurines and urns, as well as all of the wasters from these forms, were found on the lower terraces. Yet over half of the figurines and urns at the site were recorded on the three upper terraces.

In sum, production at El Palmillo was situated in domestic contexts, with most domestic units engaging in more than one craft activity. Most households utilized xerophytic plants such as maguey and local sources of stone, yet the particular suite of economic activities varied from house to house. Lapidary production was a more significant activity on Terrace 507 than on any other terrace. The spinning (and weaving) of fine cloth was more important on Terraces 507 and 335. Some spinning of fine thread also was conducted on Terrace 925 but to a limited degree. Weaving tools, however, were common on that terrace, and it is likely that the residents manufactured cloth or other perishable maguey (or other fiber) products such as rope, sandals, or baskets. Residents on the lower terraces (1147, 1148, 1162, and 1163) specialized in stoneworking and likely supplied neighboring households with tools, preforms, and possibly cores from which stone tools could be flaked. The occupants of these terraces also appear to have produced higher quantities of ceramic vessels, figurines, and urns. As the finished products from several of these specialized craft industries were found in other El Palmillo residences, a significant volume of house-to-house exchange in utilitarian and likely some finer craft products seems to have been a key aspect of the intrasite economy.

SYNTHESIS

Similar to the case today, most craft production in the Classic period Valley of Oaxaca took place in domestic settings. At both Ejutla and El Palmillo we found large quantities of craft debris mixed in with domestic rubbish. The archaeologically observed craft activities at these communities include the manufacture of a range of ceramic forms, the production of ground- and chipped-stone tools from both local and exotic (obsidian) materials, the crafting of lapidary objects and marine shell ornaments, and the processing of a range of xerophytic plants, especially maguey, for fiber and other raw materials. These fibers likely were spun and processed into various goods including cloth, rope, and baskets. Here it is worth noting that "Matatlán" (the name of the community where El Palmillo is situated) refers to "place of the nets" in Nahuatl, the language of the Aztecs.

Some of the crafts documented at Ejutla and El Palmillo are still practiced in Oaxaca today (ceramic production, weaving) or were until very recently (processing xerophytic plants for fiber). At the same time, to date, archaeologists have yet to find any solid indications that economic specialization occurred in larger-scale, nondomestic contexts during the Classic period in the Valley of Oaxaca.

No single craft industry prevailed at either Ejutla or El Palmillo. The pattern of production appears more similar to that of the early part of the twentieth century, when multiple crafts were practiced within settlements and households, in contrast to the situation in contemporary Oaxaca, where many communities are known largely for a specific craft. Shell working and ceramic production, especially of figurines, were important at Ejutla. The craft specialists also made the chert microdrills used to perforate shell ornaments and applied a similar procedure (use of tubular drills) to work both shell and ornamental stone. Stoneworking and processing xerophytic plants for fiber were practiced by most households at El Palmillo, although sometimes they specialized in different aspects of these crafts. Other activities, including lapidary crafting and ceramic production, were carried out by fewer households. Of course, our sample size at both archaeological communities remains rather small, and it is probable that further excavations would provide indications of additional artisan activities at each settlement.

Not only were several crafts practiced at both sites, but individual households often engaged in two or more craft pursuits, although sometimes these crafts were practiced at different levels of intensity. Within individual households, some production could be for exchange beyond the community (e.g., shell ornaments and ceramic figurines at Ejutla), whereas other goods were produced largely for local consumption (e.g., ceramics at El Palmillo). In Classic period Oaxaca, it is evident that domestic production could produce craft goods at very different volumes or intensities.

How much the households in each Classic period community specialized in the same suite of activities may have been more variable. At El Palmillo most households engaged in some aspect of the dominant crafts, stoneworking and processing xerophytic plants, although with excavations unearthing only seven out of 1,400 terraces, our sample is small. Because we were unable to excavate more than one house at Ejutla, we cannot make such house-to-house comparisons there. Yet, given the small area of the site with heavy shell debris on the surface, shell working may have been concentrated in a single barrio within the larger community.

Clearly, the organization of craft producers and different specializations across the valley in the Classic period was complex. Even though production was based in houses, there was variation in the intensity or volume of production, not only between houses and communities but for specific crafts within single households. Although we cannot say which crafts, if any, were practiced full-time, most crafts clearly were produced only on a part-time basis. There also was significant spatial diversity in the crafts that were produced in different parts of the valley and in upland versus valley floor locations.

Pre-Hispanically, the predominance of part-time craft manufacture in the Valley of Oaxaca likely was a consequence of extant demographic densities and a transport system that relied entirely on human carriage. Most Oaxacan settlements other than Monte Albán had fewer than 10,000 inhabitants, and there were no beasts of burden or water travel possibilities. As a consequence, full-time domestic production of even basic goods rarely could be sustained (especially if multiple households in the same community manufactured a similar set of products) because the reachable demand for those goods was not sufficiently large. In some valley communities, pre-Hispanic households likely supplemented agricultural production with a single craft. But the location of many Classic period terrace sites in agriculturally marginal settings (at least for maize) likely meant that even such an economic regime was risky for many households living in these upland locations (Nicholas 1989). Multicrafting, in which a residential unit was involved in more than one craft activity (possibly also supplemented by some farming), was a plausible domestic response to these economic circumstances.

BROADER IMPLICATIONS AND CONCLUSION

Because most, if not all, Classic period production appears to have been household based, there are key implications for understanding the modes of circulation and other aspects of the political economy during the Classic period hegemony of the Monte Albán polity. With so many households producing a range of goods at varying levels of intensity, the direct political or administrative control of production would have been an expensive as well as near impossible undertaking for those in power at Monte Albán or even at secondary centers. At the same time, no giant, centralized storage facilities have ever been found at major centers in pre-Hispanic Oaxaca, thereby indicating that the centralized collection and redistribution of large quantities of goods from valley households likely was not a key aspect of the Classic period economy. Yet, at the same time, there does seem to have been a significant amount of house-to-house, community-to-community, and even regional sector–to–regional sector circulation of goods, some of which were basic to household maintenance. The implied volume of exchange likely could not have been handled entirely through simple face-to-face reciprocity. Rather, as in the present and at the time of Spanish conquest, some kind of marketplace exchange likely was in place to link producers and consumers across a broad valley. Large open places (without monumental architecture) have been recorded at the margins of large Classic period settlements, including Monte Albán (Blanton 1978) and El Palmillo (Feinman and Nicholas 2004). Such plazas, which were found at the nexus of several roads and pathways at both sites, potentially could have served as loci of market activities. Marketplaces were noted as having been economically significant in various regions of the State of Oaxaca during the sixteenth century (Appel 1982; Pohl et al. 1997; Spores 1965).

If the direct control of (or exaction from) production was not a major revenue source for Classic period Monte Albán (and the other smaller valley centers), then how were resources procured by governing institutions? Numerous potential avenues remain. These include taxes for marketplace participation (a revenue source for the later Aztecs [Blanton 1996]) and the procurement of taxes or tribute related to the maintenance of peace and intraregional passage. The role of offense/defense as a basis for the origin and longevity of the Monte Albán state has long been maintained (e.g., Blanton 1978; Marcus and Flannery 1996). Of course, various kinds of conscriptions or labor taxes also may have been a source of support for Classic period Zapotec polities.

In sum, with many more pre-Hispanic houses excavated, analyzed, and reported than there were 20 years ago, we now have a far richer and more empirically grounded understanding of the Mesoamerican economy than we did several decades ago. At the same time, studies of contemporary household-based production in Mexican villages have served to guide and provoke important insights in regard to the nature, organization, and technologies of production in past communities. Yet we also have begun to learn how craft production in the past was both different from that of the present and diverse and complex in its own right. Clearly we need additional and even more ingenious archaeological studies at several analytical scales to clarify and amplify our understanding of past practices of production, and the larger web of economic relations of which it was part, even further.

Acknowledgments.

We gratefully thank Izumi Shimada for inviting us to contribute to this volume. We gratefully acknowledge the National Science Foundation support given to Gary M. Feinman for the excavations at El Palmillo (SBR-9805288, BCS-0349668) and Ejutla (BNS 89-19164, BNS 91-05780, SBR-9304258). We also appreciate the valuable support received from the National Geographic Society, the H. John Heinz III Fund of the Heinz Family Foundation, the Field Museum, the Graduate School of the University of Wisconsin–Madison, and Arvin B. Weinstein. This study would not have been possible without the assistance of our Oaxacan and North American field and laboratory crews. We especially thank Helen Haines, William Middleton, and Jennifer Clark for assisting us with the analysis of the lithic and faunal collections at both sites. We also are grateful to Jill Seagard, who helped prepare the graphics. We profoundly thank the Instituto Nacional de Antropología e Historia of Mexico, the Centro Regional de Oaxaca, and the local authorities of Santiago Matatlán and Ejutla de Crespo for the necessary permissions to implement these field studies, as well as for their essential support. Finally, we dedicate this work to the people of Matatlán and Ejutla. Their untiring assistance, interest, and encouragement in various stages of this study have made our effort rewarding in so many ways.

Notes

1. Yet one of the important occupations in 1968, making rope from maguey fibers, is no longer practiced in Mitla today.
2. Presumably, they are remnants from temporary earthen roofs placed over the pits during firing.

References Cited

Adams, Jenny L.

2002 *Ground Stone Analysis: A Technological Approach*. University of Utah Press, Salt Lake City.

Aldenderfer, Mark

1991 The Structure of Late Classic Lithic Assemblages in the Central Petén Lakes Region, Guatemala. In *Maya Stone Tools*, edited by Thomas R. Hester and Harry J. Shafer, pp. 119–142. Monographs in World Archaeology, No. 1. Prehistory Press, Madison, Wisconsin.

Appel, Jill

1982 A Summary of the Ethnohistoric Information Relevant to the Interpretation of Late Postclassic Settlement Pattern Data, the Central and Valle Grande Survey Zones. In *Monte Albán's Hinterland, pt. 1: The Prehispanic Settlement Patterns of the Central and Southern Parts of the Valley of Oaxaca, Mexico*, by Richard E. Blanton, Stephen A. Kowalewski, Gary M. Feinman, and Jill Appel, pp. 139–148. Memoirs No. 15. Museum of Anthropology, University of Michigan, Ann Arbor.

Arnold, Phillip, III

2000 Working without a Net: Recent Trends in Ceramic Ethnoarchaeology. *Journal of Archaeological Research* 8:105–133.

Balkansky, Andrew K., Gary M. Feinman, and Linda M. Nicholas

1997 Pottery Kilns of Ancient Ejutla, Oaxaca, Mexico. *Journal of Field Archaeology* 24:139–160.

Baskes, Jeremy

2000 *Indians, Merchants, and Markets: A Reinterpretation of the Repartamiento and Spanish–Indian Economic Relations in Colonial Oaxaca 1750–1821*. Stanford University Press, Stanford.

Beals, Ralph L.

1975 *The Peasant Marketing System of Oaxaca, Mexico*. University of California Press, Berkeley.

1979 Economic Adaptations in Mitla, Oaxaca. In *Mesoamerica: Homenaje al Doctor Paul Kirchoff*, edited by Barbro Dahlgren, pp. 165–193. Instituto Nacional de Antropología e Historia, Mexico City.

Berdan, Francis F.

1987 Cotton in Aztec Mexico: Production, Distribution, and Uses. *Mexican Studies/Estudios Mexicanos* 3:235–262.

Blanton, Richard E.

1978 *Monte Albán: Settlement Patterns at the Ancient Zapotec Capital*. Academic Press, New York.

1996 The Basin of Mexico Market System and the Growth of Empire. In *Aztec Imperial Strategies*, edited by Frances F. Berdan, Richard E. Blanton, Elizabeth Boone, Mary Hodge, Michael E. Smith, and Emily Umberger, pp. 47–84. Dumbarton Oaks, Washington, D.C.

Blanton, Richard E., Stephen Kowalewski, Gary Feinman, and Jill Appel
1982 *Monte Albán's Hinterland, pt. 1: The Prehispanic Settlement Patterns of the Central and Southern Parts of the Valley of Oaxaca, Mexico.* Memoirs No. 15. Museum of Anthropology, University of Michigan, Ann Arbor.

Boksenbaum, Martin W.
1980 Basic Mesoamerican Stone-Working: Nodule Smashing? *Lithic Technology* 9: 12–26.

Carpenter, Andrea, and Gary M. Feinman
1999 The Effects of Behavior on Ceramic Composition: Implications for the Definition of Production Locations. *Journal of Archaeological Sciences* 26:783–796.

Caso, Alfonso
1938 *Exploraciones en Oaxaca, quinta y sexta temporadas 1936–1937.* Publicación 34. Instituto Panamericano de Geografía e Historia, Mexico City.

Charlton, Cynthia O.
1993 Obsidian as Jewelry: Lapidary Production in Aztec Otumba, Mexico. *Ancient Mesoamerica* 4:231–243.

Charlton, Cynthia O., Thomas H. Charlton, and Deborah L. Nichols
1993 Aztec Household-Based Craft Production: Archaeological Evidence from the City-State of Otumba, Mexico. In *Prehispanic Domestic Units in Western Mesoamerica: Studies of the Household, Compound, and Residence,* edited by Robert S. Santley and Kenneth G. Hirth, pp. 147–171. CRC Press, Boca Raton.

Chibnik, Michael
2000 The Evolution of Market Niches in Oaxacan Woodcarving. *Ethnology* 39:225–242.

Cohen, Jeffrey H.
1999 *Cooperation and Community: Economy and Society in Oaxaca.* University of Texas Press, Austin.

Cook, Scott
1982 *Zapotec Stoneworkers: The Dynamics of Rural Simple Commodity Production in Modern Mexican Capitalism.* University Press of America, Lanham, Maryland.

Cook, Scott, and Martin Diskin (editors)
1976 *Markets in Oaxaca.* University of Texas Press, Austin.

Costin, Cathy L.
1991 Craft Specialization: Issues in Defining, Documenting, and Explaining the Organization of Production. *Archaeological Method and Theory* 3:1–56.

Drucker-Brown, Susan
1982 *Malinowski in Mexico: The Economics of a Mexican Market System.* Routledge and Kegan Paul, London.

Druijven, Peter, and Annette Kruithof
1992 Tlacolula Valley, Oaxaca, Mexico. In *Coping with Semiaridity: How the Rural Poor Survive in Dry-Season Environments,* edited by Henk Reitsma, Ton Dietz, and Leo de Haan, pp. 132–152. Netherlands Geographic Studies 146. Department of Human Geography, University of Amsterdam, Amsterdam.

Feinman, Gary M.
1980 The Relationship between Administrative Organization and Ceramic Production in the Valley of Oaxaca, Mexico. Unpublished Ph.D. dissertation, Department of Anthropology, Graduate Center, City University of New York.
1986 The Emergence of Specialized Ceramic Production in Formative Oaxaca. *Research in Economic Anthropology, Supplement* 2:347–373.
1997 Macro-Scale Perspectives on Settlement and Production in Ancient Oaxaca. In *Economic Analysis beyond the Local System,* edited by Richard E. Blanton, Peter N. Peregrine, Deborah Winslow, and Thomas D. Hall, pp. 13–42. University Press of America, Lanham, Maryland.
1999 Rethinking Our Assumptions: Economic Specialization at the Household Scale in Ancient Ejutla, Oaxaca, Mexico. In *Pottery and People: A Dynamic Interaction,* edited by James M. Skibo and Gary M. Feinman, pp. 81–98. University of Utah Press, Salt Lake City.

Feinman, Gary M., and Andrew K. Balkansky
1997 Ceramic Firing in Ancient and Modern Oaxaca. In *Prehistory and History of Ceramic Kilns,* edited by Prudence Rice, pp. 129–147. American Ceramic Society, Westerville, Ohio.

Feinman, Gary M., and Linda M. Nicholas
1990 At the Margins of the Monte Albán State: Settlement Patterns in the Ejutla Valley, Oaxaca, Mexico. *Latin American Antiquity* 1:216–246.
1991 The Monte Albán State: A Diachronic Perspective on an Ancient Core and Its

Periphery. In *Core/Periphery Relations in Precapitalist Worlds*, edited by Christopher Chase-Dunn and Thomas D. Hall, pp. 240–276. Westview Press, Boulder.

1993 Shell Ornament Production in Ejutla: Implications for Highland–Coastal Interaction in Ancient Oaxaca. *Ancient Mesoamerica* 4:103–119.

1995 Household Craft Specialization and Shell Ornament Manufacture in Ejutla, Mexico. *Expedition* 37(2):14–25.

2000 High-Intensity Household-Scale Production in Ancient Mesoamerica: A Perspective from Ejutla, Oaxaca. In *Cultural Evolution: Contemporary Viewpoints*, edited by Gary M. Feinman and Linda Manzanilla, pp. 119–142. Kluwer Academic/Plenum Publishers, New York.

2001 Excavations at El Palmillo: A Hilltop Terrace Site in Oaxaca, Mexico. *In the Field* 72(2):2–5.

2003 Top-Notch Tomb. *Archaeology* 56(5): 14.

2004 *Hilltop Terrace Sites of Oaxaca, Mexico: Intensive Surface Survey at Guirún, El Palmillo, and the Mitla Fortress*. Fieldiana, Anthropology, New Series No. 37. Field Museum of Natural History, Chicago.

Feinman, Gary M., Linda M. Nicholas, and Scott Fedick

1991 Shell Working in Prehispanic Ejutla, Oaxaca (Mexico): Findings from an Exploratory Field Season. *Mexicon* 13:69–77.

Feinman, Gary M., Linda M. Nicholas, and Helen R. Haines

2002 Houses on a Hill: Classic Period Life at El Palmillo, Oaxaca, Mexico. *Latin American Antiquity* 13:251–277.

2006 Socioeconomic Inequality and the Consumption of Chipped Stone at El Palmillo, Oaxaca, Mexico. *Latin American Antiquity* 17:151–175.

Feinman, Gary M., Linda M. Nicholas, and William D. Middleton

1993 Craft Activities at the Prehispanic Ejutla Site, Oaxaca, Mexico. *Mexicon* 15:33–41.

2001 Domestic Life at Classic Period Hilltop Terrace Sites: Perspectives from El Palmillo, Oaxaca. *Mexicon* 23:42–48.

Finsten, Laura

1995 *Jalieza, Oaxaca: Activity Specialization at a Hilltop Center*. Publications in Anthropology 48. Vanderbilt University, Nashville.

Fish, Suzanne K., Paul R. Fish, and John H. Madsen

1992 Evidence for Large-Scale Agave Cultivation in the Marana Community. In *The Marana Community in the Hohokam World*, edited by Suzanne K. Fish, Paul R. Fish, and John H. Madsen, pp. 73–87. Anthropological Papers No. 56. University of Arizona, Tucson.

Flannery, Kent V.

1983 The Legacy of the Early Urban Period: An Ethnohistoric Approach to Monte Albán's Temples, Residences, and Royal Tombs. In *The Cloud People: Divergent Evolution of the Zapotec and Mixtec Civilizations*, edited by K. V. Flannery and J. Marcus, pp. 132–136. Academic Press, New York.

González Licón, Ernesto

2003 Social Inequality at Monte Albán, Oaxaca: Household Analysis from Terminal Formative to Early Classic. Unpublished Ph.D. dissertation, Department of Anthropology, University of Pittsburgh.

Haines, Helen R., Gary M. Feinman, and Linda M. Nicholas

2004 Household Economic Specialization and Social Differentiation: The Stone-Tool Assemblage at El Palmillo, Oaxaca. *Ancient Mesoamerica* 15:251–266.

Haller, Mikael, Gary M. Feinman, and Linda M. Nicholas

2006 Socioeconomic Inequality and Differential Access to Faunal Resources at El Palmillo, Oaxaca, Mexico. *Ancient Mesoamerica* 17:39–56.

Hayden, Brian

1987 Past to Present Uses of Stone Tools and Their Effects on Assemblage Characteristics in the Maya Highlands. In *Lithic Studies among the Contemporary Maya Highlands*, edited by Brian Hayden, pp. 160–234. University of Arizona Press, Tucson.

Hendry, Jean Claire

1992 *Atzompa: A Pottery Producing Village of Southern Mexico in the Mid-1950s*. Publications in Anthropology 40. Vanderbilt University, Nashville.

Hester, Thomas R., and Robert F. Heizer

1972 Problems in the Functional Interpretation of Artifacts: Scraper Planes from Mitla and Yagul, Oaxaca. *University of*

California Archaeological Research Facility 14:107–123.

Kowalewski, Stephen A.

2003 Scale and the Explanation of Demographic Change: 3,500 Years in the Valley of Oaxaca. *American Anthropologist* 105:313–325.

Kowalewski, Stephen A., Gary M. Feinman, Laura Finsten, Richard E. Blanton, and Linda M. Nicholas

1989 *Monte Albán's Hinterland, pt. 2: Prehispanic Settlement Patterns in Tlacolula, Etla, and Ocotlán, the Valley of Oaxaca, Mexico*. Memoirs No. 23. Museum of Anthropology, University of Michigan, Ann Arbor.

Lind, Michael, and Javier Urcid

1983 The Lords of Lambityeco and Their Nearest Neighbors. *Notas Mesoamericanas* 9:78–111.

Marcus, Joyce, and Kent V. Flannery

1996 *Zapotec Civilization: How Urban Society Evolved in Mexico's Oaxaca Valley*. Thames and Hudson, London.

McGovern, Patrick E.

1989 Ceramics and Craft Interaction: A Theoretical Framework, with Prefactory Remarks. In *Cross-Craft and Cross-Cultural Interactions in Ceramics*, edited by Patrick E. McGovern, pp. 1–11. American Ceramic Society, Westerville, Ohio.

Middleton, William D.

1998 Craft Specialization at Ejutla, Oaxaca, Mexico: An Archaeometric Study of the Organization of Household Craft Production. Unpublished Ph.D. dissertation, Department of Anthropology, University of Wisconsin, Madison.

Middleton, William D., Gary M. Feinman, and Guillermo Molina Villegas

1998 Tomb Use and Reuse in Oaxaca, Mexico. *Ancient Mesoamerica* 9:297–307.

Middleton, William D., Gary M. Feinman, and Linda M. Nicholas

2001 An Investigation of the Use of Xerophytic Plant Resources in the Economy and Subsistence of El Palmillo, Oaxaca, Mexico. Project report submitted to the Heinz Family Foundation, Pittsburgh.

2002 Domestic Faunal Assemblages from the Classic Period Valley of Oaxaca, Mexico: A Perspective on the Subsistence and Craft Economies. *Journal of Archaeological Science* 29:233–249.

Middleton, William D., and T. Douglas Price

1996 Identification of Activity Areas by Multi-Element Characterization of Sediments from Modern and Archaeological House Floors Using Inductively Coupled Plasma–Atomic Emission Spectroscopy. *Journal of Archaeological Science* 23: 673–687.

Murphy, Arthur D., and Alex Stepick

1991 *Social Inequality in Oaxaca: A History of Resistance and Change*. Temple University Press, Philadelphia.

Murphy, Arthur D., Mary Winter, and Earl W. Morris

1997 Household Adaptations in a Regional Urban System: The Central Valleys of Oaxaca, Mexico. In *Economic Analysis beyond the Local System*, edited by Richard E. Blanton, Peter M. Peregrine, Deborah Winslow, and Thomas D. Hall, pp. 235–254. University Press of America, Lanham, Maryland.

Nicholas, Linda M.

1989 Land Use in Prehispanic Oaxaca. In *Monte Albán's Hinterland, pt. 2: Prehispanic Settlement Patterns in Tlacolula, Etla, and Ocotlán, the Valley of Oaxaca, Mexico*, by Stephen A. Kowalewski, Gary M. Feinman, Laura Finsten, Richard E. Blanton, and Linda M. Nicholas, pp. 449–505. Memoirs No. 23. Museum of Anthropology, University of Michigan, Ann Arbor.

Paddock, John, Joseph R. Mogor, and Michael D. Lind

1968 *Lambityeco Tomb 2: A Preliminary Report*. Boletín de Estudios Oaxaqueños 25. Museo de Frissell de Arte Zapoteca, Mitla, Oaxaca.

Palerm, Angel, and Eric Wolf

1957 Ecological Potential and Cultural Development in Mesoamerica. *Pan American Union Social Science Monograph* 3:1–37.

Parry, William J.

1987 *Chipped Stone Tools in Formative Oaxaca, Mexico: Their Procurement, Production, and Use*. Memoirs No. 20. Museum of Anthropology, University of Michigan, Ann Arbor.

Parsons, Elsie Clews

1936 *Mitla: Town of Souls*. University of Chicago Press, Chicago.

Parsons, Jeffery R., and Mary H. Parsons

1990 *Maguey Utilization in Highland Cen-*

tral Mexico: An Archaeological Ethnography. Anthropological Papers No. 82. Museum of Anthropology, University of Michigan, Ann Arbor.

Parsons, Mary H.

1972 Spindle Whorls from the Teotihuacan Valley, Mexico. In *Miscellaneous Studies in Mexican Prehistory*, by Michael W. Spence, Jeffrey R. Parsons, and Mary H. Parsons, pp. 45–79. Anthropological Papers No. 45. Museum of Anthropology, University of Michigan, Ann Arbor.

Pohl, John M. D., John Monaghan, and Laura R. Stiver

1997 Religion, Economy, and Factionalism in Mixtec Boundary Zones. In *Códices y documentos sobre México, segundo simposio, volumen I*, edited by S. R. Smither, V. Sosa, and R. M. Baracs, pp. 205–232. Instituto Nacional de Antropología e Historia, Mexico City.

Redmond, Elsa M.

1979 A Terminal Formative Ceramic Workshop in the Tehuacan Valley. In *Prehistoric Social, Political, and Economic Development in the Area of the Tehuacan Valley: Some Results of the Palo Blanco Project*, edited by Robert D. Drennan, pp. 111–125. Technical Reports No. 11. Museum of Anthropology, University of Michigan, Ann Arbor.

Robles G., Nelly M.

1994 *Las Canteras de Mitla, Oaxaca: Tecnología para la Arquitecture Monumental*. Publications in Anthropology No. 47. Vanderbilt University, Nashville.

Rothstein, Arden A., and Anya L. Rothstein

2002 *Mexican Folk Art from Oaxacan Artist Families*. Schiffer Publishing, Atglen, Pennsylvania.

Shepard, Anna O.

1963 *Beginnings of Ceramic Industrialization: An Example from the Valley of Oaxaca*. Notes from a Ceramic Laboratory 2. Carnegie Institution, Washington, D.C.

Shimada, Izumi

1996 Sicán Metallurgy and Its Cross-Craft Relationships. *Boletín del Museo del Oro* 41:27–61.

Spores, Ronald

1965 The Zapotec and Mixtec at Spanish Contact. In *Handbook of Middle American Indians*, edited by Gordon R. Willey, Vol. 3:962–987. University of Texas Press, Austin.

Stark, Barbara L.

1985 Archaeological Identification of Pottery Production Locations: Ethnoarchaeological and Archaeological Data in Mesoamerica. In *Decoding Prehistoric Ceramics*, edited by Ben A. Nelson, pp. 158–194. Southern Illinois University Press, Carbondale.

Stark, Miriam

2003 Current Issues in Ceramic Ethnoarchaeology. *Journal of Archaeological Research* 11:193–242.

Starr, Frederick

1900 *Notes upon the Ethnography of Southern Mexico*. Proceedings of the Davenport Academy of Natural Sciences 8. Davenport, Iowa.

Stephen, Lynn

1991 *Zapotec Women*. University of Texas Press, Austin.

Taylor, William B.

1972 *Landlord and Peasant in Colonial Oaxaca*. Stanford University Press, Stanford.

Van de Velde, Paul, and Henriette R. Van de Velde

1939 *The Black Pottery of Coyotepec, Oaxaca, Mexico*. Southwest Museum Papers 13. Los Angeles.

Wasserspring, Lois

2000 *Oaxacan Ceramics: Traditional Folk Art by Oaxacan Women*. Chronicle Books, San Francisco.

Winter, Marcus C.

1974 Residential Patterns at Monte Albán, Oaxaca, Mexico. *Science* 186:981–987.

1984 Exchange in Formative Highland Oaxaca. In *Trade and Exchange in Early Mesoamerica*, edited by Kenneth G. Hirth, pp. 179–214. University of New Mexico Press, Albuquerque.

Winter, Marcus C. (editor)

1995 *Entierros humanos de Monte Albán: Dos estudios*. Contribución No. 7 del Proyecto Especial Monte Albán 1992–1994. Centro Regional de Oaxaca, Instituto Nacional de Antropología e Historia, Oaxaca.

5

Classic Maya Elite Competition, Collaboration, and Performance in Multicraft Production

TAKESHI INOMATA

It has long been argued, and is now well established, that in Classic Maya society (A.D. 250–900) many of the art objects of the finest quality were produced by elites (Coe 1977; Stuart 1989; Webster 1989). Recent developments in archaeology, epigraphy, and iconography have added more detailed information on craftspersons' identities, the process of their work, and the types of products (Coe and Kerr 1997; Inomata 2001b; Reents-Budet 1994, 1998). Such studies demonstrate that the work of these individuals on many occasions can be characterized adequately as multicraft production (Shimada, this volume). Classic Maya elite craftspersons were at once skilled artists and knowledgeable scribes, who worked on various media and materials.

In this chapter I examine craft production by the Classic Maya elite from the perspectives of multicraft production and performance theory. The practice of multicraft production forces us to examine the social and cognitive aspects of craft activities. Not only does the acquisition of various kinds of raw material and of necessary tools demand a reliance on extensive social networks, but the maintenance and transmission of the skills and knowledge required for multicraft production entail elaborate and sustained coordination. In addition, craft production by elites was not purely an economic activity. I (2001b) have argued elsewhere that elite engagements in artistic creation were profoundly political and ideological. The work of Maya elite artists had significant implications and consequences for the competition among courtiers and for the distinction between elites and subordinates. Thus, collaboration and competition were coexisting aspects of this production system. Shimada (this volume) argues that the "top-down" approach to craft production, with the emphasis on perspectives from consumers or on the overall organization of exchange and sponsorship, is not enough. For the study of the social and cognitive aspects of craft production by Maya elites, what he calls the "bottom-up" approach is important. We need to incorporate a perspective from producers and examine the social implications of craft activities themselves in addition to their material products and their circulation (see Costin and Wright 1998).

In this regard, the theory of performance provides a useful insight. By *performance* I mean communicative physical acts that involve an actor (or actors) and an audience. Performance is about showing and seeing, and it inevitably has political consequences (Bell 1992:197–223; Kertzer 1988; Turner 1986). Craft production often has a quality of performance (Carter 2007). Acts of production are recognized and evaluated by others who witness them. Activities that take place in such social matrices shape the actors' identities and political relations with others. The notion of performance is germane to practice theory developed by Bourdieu, Ortner, Sahlins, and others, which has had profound effects on archaeological thoughts, particularly in the study of production and specialization. Practice theory, with the central interest in political relations, implicitly concerns those who see or perceive the acts of agents. Performance studies help us to go further by more explicitly addressing the relationships between actors and audience and the political implications of acting and seeing (or not seeing).

The performance perspective, in other words, mediates the top-down and bottom-up approaches in the study of craft production. Performance theory explicitly criticizes the view that privileges the preexistence and primacy of abstract structures, generative rules, and mental processes that one-directionally determine people's action. Instead, it addresses how social agents' physical acts define social reality and its meaning (Austin 1962; Bell 1997; Bloch 1986; Kertzer 1988; Tambiah 1979). This focus on action, however, does not mean that the theory privileges it over abstract meanings or emphasizes unrelated individual events. On the contrary, the performance approach recognizes and examines how such acts and their perception by others are embedded in specific historical and social contexts. Rather than dichotomizing the top-down and bottom-up processes in social constitution, it addresses how the two are indissolubly intertwined and how they shape each other.

The issue of performance relates to the main themes of the volume: the identity of craftspersons and power relations, as well as competition and collaboration associated with craft production of multiple media. Did master artists work in open public view or in seclusion? Who worked together in what contexts? Who were allowed to share the knowledge and skills of production? At issue are not only the questions of craftspersons' identities and their political relations with others but also the processes in which their identities were shaped and power relations were asserted, negotiated, and transformed.

MULTICRAFT PRODUCTION BY AGUATECA ELITES

Excavations at the Classic Maya center of Aguateca, Guatemala, have provided a significant data set on multicraft production by elites. The central part of this fortified center was burned around A.D. 810, most likely during an enemy attack. The rapid abandonment of elite residences resulted in remarkable assemblages of numerous complete and reconstructible objects (Inomata 1997, 2003; Inomata and Triadan 2000; Inomata et al. 2001). All the excavated elite residences located outside of the royal palace contained remains of craft production. These buildings, Structures M7-35, M8-4, M8-8, and M8-10, were found close together in the elite residential area along the causeway that connected the royal palace and the main plaza (Figure 5.1). Numerous utilitarian objects, including storage jars, cooking vessels, and grinding stones, indicate that they served as residences. The high status of their occupants is suggested by numerous jade ornaments and, in the case of Structures M8-4 and M8-10, objects carved with the Aguateca emblem glyph, that is, the royal title reserved for individuals of the highest rank. Considering the relatively well-structured distribution patterns of artifacts in these buildings and the placement of burials in Structure M8-10, I have argued that each of these residences was occupied by a relatively small number of individuals, possibly including a male household head, who was a scribe-artist, his wife or wives, and their children. The work of these master artists may have been helped by assistants and apprentices who lived in different buildings (Inomata 2001b; Inomata and Stiver 1998; Inomata and Triadan 2000; Inomata et al. 2002; Triadan 2000).

One of the clearest data sets on multicraft production came from Structure M8-10, which we nicknamed the "House of the Scribe." This building contained eight small stone mortars

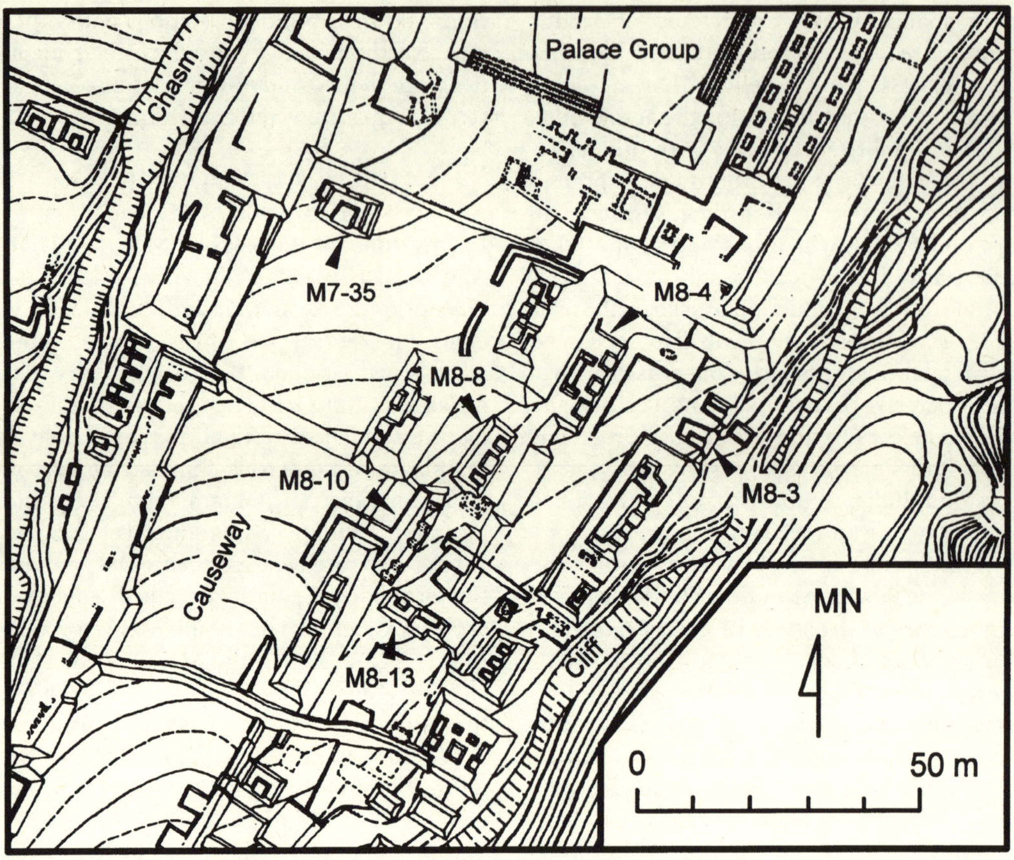

FIGURE 5.1. Central part of Aguateca, with 1 m contour lines in solid lines.

and six stone pestles used for pigment preparation (Inomata 1997: fig. 11). Three halved shells associated with them most likely served as ink pots, as Maya ceramic paintings often depict them in the hands of scribes painting codices (Inomata 1997: fig. 12). In addition, a shell ornament found in this structure was carved with a personal name followed by *its'aat*, a title for scribes, which further corroborates the scribal status of a resident (Inomata 1997: fig. 14). The work of the elite occupant of the House of the Scribe probably included the painting of codices, although such products did not preserve in the tropical environment. Among the most impressive objects unearthed in this residence are numerous shell and bone ornaments (Inomata and Stiver 1998: figs. 8–9). Some of these items may have been carved by the same scribe who lived there. The presence of abundant bone work debitage and a small amount of shell debitage in a dense midden south of the building supports this idea. Unworked river clams found in this structure may have been raw materials for such work, although it is also possible that they served as ink pots. Some of the numerous chipped-stone tools associated with this building, including retouched obsidian blades with a pointed end, may have been used for shell and bone carving. In addition, excavators uncovered four polished stone axes. Because the Maya at Aguateca appear to have used chert bifaces to work on stone blocks for construction (Aoyama 2003), these finely polished axes may have served for more elaborate work, namely, stone or wood sculptures. Thus, the scribe-artist of the House of the Scribe appears to have produced various

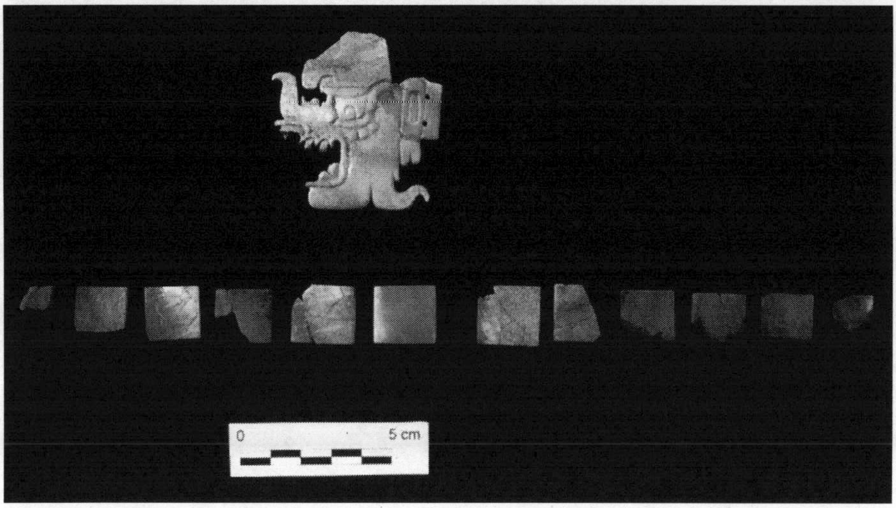

Figure 5.2. Jester God headband found in the south room of Structure M8-4 (the House of Mirrors).

types of objects, potentially including codices, shell and bone ornaments, and wood or stone sculptures.

Structure M8-13, a less elaborate building, was located to the south of the House of the Scribe and faces the same patio area. Kitty Emery (2003) has identified numerous bone work debitage in the area behind this building. Kazuo Aoyama (2003), who examined use-wear on the lithics, found that numerous stone tools from the same area were used for carving bone or shell, resulting in a remarkable accordance of the two lines of evidence. The residents of Structure M8-13, who appear to have been of lower status than those of the House of the Scribe, most likely engaged in bone working, which may have been related to the artistic production by the scribe-artist of the House of the Scribe.

Structure M8-4, which we called the "House of Mirrors," has provided equally compelling evidence of multicraft production. As in the case of the House of the Scribe, ten stone mortars and two pestles indicate the preparation of pigments, which may have been used for painting codices, ceramics, or other materials (Inomata 2001b: fig. 6). On the rear bench of the south room, excavators unearthed a Jester God headband, which was a symbol of rulership (Figure 5.2). The scribe-artist of this building appears to have produced or refurbished royal headdresses and other ceremonial attire, which were meant to be used with the Jester God diadem. Prior to the abandonment of Aguateca, this artisan was recycling mosaic pieces of pyrite mirrors and was cutting and polishing them into rectangular-shaped ornaments. Marcelo Zamora and I identified numerous pyrite pieces in the process of recycling, along with finished ornaments and original mosaic mirror pieces (Figure 5.3). These rectangular pyrite pieces, originally shining in a golden color, were probably going to be embedded in a headdress or other kinds of ceremonial gear. Small bone pieces found in the same area may also have been parts of the same ceremonial attire. These objects were rounded and well polished on one end, whereas the other end was simply snapped along an incision and had small side notches (Figure 5.4). Interestingly, the rounded ends seem have been exposed to more severe heat than the snapped ends when the building was burned. It is likely that they were embedded in an object made of leather or wood at the snapped ends, leaving the rounded ends visible. Use-wear analysis by Aoyama (2003) shows a significantly higher ratio of stone tools used for shell or bone working in this area than in other parts of the building.

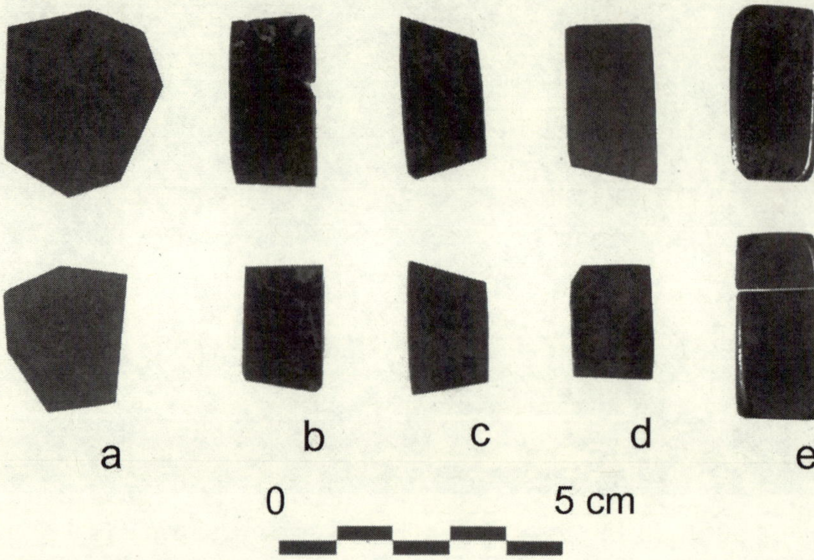

FIGURE 5.3. Pyrite pieces found in the south room of Structure M8-4 (the House of Mirrors): (a) mosaic pieces of a pyrite mirror before recycling; (b–d) pyrite pieces in the process of recycling; (e) finished pyrite ornaments.

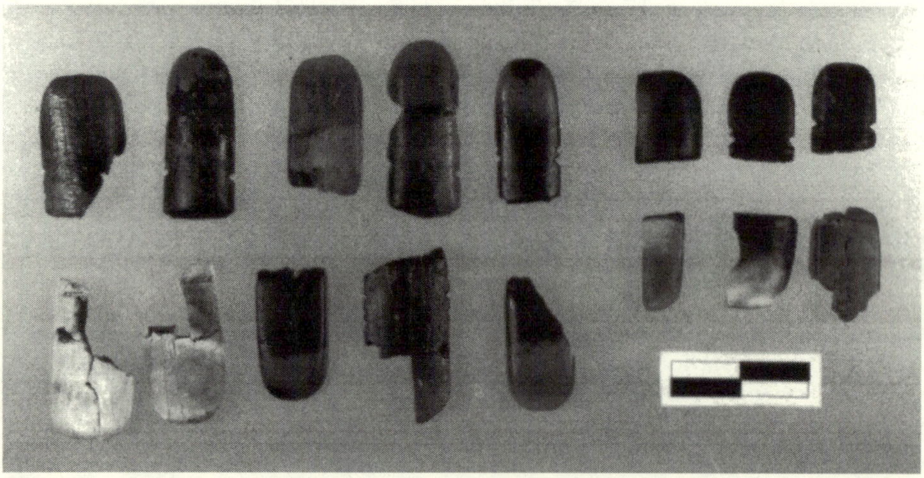

FIGURE 5.4. Small bone ornaments found in the south room of Structure M8-4 (the House of Mirrors). Note that the rounded ends are more severely burned.

Fifteen round reworked ceramic sherds measuring 10–15 cm in diameter were found on the exterior floor south of the House of Mirrors and in the north annex (Inomata et al. 2002: fig. 15). Burned clay was caked on the exteriors of some of these disks, and excavators unearthed a chunk of unfired clay on the southern side of the building. It appears that the residents of the House of Mirrors or their assistants used these objects to work on clay, although it is not clear whether they made ceramic vessels or other types of artifacts. The building also contained seven polished stone axes. The results of the use-wear analysis suggested to Aoyama (2003) that they were used for carving stone. In sum, the craft work conducted by the residents of the House of Mirrors involved various types of material, including pigments, clay, and stone. The production of ceremonial gear, in particular, combined different types of material, such as pyrite, bone, and stone. Such attire probably consisted also of feather and leather, as depicted in stelae and ceramic paintings.

Structure M8-3, a small one-room building located in front of the House of Mirrors, contained a ball of specular hematite pigment, as well as four flat limestone slabs and three manos (handheld grinding stones), which appear to have been used for craft production. A soil chemistry analysis by Richard Terry (Terry et al. 2004) showed high concentrations of iron and copper in and around this structure, pointing to craft work involving the use of pigments. Structure M8-3 may have been a workshop associated with the House of Mirrors or the unexcavated Structure M8-5.

Structures M8-8 and M7-35 also exhibited evidence of craft production by elites, but the materials involved in their work may have been less diverse. Structure M8-8, or the "House of Axes," was associated with 22 polished axes, a number particularly high in comparison to the other buildings. Seventeen of them were found in two concentrations in and near the north annex and consisted of axes of various sizes and shapes (Inomata 2001b: fig. 5). Based on the results of his use-wear analysis, Aoyama (2003) concludes that they were used for carving stone. They probably formed a tool kit of a stelae carver. The nature of the craft work conducted in Structure M7-35, or the "House of the Niche," remains unclear, but it involved a small stone disk and a pestle, as well as three polished axes.

Epigraphic and iconographic data indicate that many of the elite scribes and artists who painted exquisite ceramics and carved stone monuments were males. The primary craft product of elite females was textiles. In the Aguateca elite residences, we can recognize a relatively clear division of male and female spaces. In the House of Mirrors and House of the Scribe, scribal implements were found in the central room and in one of the side rooms of the building, whereas textile production tools, such as spindle whorls and bone needles, were concentrated in the other side room (Figure 5.5; see Inomata and Stiver 1998: figs. 6, 12; Inomata et al. 2002: fig. 19). The rooms associated with textile production also contained numerous storage vessels and implements for food preparation, which accords with the notion that food preparation was a female domain in Classic Maya society. Such a pattern is less clear in the House of Axes, probably because the stelae carving by a resident of this building took place in the Main Plaza. Still, textile production tools were found associated with implements for food preparation around the north room. However, one should not overemphasize the level of gender-based spatial division. Male and female artists may often have worked together on many occasions. Numerous shell ornaments found in the House of the Scribe, which were possibly made by a male artist, had small perforations, suggesting that they were attached to textiles, products of females.

KNOWLEDGE, COMPETITION, AND COLLABORATION

These data on elite Maya craft specialists urge us to go beyond the static divisions of sponsors and producers and of attached and independent specialists (Ames 1995; Costin and Wright 1998; Inomata 2001b). Elite Maya scribes and artists made precious objects for rulers and other elites, at least on certain occasions. It is, however, most likely that they sometimes produced art pieces on their own initiative for such purposes as gift exchange and dedications to the supernatural. Thus, they engaged in both independent and attached production. They did not merely follow the orders of their lords but often used their abilities and products for their own purposes.

The esoteric knowledge of artistic production is a critical factor in this production system. Elite art objects served primarily as media for recording and expressing dynastic events and mythological notions in writing, painting, and carving with appropriate calendrical notations. Thus, the work of Maya elite artists required not only technical skills and familiarity with material properties but also the knowledge of writing, history, myths, the

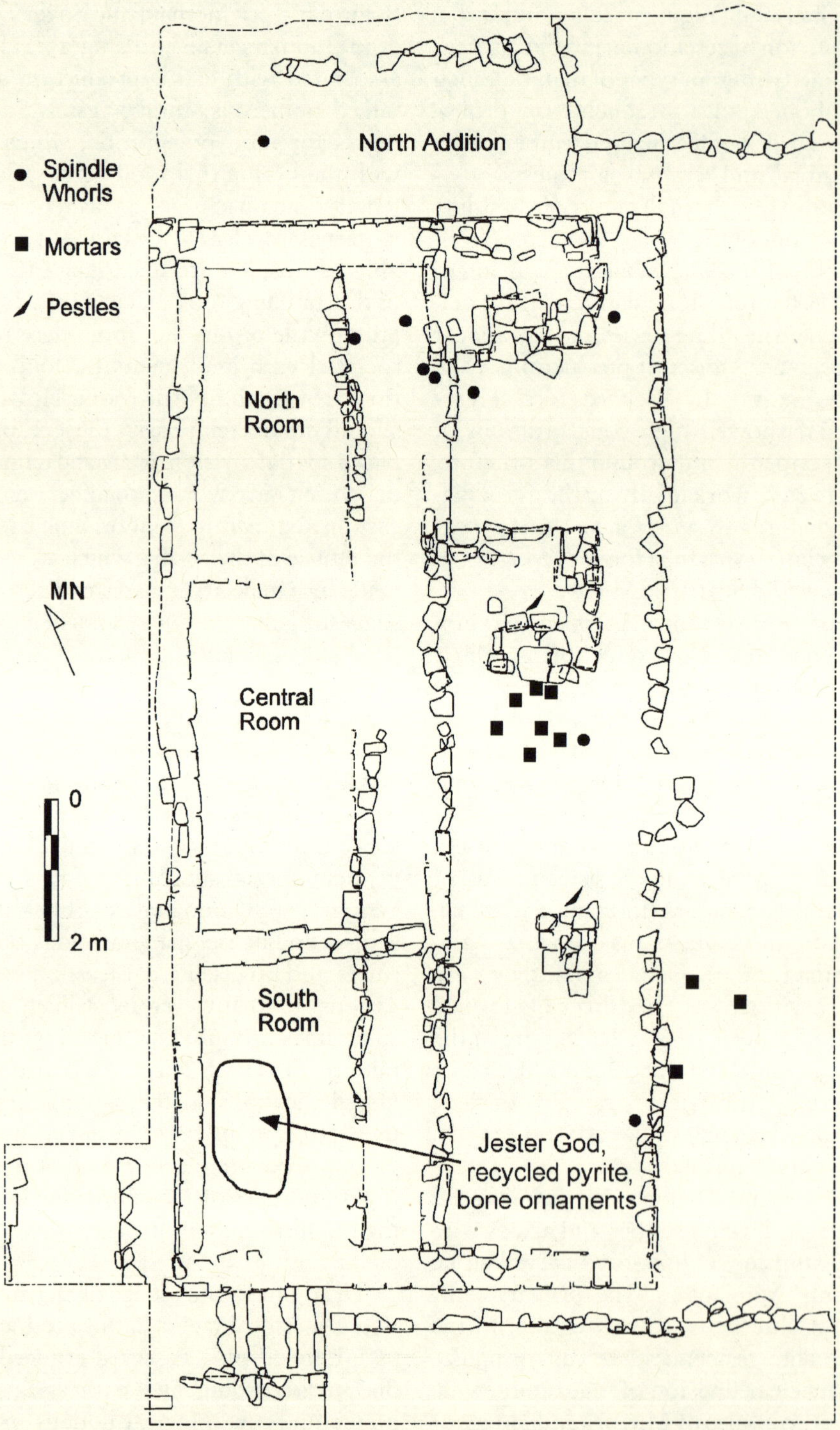

FIGURE 5.5. The distribution of spindle whorls, stone mortars, and pestles in Structure M8-4 (the House of Mirrors). Drawing also indicates the area in the south room where possible pieces for ceremonial attire (a Jester God headband, recycled pyrite pieces, and small bone ornaments) were found.

calendar, astronomy, and mathematics. Once mastered, such knowledge could be applied to various types of craft production. Reents-Budet (1994:218) has noted that Maya artists participated in many different projects and worked on different media. Direct evidence of such practice has been presented by Grube (1990:320), who has found the same scribe's signatures on a carved stone lintel from Xkalumkin and on a carved vessel. We should note that at Aguateca a certain level of specialization existed among elite scribes and artists. The resident of the House of Axes appears to have been more focused on stone carving, whereas activities in the House of the Scribe involved primarily the use of pigments. Still, the results of excavations at Aguateca accord well with the observations of multicraft production. The data from each of the Aguateca elite residences reflect a wide range of craft activities. An individual scribe-artist may have painted codices, carved shell ornaments, and painted ceramic vases.

The recognition of such knowledge probably contributed significantly to the reputation and honor that elite artists enjoyed. As Bourdieu (1984) and Helms (1993) have argued forcibly, the prestige and reputation of high achievements and ability in education, art, and culture were closely tied to political power in certain historical contexts. Epigraphic and iconographic evidence indeed suggests that scribes and artists were highly respected in Classic Maya society (Coe and Kerr 1997; Reents-Budet 1994). There are indications, including scribal signatures, that the artistic achievements of scribes and artists were clearly recognized by many (Houston 2000). Some images show deities conducting such activities, indicating that the creative acts of scribes and artists were embodiments of the power and wisdom of gods (Coe 1977; Reents-Budet 1998). Their products were endowed with a certain level of inalienability, with the producers' identities always associated with the value and meaning of the objects (see Weiner 1992). The skilled production of art objects, thus, contributed significantly to the power and prestige of scribes and artists. In this regard, the production of art objects by Maya elites was not purely aesthetic or economic work but, rather, a highly political act. Many scribes and artists were at once prominent courtiers, administrators, and diplomats, who actively engaged in power struggles at the royal court. The production of art objects had important implications and consequences in competition among elites (Inomata 2001b). This observation suggests to me that artistic production itself was also a highly competitive process.

The other side of the coin is a highly collaborative process. Learning writing, the calendar, and other kinds of esoteric knowledge undoubtedly required a long period of training and close tutelage by master scribes and artists (Coe and Kerr 1997; Reents-Budet 1994). The maintenance and transmission of the knowledge necessary for artistic production entailed sustained collective effort. Although we know little about its specific mechanisms and institutions, such collective effort pertained most likely to the elite rather than to Maya society as a whole. Bishop Landa wrote that during the contact period high priests taught writing, the calendar, medicine, and religion to the sons of lords and other priests (Tozzer 1941:27–28). In a similar manner, most scribes and artists appear to have been recruited from the elite during the Classic period. The esoteric knowledge kept among the elite served as a core element of their high culture, which enhanced their shared identities of privileged status and their cohesion for common political purposes (see Baines and Yoffee 1998). A question remains as to whether nonelites shared similar cultural values associated with the production and products of elite artists. I speculate that they did at least partially because nonelites used objects with pseudoglyphs and motifs similar to those of elite objects, but in lower qualities, assimilating to certain aspects of elite culture. In this sense, artistic production by elites themselves might have served to portray elite high culture as something superior and something to be desired, and thus it might have helped to distinguish the elite from the rest of society and to legitimize their political power at least to a certain degree (see Helms 1993).

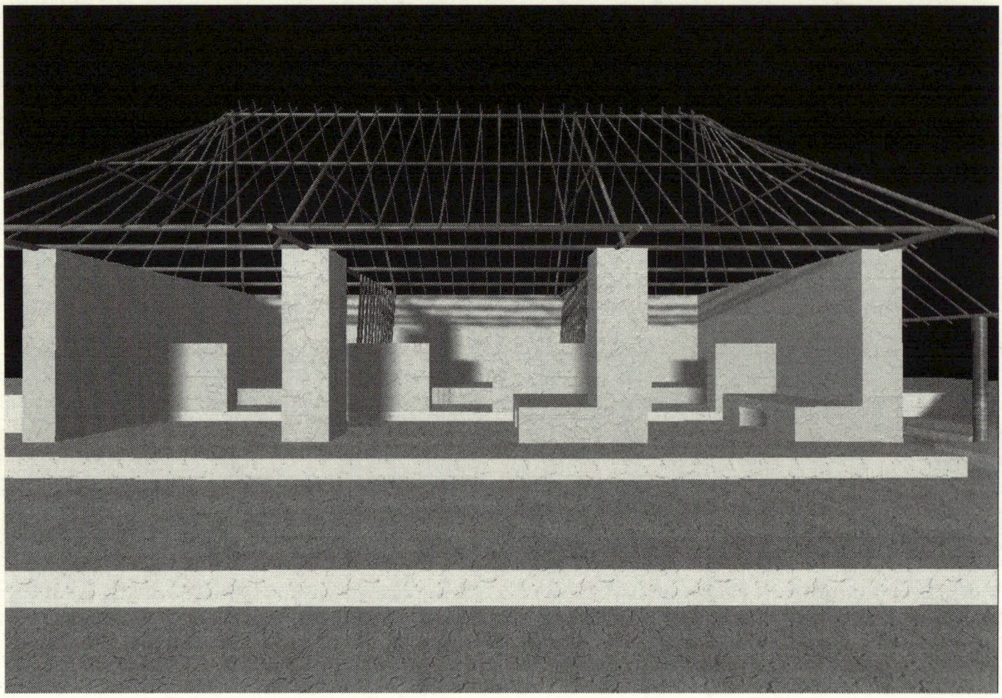

FIGURE 5.6. Reconstruction computer model of Structure M8-4 (the House of Mirrors) as viewed from the patio area. The male artists may have worked in and in front of the center room and the south room (the room on the left), whereas a female appears to have spun and woven in front of the north room (the room on the right).

In addition to the learning mechanisms, production processes were also collaborative. Although the production of certain types of objects may have been essentially individual work, others involved the collaboration of multiple individuals. The stelae of Piedras Negras, for example, bear up to eight different scribal signatures (Houston 2000). Reents-Budet (1994:218–222) suggests that some of the exquisitely painted ceramics were produced in workshop settings, where multiple individuals responsible for different stages of production worked side by side. In addition, according to Bishop Landa, Maya women in colonial Yucatán often collaborated in the production of textiles, and similar practices probably existed during the Classic period (Tozzer 1941:127). As mentioned above, males and females who often worked on different materials and in different spaces within a house must have collaborated on many occasions.

THE PERFORMANCE OF CRAFT PRODUCTION

Focusing on the aspect of performance in this system of production is an effective approach. I (2001a; Inomata et al. 2002) have argued elsewhere that Classic Maya elite residences, particularly their central rooms, were theatrical stages for political performance. The relatively open arrangement of these buildings facing patios facilitated the high visibility of meetings, negotiations, and receptions that were held there. Such political acts were probably witnessed, and had to be witnessed, by audiences of varying sizes that occupied the patios (Figure 5.6). Some such scenes were then recorded in ceramic paintings. The same may be true of craft production. In the House of the Scribe and the House of Mirrors many of the scribal instruments were found in the central rooms that were used for meetings and receptions. In the House of Mirrors, in particular, many scribal tools were probably placed on the front bench, an area highly visible from outside. Numerous pyrite pieces in the process of reworking were found in the south room of the House of Mirrors. However, we did not recover any production debitage in the room, and the levels of heavy metals in the soil were relatively low. An unexcavated area in the patio, where soil analysis revealed high metal concentrations, may have been a main pyrite-working area. The scribe probably brought pyrite pieces back to the room after the day's work. His work in the open area would have been highly visible. The high visibility of scribal work appears to be reflected in its depiction as a common theme in ceramic paintings.

Females appear to have worked generally in different areas than males, but their work was equally visible and theatrical. In modern Maya villages, where women typically work on textiles in front of buildings or patio areas, the performance of weaving has become even more emphasized with the surge of tourism. Many modern Maya women weave in front of their houses, highly conscious of the eyes of tourists. Such performance not only attracts potential buyers of textiles but also demands respect for their work and enhances the authenticity of the products that they hope to sell. Like their modern descendants, Classic Maya women must have spun and woven in visible locations, possibly in front of the side rooms in the case of the Aguateca elite residences. Although women's performance probably attracted attention and respect, women's work may have been perceived as secondary, at least in the eyes of elite males, in relation to the work of male artists and scribes that took the central stage, that is, the central room that was the most prominent space of political manifestation for an elite household. Depictions of the work of male artists are far more common in Classic Maya pictorial media than those of female weavers, which appears to reflect the higher prestige that the former enjoyed in the male-dominated court culture.

The religious beliefs of the Classic Maya add to the meanings of the performance of artistic creation. Ethnohistorical accounts of the colonial Maya indicate that the production of certain objects was accompanied by various types of ritual performance. This implies the intense interest of various parties in acts of production. The importance of religious beliefs surrounding artistic production may also have required the seclusion of such work in certain contexts. According to Bishop Landa (Tozzer 1941:160), carvers of wood statues, along with priests, shut themselves up in a temporary hut while they worked. This involved strict religious observance of fasting, bloodletting, incense burning, and abstinence. Although the productive acts of artists were shielded from public view in such cases, I should point out that temporary huts, along with associated rituals, probably had an additional effect of making people strongly aware of the ongoing sacred work of artists.

The primary viewers of the performance of craft production by elite artists were probably fellow courtiers who lived and worked in the same area. Some of them were craftspersons themselves who produced similar or related objects. The data from Aguateca demonstrate that various elite craftspersons lived and worked in close proximity to each other. Given the relatively high visibility of their productive activities, these artists were probably well aware of how others worked. Thus, the performance of artistic creation may have been a way of demonstrating and asserting the ability and knowledge of elite artists to rival courtiers. It may also have facilitated collaborative processes among them.

The work of stone monument carvers may have been unique in this regard. They worked mainly in large public plazas where such stone monuments were placed. Could many people see carvers work, or were the plazas closed off to the public while they worked? Data from Aguateca give us a rare glimpse of their work process. Structure L8-8 facing the main plaza was abandoned while it was in the process of construction (Inomata et al. 2004). It appears that many workers were carrying stones to build up the masonry core of the structure, and masons were covering some walls with cut stones. In front of this unfinished temple was an unfinished stone monument, Altar M (Figure 5.7). In other words, the construction of a temple and the carving of an

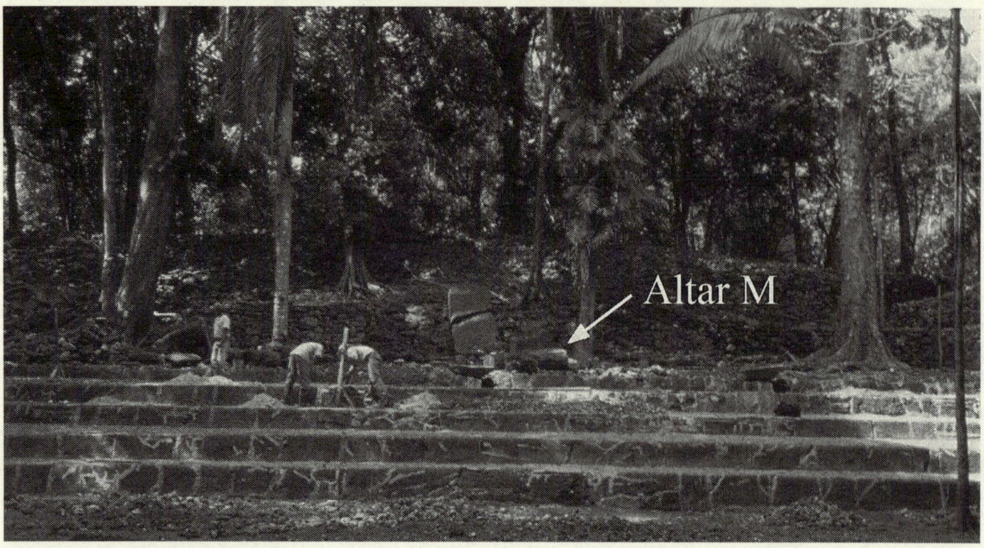

FIGURE 5.7. Altar M on the front terrace of the unfinished temple, Structure L8-8.

associated monument were proceeding simultaneously. Numerous workers, elites and nonelites alike, could probably witness the work of stone monument carvers, unless their work was shielded by a temporary structure built around the monument. Again, such a building for seclusion would only have made the general public even more conscious of the fact that carvers were currently working.

CONCLUSIONS

Although archaeological inquiries into craft production have often used "top-down" approaches addressing the overall organization of exchange and the economic role of material products, recent studies demonstrate the importance of a "bottom-up" approach with a specific focus on producers and the social meaning of their activities (Costin and Wright 1998; Inomata 2001b; Shimada, this volume). Such an approach to craft production by the Classic Maya elite provides significant insights, particular regarding the following points.

First, we need to examine the nature of multicraft production by Maya artists and scribes. An overemphasis on the "top-down" approach may result in the imposition of classificatory schemes created by archaeologists, which may not fit the reality of past practices. For Maya craftspersons, the socially constructed values and meanings of objects may have shaped more important categorizations of products than those determined by material types. Many objects produced by Maya elite artists combine multiple types of material. Although a certain level of division of labor existed among elite craftspersons, many of them worked on various media. We should be aware that the pattern of specialization among archaeologists, which typically follows the division of ceramics, lithics, faunal remains, and so forth, does not necessarily correspond with that of the past producers whom we study. If archaeologists sort artifacts by material types in the field and simply leave them to respective specialists, they may overlook fundamental information from the overall assemblage. Though we will never learn the exact emic perception of craft production, we need to try to access the perspective of past producers.

Second, the knowledge and belief associated with productive activities were important factors related to multicraft production. Maya scribe-artists were not so much specialists in ceramics, stone, or other materials as they were experts in the esoteric knowledge of Maya elite culture, which encompassed writing, the calendar, mathematics, astronomy, history, and religion. Once they learned such esoteric knowledge, it could potentially be applied to various material media. This may also mean that Maya elite artists had to familiarize themselves with various types of materials and to master different kinds of techniques in order to give the desired material forms to their esoteric knowledge. Such knowledge was closely tied to beliefs in the cultural values of productive activities. The study of this production system requires a holistic approach that pays attention not only to the visible and tangible domain of materiality but also to the abstract and cognitive domains of knowledge and belief.

Third, craft production by Maya elites involved both collaborative and competitive processes, which shaped, and were shaped by, the identities of producers and their power relations with others. It is likely that many Maya highly respected the esoteric knowledge of scribe-artists and placed significant cultural value on their art products. Thus, the production of art objects—the act of production itself in addition to the economic and political gains from the exchange of products—probably contributed considerably to the prestige and power of the artists. Artistic production was doubtless a highly competitive process with substantial political implications. At the same time, production at a high standard, as well as the maintenance of the necessary knowledge, certainly required close collaboration among multiple individuals. The competition and collaboration involved in elite craft production were reflections of similar processes that took place in the royal court. Internally, the court was an inherently competitive organization saturated with political struggles among various individuals, groups, and factions. In relation to the rest of society, however, the court distinguished itself with a shared subculture and values, which led to, and were sustained by, a certain level of collaboration among its members (Inomata and Houston 2001).

Fourth, to understand the social and political implications of artistic production, we need to examine the aspect of performance. A strong emphasis on the visibility of artists' work underlay the ideological values of such acts. What mattered for the Classic Maya elite was not only the aesthetic and practical values of the final products but the act of artistic creation itself, the act that was the embodiment of god's power, the struggle toward the realization of cultural ideals, and the confirmation of ideological values. An identity and a reputation as a skilled craft producer were in a sense a practical achievement accumulated through performance in front of viewers. The performance of elite craft production was enlaced with competition and collaboration. On the one hand, elite artists as active players in court politics demanded attention, respect, and recognition through such acts. The open nature of their activities, on the other hand, facilitated the collaborative maintenance of knowledge and the creation of a system of shared values.

Our goal is not to decipher religious and cultural beliefs associated with craft production completely or to reconstruct every detail of performance. It is simply impossible in most archaeological situations. Yet, by examining not only the final products but also the acts of production, underlying knowledge, and associated beliefs, we can perhaps gain a richer and deeper appreciation of the social processes surrounding craft production.

Acknowledgments

I am indebted to the personnel of the Instituto de Antropología e Historia de Guatemala for permits and generous support for our fieldwork at Aguateca. The Aguateca Archaeological Project was funded by the National Science Foundation (SBR-9707950, SBR-9974302, BCS-9910594), the Mitsubishi Foundation, the National Geographic Society, the Foundation for the Advancement of Mesoamerican Studies Inc., the H. John Heinz III Charitable Trust, Yale University, and the University of Arizona. I thank Izumi Shimada for inviting me to contribute to his Society for American Archaeology session and to this volume.

References Cited

Ames, Kenneth M.
1995 Chiefly Power and Household Production on the Northwest Coast. In *Foundations of Social Inequality*, edited by T. Douglas Price and Gary M. Feinman, pp. 155–188. Plenum Press, New York.

Aoyama, Kazuo
2003 La producción artesanal de la elite en la sociedad clásica maya: Evidencia de la lítica de los grupos domésticos en Aguateca, Guatemala. In *XVI Simposio de Investigaciones Arqueológicas en Guatemala*, edited by Juan Pedro Laporte, Bárbara Arroyo, Héctor L. Escobedo, and Héctor E. Mejía, pp. 519–531. Ministerio de Cultura y Turismo, Instituto de Antropología e Historia, and Asociación Tikal, Tikal.

Austin, J. L.
1962 *How to Do Things with Words*. Oxford University Press, Oxford.

Baines, John, and Norman Yoffee
1998 Order, Legitimacy, and Wealth in Ancient Egypt and Mesopotamia. In *Archaic States*, edited by Gary M. Feinman and Joyce Marcus, pp. 199–260. School of American Research Press, Santa Fe.

Bell, Catherine
1992 *Ritual Theory, Ritual Practice*. Oxford University Press, Oxford.
1997 *Ritual: Perspectives and Dimensions*. Oxford University Press, Oxford.

Bloch, Maurice
1986 *From Blessing to Violence: History and Ideology in the Circumscription Ritual of the Merina of Madagascar*. Cambridge University Press, Cambridge.

Bourdieu, Pierre
1984 *Distinction: A Social Critique of the Judgement of Taste*. Harvard University Press, Cambridge.

Carter, Tristan
2007 The "Necrolithic": Death, Obsidian, and Performance in the Early Cycladic Burial Arena. In *Rethinking Production in Archaeological Contexts*, edited by Zachary X. Hruby and Rowan Flad, pp. 88–107. Archaeological Papers of the American Anthropological Association No. 17. Arlington, Virginia.

Coe, Michael D.
1977 Supernatural Patrons of Maya Scribes and Artists. In *Social Process in Maya Prehistory: Essays in Honor of Sir Eric Thompson*, edited by Norman Hammond, pp. 327–347. Academic Press, London.

Coe, Michael D., and Justin Kerr
1997 *The Art of the Maya Scribe*. Harry N. Abrams, New York.

Costin, Cathy Lynne, and Rita P. Wright (editors)
1998 *Craft and Social Identity*. Archaeological Papers of the American Anthropological Association No. 8. Arlington, Virginia.

Emery, Kitty F.
2003 The Noble Beast: Status and Differential Access to Animals in the Maya World. *World Archaeology* 34:498–515.

Grube, Nikolai
1990 The Primary Standard Sequence on Chocholá Style Ceramics. In *The Maya Vase Book*, edited by Justin Kerr, Vol. 2:320–330. Kerr Associates, New York.

Helms, Mary W.
1993 *Craft and the Kingly Ideal: Art, Trade, and Power*. University of Texas Press, Austin.

Houston, Stephen D.
2000 Into the Minds of Ancients: Advances in Maya Glyph Studies. *Journal of World Prehistory* 14:121–201.

Inomata, Takeshi
1997 The Last Day of a Fortified Classic Maya Center: Archaeological Investigations at Aguateca, Guatemala. *Ancient Mesoamerica* 8:337–351.
2001a The Classic Maya Royal Palace as a Political Theater. In *Reconstruyendo la ciudad maya: El urbanismo en la sociedades antigua*, edited by Andrés Ciudad Ruiz, María Josefa Iglesias Ponce de León, and María del Carmen Martínez Martínez, pp. 341–362. Sociedad Española de Estudios Mayas, Madrid.
2001b The Power and Ideology of Artistic Creation: Elite Craft Specialists in Classic Maya Society. *Current Anthropology* 42:321–333.
2003 War, Destruction, and Abandonment: The Fall of the Classic Maya Center of Aguateca, Guatemala. In *The Archaeology of Settlement Abandonment in Middle America*, edited by Takeshi Inomata and Ronald W. Webb, pp. 43–60. University of Utah Press, Salt Lake City.

Inomata, Takeshi, and Stephen D. Houston
2001 Opening the Royal Maya Court. In *Royal Courts of the Ancient Maya, vol. 1: Theory, Comparison, and Synthesis*,

edited by Takeshi Inomata and Stephen D. Houston, pp. 3–23. Westview Press, Boulder.

Inomata, Takeshi, Erick Ponciano, Oscar Santos, Oswaldo Chinchilla, Otto Román, and Véronique Breuil-Martínez

2004 An Unfinished Temple at the Classic Maya Center of Aguateca, Guatemala. *Antiquity* 78:798–811.

Inomata, Takeshi, Erick Ponciano, Richard Terry, Daniela Triadan, and Harriet F. Beaubien

2001 In the Palace of the Fallen King: The Excavation of the Royal Residential Complex at the Classic Maya Center of Aguateca, Guatemala. *Journal of Field Archaeology* 28:287–306.

Inomata, Takeshi, and Laura R. Stiver

1998 Floor Assemblages from Burned Structures at Aguateca, Guatemala: A Study of Classic Maya Households. *Journal of Field Archaeology* 25:431–452.

Inomata, Takeshi, and Daniela Triadan

2000 Craft Production by Classic Maya Elites in Domestic Settings: Data from Rapidly Abandoned Structures at Aguateca, Guatemala. *Mayab* 13:57–66.

Inomata, Takeshi, Daniela Triadan, Erick Ponciano, Estela Pinto, Richard E. Terry, and Markus Eberl

2002 Domestic and Political Lives of Classic Maya Elites: The Excavation of Rapidly Abandoned Structures at Aguateca, Guatemala. *Latin American Antiquity* 13:305–330.

Kertzer, David I.

1988 *Ritual, Politics, and Power*. Yale University Press, New Haven.

Reents-Budet, Dorie

1994 *Painting the Maya Universe: Royal Ceramics of the Classic Period*. Duke University Press, Durham.

1998 Elite Maya Pottery and Artisans as Social Indicators. In *Craft and Social Identity*, edited by Cathy Lynne Costin and Rita P. Wright, pp. 71–92. Archaeological Papers of the American Anthropological Association, Vol. 8. Arlington, Virginia.

Stuart, David

1989 The Maya Artists: An Epigraphic and Iconographic Study. Unpublished B.A. thesis, Princeton University.

Tambiah, Stanley J.

1979 A Performative Approach to Ritual. *Proceedings of the British Academy* 65:113–169.

Terry, Richard E., Fabian G. Fernandez, J. Jacob Parnell, and Takeshi Inomata

2004 The Story in the Floors: Chemical Signatures of Ancient and Modern Maya Activities at Aguateca, Guatemala. *Journal of Archaeological Science* 31:1237–1250.

Tozzer, Alfred M.

1941 *Landa's Relación de las cosas de Yucatan*. Papers of the Peabody Museum of American Archaeology and Ethnology 18. Harvard University, Cambridge.

Triadan, Daniela

2000 Elite Household Subsistence at Aguateca, Guatemala. *Mayab* 13:46–56.

Turner, Victor W.

1986 *The Anthropology of Performance*. Performing Arts Journal Publications, New York.

Webster, David

1989 *The House of the Bacabs, Copan, Honduras*. Studies in Pre-Columbian Art and Archaeology 29. Dumbarton Oaks Research Library and Collection, Washington, D.C.

Weiner, Annette B.

1992 *Inalienable Possessions: The Paradox of Keeping-while-Giving*. University of California Press, Berkeley.

II

Regional and Diachronic Syntheses

6

Structures of Craft Production, Society, and Political Control: Late Prehistoric and Early Roman Temperate Europe

PETER S. WELLS

Manufacturing—at all levels—is driven by wants. Whether we consider the making of a Middle Paleolithic flint scraper or the latest SUV, the choices of raw material, techniques of manufacture, and style and ornamentation of the product are ultimately determined by the person who uses it. Manufacturing, or crafting, is an essential part of being human. All human groups make things. The earliest tools known date from 2.6 million years ago on, and for at least the past 50,000 years, humans have been making ornaments of different kinds.

The products of human manufacturing, whether we call them crafts or industrial goods, play complex roles in social interaction (Appadurai 1986; Eicher 1998; Gosden and Marshall 1999; Helms 1993). Among those roles is the communication of information about the identity, status, and beliefs of the user. This communicative role is especially apparent among objects commonly classified as "ornaments," but it is equally discernible, if more subtle, among those designated "tools" and "weapons." All made objects communicate messages between users and others.

In modern Western society, we have become so inundated with material goods that we are overwhelmed by the information that objects convey. Some categories of material goods still carry powerful social meanings of which we are directly aware, such as clothing (Eicher 1995) and cars, whereas others that seem less loaded with social and cultural implications nonetheless bear significant messages as well (Csikszentmihalyi and Rochberg-Halton 1981). It seems to be the case that in societies that are not as deluged with material goods as the modern West, all elements of material culture can have complex social and spiritual meanings (Wilson 2000). An examination of the desires for goods and of the crafting processes that provide those goods leads to significant insights about the societies involved.

MATERIAL GOODS IN LATE PREHISTORIC AND EARLY ROMAN EUROPE

Though considerably richer in material culture than communities of the Paleolithic, Neolithic, and Bronze Ages, the peoples who lived in temperate Europe between 3,000 and 1,500 years ago (the final stage of what we call "prehistory" and the beginning of "historical" times) had quite circumscribed material cultures by our twenty-first-century standards. It was not until the Industrial Revolution and the accompanying explosion of consumer goods (McKendrick et al. 1982) that material culture began to approach the richness and diversity that we known today.

In this essay, I develop the discussion around two categories of material culture that are well represented archaeologically—pottery and personal ornaments made of metal, especially bronze. These two categories involve different materials and thus provide two different sets of evidence pertaining to craft production. Pottery manufacture left rich archaeological evidence, particularly the remains of kilns and quantities of misfired sherds. The manufacture of metal ornaments, on the other hand, left only scanty traces. The scale of production of bronze and iron ornaments required only small furnaces that are rarely identifiable archaeologically. Little scrap remains from the casting and forging processes, because unlike failed pots, which are commonly disposed of near the kiln, miscast bronze objects can be tossed back into the crucible to be remelted, and damaged iron ornaments can be reforged into other objects.

I begin with a brief summary of the character of these categories of material culture in late prehistoric and Roman times in temperate Europe. Second, I review what we know about why these objects were desired—the roles they played in people's physical, social, and spiritual lives. Third, I address the issue of production—how the manufacture of these objects was organized and how that organization relates to the social roles that the objects played. The chronological range covered here is from the Late Bronze Age (beginning 1200 B.C.) to the end of the Early Roman period (A.D. 300). The geographical focus is the central regions of the European continent—today southern Germany and neighboring areas.

Pottery

During the Late Bronze Age (1200–800 B.C.), the range of ceramic forms increased (Harding 2000). At settlement sites we find sherds from a wide range of vessel types, including large, thick-walled storage vessels; bowls and dishes used for preparing and serving food; and finely made, thin-walled, often highly decorated cups for drinking liquids (Wells 1983). In well-outfitted burials, sets of ceramic vessels represent the different forms used in feasting, including large serving vessels, dishes, plates, and cups (Müller-Karpe 1955). Pottery in this period was handmade and for the most part monochrome. Many large vessels were decorated with simple fingertip impressions on the rim and with raised bands around the shoulder. The small, thin-walled cups are the only category consistently ornamented, with complex patterns of incised lines and strokes (Wells 1985). Graphite was often applied to the surface of vessels—especially the ornate cups—to give them a shiny black appearance.

In the Early Iron Age (800–450 B.C.), pottery assemblages became more complex. A wider range of vessel types was decorated with incised ornament, mainly in geometrical patterns of squares, triangles, and circles. Other categories of vessels were painted in different colors (Kimmig 1983a:101–117). The great majority were handmade, but at the few trade centers beginning around the end of the sixth century B.C., such as the Heuneburg in southwest Germany and Mont Lassois in eastern France (Figure 6.1), some vessels were made on the potter's wheel (Béarat and Bauer 1994). This technological innovation is understood in connection with the well-documented interactions with the Greek world of the Mediterranean Basin during the period 600–450 B.C. (Kimmig 2000). When those interactions declined, the commercial centers in temperate Europe did as well, though the relationship between those changes is not well understood. The potter's wheel remained in use in some places but not generally for fine wares comparable to those produced at the Early Iron Age centers.

In the first part of the Late Iron Age (450–150 B.C.), again pottery was generally plainer in color and ornamentation and less differentiated in form than that of the preceding era. During this period of roughly 300 years, few trade centers are known in temperate Europe, and the trend toward the centralization of crafts that was evident at the centers of the Early Iron Age disappeared. In the course of the second century B.C., the *oppida* emerged as major centers of population and economic organization, with production on a much larger scale than earlier (Collis 1995). The Latin term *oppidum* (plural *oppida*), used by Julius Caesar in his commentaries on his mili-

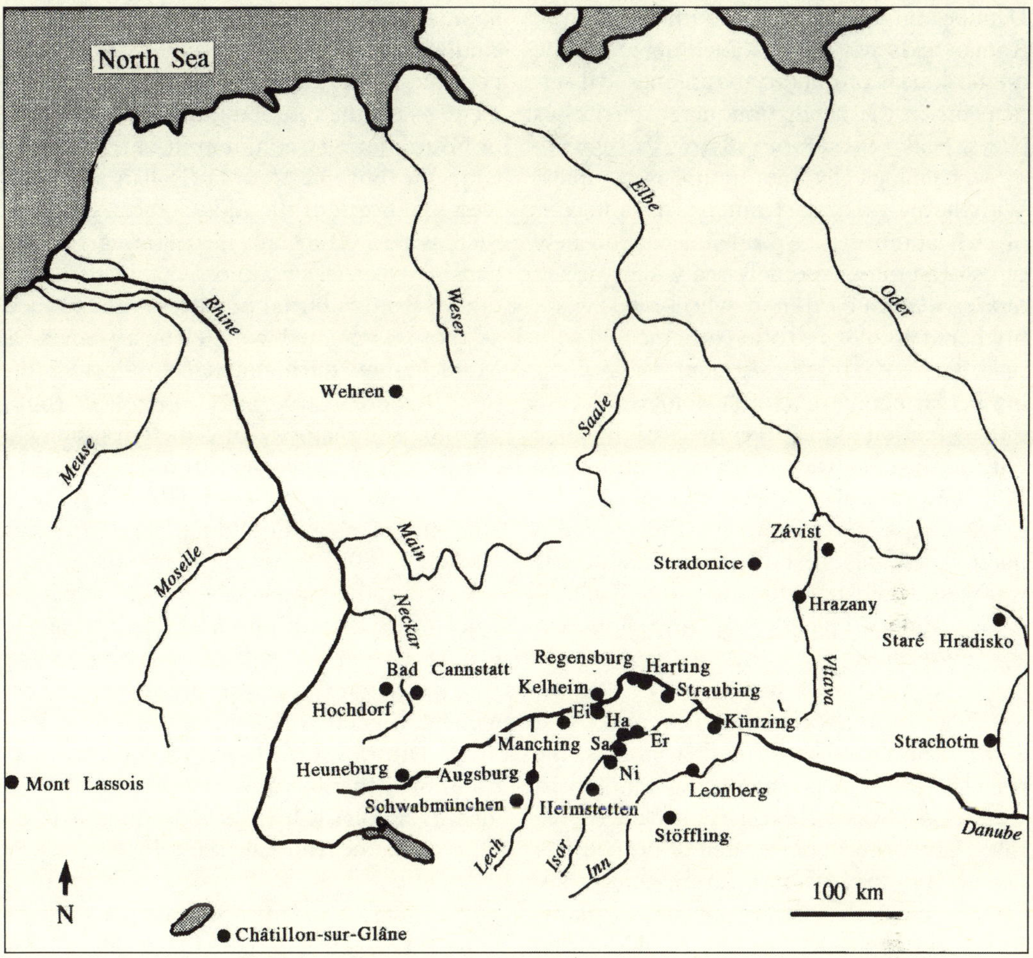

FIGURE 6.1. Map showing the locations of sites mentioned in the text. Ei = Eining, Er = Ergolding, Ha = Hascherkeller, Ni = Niedererlbach, Sa = Sallmannsberg.

tary campaigns of 58–51 B.C. in Gaul, is used by modern investigators to refer to the large (some over 1,000 acres) walled settlements of Late Iron Age temperate Europe, some 150 of which have been identified. At this time the fast-turning potter's wheel came into regular use, and at the principal oppida the great majority of the pottery was wheel made (some 75 percent at Manching—see below), and we can speak of series or mass production. Significant diversity in ceramics developed, both in material and in form (Lasch 1999). Among the principal types are large, coarse ware storage vessels; thick-walled cooking pots of a graphite–clay mix (the graphite is thought to have served to spread heat more evenly and thus prevent cracking); finely made wheel-turned tableware; and thin-walled, often tall, narrow vessels with polychrome painted patterns (Maier 1970).

With respect to the ways in which research traditions have developed, the Roman period represents a break with the prehistoric Iron Age. For the first time, texts inform us directly about events in temperate Europe. Latin texts tell about the process of the Roman conquest of much of temperate Europe. For example, Julius Caesar's commentary about his conquest of Gaul relates details of that military venture, completed in 51 B.C. In 15 B.C., Generals Tiberius and Drusus led their armies in the conquest of the lands between the Upper

Danube and the Alps, now southern Bavaria. Roman texts also tell us something about the establishment of military bases and civil settlements in the newly conquered territories (Czysz et al. 1995; Horn 1987; Wells 1999).

Archaeologically, the picture is complex. With the arrival of the Roman army and later of civil authorities, we do indeed find new types of pottery, especially the widely prized *terra sigillata*, a hard-fired, wheel-made ware, bright red in color and often ornamented with relief figures (Garbsch 1982). But, as is becoming increasingly clear as archaeologists devote more attention to them, ceramics of the same types as the Late Iron Age wares continued to be made and used at Roman sites for two and in some cases three centuries after the conquest. These include both coarse handmade wares and fine polychrome-painted wheel-made ceramics (Flügel 1996; von Schnurbein 1993; Wieland 1993).

Metal Ornaments

During the Late Bronze and Iron ages, the principal kinds of personal ornaments worn were ring jewelry (bracelets, necklaces, leg rings, finger rings), most often of bronze; fibulae (bronze or iron and, rarely, gold or silver); beads of glass or amber; and pendants (mostly of bronze). During the middle part of the Late Iron Age, ornate bronze link belts became common in women's graves, as did glass bracelets. In the final phase of the Late Iron Age, ornamental bronze belt hooks occurred in some women's graves. Throughout the Late Bronze and Iron ages, all categories of personal ornaments were more common in women's burials than in men's (Krämer 1985; Waldhauser 1987).

Personal ornamentation in men's graves often appears in the form of decoration on weapons, especially on daggers in the Early Iron Age (Sievers 1982) and on the scabbards of swords in the Late Iron Age (De Navarro 1972; Frey 1995). Bronze vessels that often occur in richly outfitted men's graves (less frequently in women's) also sometimes bear significant decoration.

The use of personal ornaments became more varied during the Early Roman period. Standard Roman burial practice in Italy included few grave goods, such as oil lamps, perfume bottles, and occasional small ornaments—not the more abundant assemblages of bronze jewelry common in Late Iron Age temperate Europe. After the Roman conquests west and north of the Alps, some cemeteries, such as Bad Cannstatt, near Stuttgart (Nierhaus 1959), reflect almost complete adoption of Roman burial practices. Others, such as those associated with Heimstetten, near Munich, show the maintenance of local traditions (Keller 1984; Volpert 2002; Wells 1999: 96–98). The majority of cemeteries indicate integration of traditional Iron Age practices with new ones introduced by Roman soldiers and settlers, as at Ergolding, near Landshut (Struck 1996). Pottery and metal ornaments are among the strongest signs of the persistence of local tradition in graves during the Roman period. Of particular interest is the creation of wholly new forms of personal ornamental metalwork during the second and third centuries A.D., based on local stylistic traditions, that became important signs of identity among Roman troops in many locations along the frontiers (see below).

DESIRES AND MEANINGS: WANTS AND USES OF THE CRAFTED OBJECTS

Pottery

On a functional level, pottery of course was desired and used to contain things—mainly for the storage, preparation, and consumption of food and drink. But pottery was much more than simply functional. Plain, undecorated pottery would have served the purposes of containing food and beverages, but much attention was devoted to making ornate wares. Pottery occurs in three principal kinds of contexts—at settlements, where it is fragmentary; in graves, where whole vessels are common; and in deposits, where both complete and partial vessels are recovered. The ways in which pottery was treated in burials and other kinds of deposits make it clear that pottery was used for a variety of social and ritual purposes. The contexts in which highly ornate pottery was used include feasting, an important medium in late prehistoric Europe for the negotiation and expression of social status (Arnold 1999; Dietler 1990; Rieckhoff 1998).

When spatial patterns of finer, more ornate ceramics at settlement sites are examined, it is apparent that finer wares tend to occur in association with other high-value materials (e.g., Wells 1997). Ornate categories of pottery are most common in otherwise richly outfitted burials (e.g., Böhme-Schönberger 1998; Haffner 1989) and are rare in more modest graves. Thus, ornate pottery was linked to high social status in both settlement and burial contexts. The role of pottery in special deposits not directly connected to domestic settlement activity or to placement in graves (e.g., Maier 1976) is not well understood at present.

Metal Ornaments

As is the case with all personal ornaments (Bogatyrev 1971; Eicher 1998; Sørensen 1997), fibulae, pendants, and attachments conveyed information about the person wearing them. Pauli (1972) has argued that in the Late Hallstatt Period (600–450 B.C.) in Württemberg, specific types of fibulae were worn by women in particular age and status categories. Gold neck rings of this period were marks of elite status displayed by elite individuals throughout west-central Europe (Biel 1985:77–91). A recent study by Schönfelder (1998) shows that gold earrings played a similar role in displaying special status in both the Early and the Late Iron ages. For the Roman province of Raetia (now much of southern Bavaria and western Austria), Gschwind (2001) makes a strong case that the fibula type known as Almgren 86 was a marker of identity worn by persons living in that particular region. It is thus apparent that such personal ornaments played important roles in manifesting in material ways the status, position, and identity of the wearers. This topic has just begun to attract systematic research attention and will certainly lead to significant results for understanding the relationships between material culture and expression of identity at different levels (see, e.g., Wells 1998).

Soldiers serving in the Roman army along the Rhine and Danube frontiers wore metal ornaments, often in sizable numbers, on their belts and on other parts of their uniforms. Some of these were attachments for the suspension of weapons or for holding armor or garments together; others were purely ornamental. Oldenstein (1976:68) reckons an average of at least ten personal ornaments per soldier. Such ornaments are common in Roman cemeteries, and they often occur in substantial quantities on military bases, such as those at Eining (Jütting 1995), Regensburg (von Schnurbein 1977), and Straubing (Prammer 1989) on the Danube River. In contrast to what we might expect as strict military regulations about uniforms, the Roman soldiers based in the centers along the frontiers seem to have had considerable latitude to wear whatever ornaments struck their fancy, thus allowing for individual creativity in what they chose to use to decorate their uniforms. The frontier troops are thought to have been largely auxiliaries, not legionaries (Davies 1989), and this factor may have allowed for greater freedom in their choice of ornaments. Legionaries were Roman citizens, whereas auxiliaries were men from the conquered territories or even from unconquered lands beyond the frontiers.

During the second and third centuries A.D., bronze fibulae and other ornaments, such as belt attachments and scabbard mounts, decorated in motifs derived from the La Tène style of Late Iron Age Europe became very popular among the military populations along the frontiers (Oldenstein 1976, 1985). This striking phenomenon has not yet been well understood but has been linked to what is known as the "Celtic renaissance"—a resurgence in the expression of traditional style and motifs among the peoples in the Roman frontier zones several generations after the conquest. These objects may be the material manifestations of revivalist movements documented in other, related situations (Webster 1999, 2001).

ORGANIZATION OF MANUFACTURING: SATISFYING DESIRES FOR CRAFTED OBJECTS

Pottery

In the Late Bronze Age, pottery kilns are often associated with the characteristic settlements of 20–30 people, typically situated just beyond the immediate settlement area, as, for example, at Hascherkeller in southern Bavaria (Wells 1983). The frequency of kilns at Late Bronze Age settlements suggests that pottery manufacture was community based, rather than purely domestic (in the individual household) or regionally centralized (Kimmig 1983b). There is no evidence for the existence of centers during the Late Bronze Age, for the production of pottery or of anything else. As study of the kiln debris at Hascherkeller has revealed, all of the different varieties of pottery in use in the Late Bronze Age were processed together in the same facilities. This level of organization of pottery production remained typical throughout later prehistoric times.

In the later part of the Early Iron Age, a number of centers emerged at which manufacturing was organized on a more specialized basis, though the great majority of pottery was still produced in smaller undertakings in the villages. Some of the new centers were regional in nature, such as that at Niedererlbach in Lower Bavaria (Nebelsick and Kohnke 1985), whereas others, such as Mont Lassois in eastern France (Brun and Chaume 1997), Châtillon-sur-Glâne in Switzerland (Schwab 1983), and the Heuneburg in southwestern Germany, were of interregional importance. As noted above, at the larger centers, such as the Heuneburg, special fine ceramics, such as the thin-walled, wheel-made, polychrome-painted wares (Dämmer 1978), were produced for limited consumption, perhaps specifically for ritual feasting and for the burials of elites (Kimmig 1983b:25). The production of such special pottery ended with the decline of the centers around the middle of the fifth century B.C. For the next three centuries, pottery production was village based, primarily handmade, and usually monochrome.

At the urban oppida of the final two centuries B.C., various different types of pottery were produced to serve special functions (Maier 1991; Sievers 2003). They include coarse handmade wares in the form of large storage vessels; graphite-and-clay, thick-walled cooking pots; fine, sharply profiled tableware; and thin-walled, polychrome-painted, usually tall and thin vessels. For the first time, the majority of the pottery was produced using the fast-turning potter's wheel.

Yet, even with this highly specialized, series production, manufacturing was still carried out on a small scale. At Manching on the Danube River in southern Bavaria—the most extensively excavated oppidum in Europe—in the 25 ha of land investigated to date, only three kilns have been found, none of them substantially larger than the kilns at Late Bronze Age settlements (Sievers 1998:633–634). Apparently, in order to produce the vastly larger quantities of pottery needed by the thousands of inhabitants of the site, and of surrounding communities as well, the potters at Manching built and used more kilns, rather than increasing the size and capacity of their kilns. The evidence of kilns and waster sherds at other oppida, such as Staré Hradisko in Moravia (Meduna 1970), is consistent with that at Manching. Although a great deal of pottery was being produced at these centers (hundreds of thousands of sherds have been recovered and studied at Manching), the organization of production remained remarkably dispersed. Instead of a concentration of large kilns in one part of a settlement, we consistently find small kilns in different locations throughout the settlement areas.

Potters working at some of the much smaller, more typical settlements of the period also produced pottery, fired in kilns similar to those at Manching. In southern Germany not far from Manching, examples at Sallmannsberg (Koch 1997) and Künzing (Irlinger 2002) have been studied recently.

After the decline of the oppida around the middle of the first century B.C., the centralized manufacture of pottery and other goods ceased. Small communities continued to make their own pottery, as, for example, at Harting (Rieckhoff 1995), but, as we might expect, handmade wares came to predominate over wheel made. At Manching, about 75 percent of the pottery was wheel made, and 25 percent was handmade (Stöckli 1979:3); at Harting the ratio is about 10 percent wheel made to 90 percent handmade.

In the Roman period, the archaeology shows a complex situation with a variety of different styles of pottery produced in different manufacturing systems. For example, the widely favored, mass-produced terra sigillata (see above) was manufactured initially in central Italy and exported throughout the greater Roman world and beyond. As demand in temperate Europe grew after the Roman conquest, new production facilities were established in southern Gaul and later in the Rhineland and in Bavaria (Bernhard 2000).

But other wares, of local origin, were also in use. Locally made coarse ware pottery is common at Early Roman sites, both military and civilian, and still abundant in some places during the second and third centuries A.D. (Flügel 1996; Wieland 1993). For example, in graves at the cemetery belonging to a Roman villa community at Ergolding near Landshut, much of the pottery placed in the burials after the middle of the second century A.D. is very similar to that produced in the area before the Roman conquest of a century and a half earlier (Struck 1996). Local fine wares of Late Iron Age type are also common at some Roman sites (von Schnurbein 1993). In addition, in every region of temperate Europe, distinctive new styles of pottery developed after the Roman conquest, different both from the mass-produced Roman wares such as terra sigillata and from the local Late Iron Age pottery. In the Lower Rhineland, the distinctive regional ceramics are known as Belgic Ware (Willems 1986). In southern Bavaria, the new styles of pottery are referred to as Raetian Ware (von Schnurbein 1977) in western and central areas and as Norican Ware (Maier 1984) in eastern areas.

The creation of specialized pottery centers is a significant sign of the increasing centralization of at least one branch of manufacturing during the Roman period. At the Roman-period village of Rapis in modern Schwabmünchen in southern Germany, over 70 kilns have been excavated, dating from the middle of the first century A.D. on (Czysz and Sorge 2000). The wide range of different types of pottery manufactured on the site and the variety of tools used in the process are striking. Products from the kilns at Schwabmünchen are found over a wide landscape, suggesting that such centers were not numerous and that they supplied many communities, some at considerable distances away.

Yet the establishment of such regional production centers for pottery may not represent as much of a departure from earlier patterns of craft organization as has been thought. From the Late Iron Age, specialized village pottery production centers have been identified at Wehren in Hessen (Frey and Laumann 1978) and at Strachotín in Moravia (Cizmar 1987). Perhaps the potteries at the oppida produced primarily for the inhabitants of those urban settlements, and production aimed more at export was carried out at many such smaller, specialized centers at locations away from the oppida.

Metal Ornaments

In many ways, the organization of the manufacture of metal ornaments is similar to that for pottery. In the Late Bronze Age, many of the small village settlements yield evidence for bronze casting, including molds and droplets of bronze, as at Hascherkeller (Wells 1996). Although it is likely that molds were manufactured by specialist craft workers, the evidence suggests that bronze casting was carried out by local individuals in many, if not most, settlements (Harding 2000). In contrast to clay for pottery, the copper and tin to make bronze had to be brought in from outside, making the whole process more complex. My focus here is on the manufacturing, not on the trade of raw materials.

In the Early Iron Age, even at the interregional centers such as the Heuneburg, the manufacture of bronze ornaments (Mansfeld 1973) was organized similarly to that at Late Bronze Age villages. Sandstone molds of types similar to those of the Late Bronze Age were used to cast several ornaments at a time, and the character of the casting debris suggests processes very similar in scale (Kimmig 1983a:120; Wells 1996). Just as for the rare wheel-made pottery at the centers (see above), special items of restricted distribution, such as the gold fibulae, bracelet, and neck ring in the Hochdorf grave, were surely produced by special craft workers at the behest of the elites of this period (Biel 1985:84, 89). But, like the specially painted wheel-made pottery produced at the centers at this time, those special crafts declined during the fifth century B.C.

At the oppida of the Late Iron Age, a significant change occurred in the production of personal ornaments such as fibulae. Most fibulae at the oppida were made in series (Büchsenschütz 2002), and a specific type, known as the Nauheim fibula (Striewe 1996; Werner 1955), which was designed especially to be mass produced (Drescher 1955; Furger-Gunti 1977), became one of the most common types. The evidence pertaining to fibulae manufacture at the oppida is primarily partly made pieces recovered in the course of settlement excavations (Furger-Gunti 1977; Gebhard 1991:50–51). Molds for casting other kinds of bronze ornaments are also well represented at the oppida (Werner 1963). Just as the evidence of the multiple small kilns at Manching suggests dispersed rather than concentrated manufacturing within the settlement, the traces of bronze casting at Manching and other sites indicate not a workshop quarter within the settlements but, rather, many small workshops operating in different parts of the complexes (Jacobi 1974:260–268; van Endert 1991:110). Similar patterns in the distribution of workshop debris have been identified at Kelheim in Bavaria (Wells 1993), Staré Hradisko in Moravia (Meduna 1970), and the sites of Hrazany, Stradonice, and Závist in Bohemia (Drda and Rybová 1995).

In the post-oppidum period, 50 B.C.–A.D. 50, we observe a trend away from the mass-produced fibulae and back to a process of individual manufacture (Wells 1995). More of the individual fibulae of this period are unique, as opposed to the large series of very similar objects that characterize the fibula assemblages at the oppida. The remains of the production process have been identified at a number of small settlements, including Leonberg (Pietsch 2002) and Stöffling (Irlinger 2002) in southern Bavaria.

In the Early Roman period, A.D. 50–300, the new factor of the Roman military and its huge material needs was added to the existing cultural landscape. By the second century A.D., some 20,000 troops were stationed in southern Germany in bases along the *limes* boundary and the Danube frontier, and they needed to be supplied with massive quantities of foodstuffs, leather and textiles for uniforms, cooking and eating utensils, weapons, and personal ornaments.[1]

Direct evidence for the production of the bronze fibulae, belt attachments, scabbard decorations, and other ornaments worn by the soldiers is meager at present (Flügel 2000). But indications are that such items were manufactured in three different kinds of contexts —in workshops at the military bases, in the towns (*vici*) that grew up around the bases to supply goods to the soldiers, and in settlements in the countryside away from the frontier (Gschwind 1997). Stone molds for casting fibulae and other ornaments, lead models for making molds, crucibles, partly made objects, and failed castings are among the indicators of such production (Burmeister 1991; Oldenstein 1976, 1985). A whole range of different kinds of arrangements existed for the manufacture of the items the soldiers wanted, but the accumulating evidence is indicating increasingly that a large proportion of this production was in the hands of small-scale local craft workers acting as entrepreneurs. It is not clear at this stage whether production was conducted more on a made-to-order basis or whether bronze workers made products they thought the soldiers might want and then peddled them at the towns next to the bases.[2] A cache of over 100 fibulae of different types, deposited in A.D. 69 or 70 at the town next to the Roman base at Augsburg in southern Bavaria (Bakker 1986), provides a view into the mix of wares that such bronze workers made to sell to the troops. As more studies focus on the evidence for the production and distribution of such materials (such as Gschwind 1997; Völling 1994), we shall gain better understanding of the roles of those goods in social processes in Roman-period Europe.

MULTICRAFTING?

The evidence currently available from temperate Europe shows no clear patterns of multicrafting in the sense of the same individuals crafting in different kinds of materials or of producers in different crafts working together on specific objects. My discussion has focused on two of the principal materials crafted—pottery and metal ornaments. I did not distinguish between bronze and iron. In some contexts, principally where metalworking was conducted on a relatively small scale and where the objects produced did not require a high degree of specialized skill, the evidence indicates that those two metals were crafted by the same individuals (as in the case of the man buried at Poysdorf, noted below). But in cases in which metalworking was carried out on a large scale or when special metallurgical techniques were needed to manufacture objects, then bronze smiths and iron smiths seem to have been different individuals. But for the categories of pottery and metal ornaments, workshop evidence indicates that these materials were processed at different parts of settlements and that there was no clear collaboration between potters and metalsmiths. At Late Bronze Age Hascherkeller, the pottery kiln was separated from all other economic activities carried out in the settlement. Traces of bronze casting were recovered from a different part of the site. At the extensively excavated Late Iron Age oppidum of Manching, the evidence for pottery production, bronze casting, and iron smithing is spatially separated. Whereas in some cultural traditions ceramic molds to cast bronze and other metals may have involved potters and metalsmiths working together, the majority of casting molds in late prehistoric Europe were of stone (frequently sandstone), though a few special forms, such as Late Bronze Age socketed bronze axes, involved ceramic molds. Most likely bronze smiths made the stone molds in which they cast their tools, weapons, and ornaments, but the molds were only devices for manufacturing the final products.

CRAFT PRODUCERS IN THEIR SOCIOPOLITICAL CONTEXTS

A detailed discussion of the place of the craft worker in the societies of late prehistoric and Roman-period Europe is beyond the scope of this chapter, but a few observations are relevant here. The relatively small scale and dispersed nature of all of the crafts would have mitigated against substantial control of craft production by powerful elites. Ongoing field research involving larger numbers of small sites is showing increasingly that earlier models of major centers as the places of production need to be revised, as many newly identified sites show evidence of such manufacturing.

In his analysis of evidence pertaining to social and political organization among communities of salt miners at the Iron Age site of the Dürrnberg at Hallein in Austria, Ludwig Pauli (1978) argues that the miners were likely something akin to independent contractors, like their counterparts in the Middle Ages. Those medieval miners sold their technical expertise to elite landowners who wanted resources extracted from their lands. There is considerable evidence that could be argued to support a similar model for craft workers in pottery and metals in late prehistoric and Roman-period Europe. If specialized craft workers could use their technical skills to operate a least partly independently within the sociopolitical systems of their societies, then perhaps many of them had similar "special" statuses. But this issue is complicated not only by changes over time and variation between larger and smaller communities but also by differences in the production of objects that required different degrees of technical knowledge and skill. Numerous smiths' graves have been identified, from the Bronze Age through the Iron Age and into early medieval times, characterized by tools used in metalworking, and these graves are often outfitted with signs of above-average status. A particularly informative burial at Poysdorf in Austria (Daim and Mehofer 2003), dating to just after the end of the Roman period, indicates an individual buried with both bronze-working and ironworking implements, as well as with weapons that include a sword, a common sign of elite status in men's graves.

The special status of some craft workers, especially metalworkers whose craft involved the use of intense fire and the transformation of materials, is underscored by the association of magic with their processes and results.

Richard Hingley (1997) has explored the significance of myths and rituals associated with ironworking, and Budd and Taylor (1995) have done the same for bronze metallurgy. These powerful associations with supernatural forces may have played a role in helping some of these craft specialists maintain a certain level of independence within the social and political context in which they worked.

Although much craft activity was carried on by permanent residents in village and town communities, there is considerable evidence in both the archaeology and, from the start of the Roman period on, texts that some craft workers moved from place to place, adopting ideas and styles as they traveled. Technological innovations such as the short-lived adoption of the potter's wheel at the end of the Early Iron Age, and its more complete adoption around the middle of the Late Iron Age, can be best understood in terms of traveling craft workers encountering these devices in lands in the Mediterranean Basin and bringing the technology into temperate Europe. We observe similar processes at work in the introduction of other technologies from the Mediterranean world, including coinage and writing.

CRAFTS, WANTS, AND POLITICAL ORGANIZATION

The small-scale and dispersed organization of manufacturing is a distinctive feature of later prehistory and the Early Roman period in temperate Europe. Even in instances where larger communities developed, at the end of the Early Iron Age and at the oppida at the end of the Late Iron Age, as well as during the Roman period, when we might anticipate more centralized structures of production, the archaeological evidence shows that manufacturing remained fundamentally dispersed. This pattern is distinctly different from that evident in more centralized societies of the past, such as those of the Uruk period in ancient Mesopotamia, the Classical period in Greece, and the Mayan period in Mesoamerica (Scarre and Fagan 2003).

The evidence pertaining to the production of craft goods—in this case pottery and metal ornaments—is coincident with that for social and political organization during the 15 centuries from the start of the Late Bronze Age to the end of the Early Roman period in temperate Europe. When social systems became more stratified, as in the later part of the Early Iron Age (Arnold 1995), some aspects of manufacturing became more specialized, as at the Heuneburg and other interregional centers. When the larger, urban populations formed at the oppida at the end of the Iron Age, an increase in the number of workshops and the development of series production in pottery and in ornaments accompanied the growth of new, larger political units.

But at no time do we see evidence for highly concentrated or tightly controlled crafts, such as existed at centers in more strictly controlled and more hierarchical political systems, as at Ur during the Early Dynastic Period and at Anyang during the Shang Bronze Age. In temperate Europe, no one center gained political supremacy over others, and in both the Early Iron Age and the Late Iron Age, centers were short-lived phenomena—thriving for less than a century in both periods. No pre-Roman kingdoms or empires developed in temperate Europe. The fundamentally dispersed character of craft production reflects this pattern of segmentary, noncentralized political organization.

The persistence of dispersed patterns of manufacturing even after the Roman conquest is particularly interesting. Though Rome was a highly centralized empire politically and economically, in the frontier regions of temperate Europe, the traditional local patterns of craft production persisted, at least well into the third century A.D.[3] The maintenance of these structures of production and supply suggests that the social and political systems of the Late Iron Age also maintained much of their integrity during the Roman period.

This pattern recognizable in the rich database of temperate Europe may also be identifiable in other imperial contexts, where the data are not yet so abundant. Such an approach to the evidence relating to craft production may help us understand the essence of social and political systems on the edges of empires elsewhere as well.

THE PRODUCERS' PERSPECTIVE

The study of material culture—tools, ornaments, weapons, and other manufactured objects—from the perspective of the makers provides an important counterpoint to more common studies of burial goods, settlement debris, and ritual offerings—the contexts in which manufactured goods typically end up. In the field of European prehistory, close attention to patterns of workshop distribution and activity is still relatively new, and much remains to be done before we can say that we have a good understanding of the production side of the equation. By later prehistoric times, most communities had long since solved the basic problems of subsistence, and a substantial proportion of economic activity was devoted to making tools, weapons, ornaments, and household goods such as pottery. In order to fully understand the social uses to which manufactured objects were put, it is essential to understand the context of their production. A research focus on workshop location, organization, and production allows us to address a variety of important questions about the economic, social, and political aspects of supplying societies with the goods they need and desire.

In the archaeology of late prehistoric and early historic Europe, excavations during the past several decades have radically changed our understanding of craft production. The biggest change has been in the realization that workshop activity was much more widespread than earlier investigators had thought and that a much wider range of individuals was crafting than imagined. The "producer's perspective" allows us to formulate a whole series of new questions about the relationships between producers and consumers and between producers and the communities and societies in which they worked.

Notes

1. The larger questions of supplying the troops are beyond the scope of this chapter (see Wells 1999:141–147).
2. Surely situations varied, and they changed over time.
3. Changes during the Late Roman period are the subject of another study.

References Cited

Appadurai, A. (editor)

1986 *The Social Life of Things*. Cambridge University Press, New York.

Arnold, B.

1995 The Material Culture of Social Structure: Rank and Status in Early Iron Age Europe. In *Celtic Chiefdom, Celtic State*, edited by B. Arnold and D. B. Gibson, pp. 43–52. Cambridge University Press, Cambridge.

1999 "Drinking the Feast": Alcohol and the Legitimation of Power in Celtic Europe. *Cambridge Archaeological Journal* 9: 71–93.

Bakker, L.

1986 Ausgrabungen an der Kornhausgasse in der Provinzialhauptstadt Augusta Vindelicum-Augsburg. *Das archäologische Jahr in Bayern 1985*:101–104.

Béarat, H., and I. Bauer

1994 Früheisenzeitliche Keramik von Baarburg ZG und Üetliberg ZH: Eine mineralogisch-petrographische und chemische Untersuchung zur Frage der Herstellungsorte scheibengedrehter Keramik in der ausgehenden Hallstattzeit. *Germania* 72:67–93.

Bernhard, H.

2000 Terra Sigillata und Keramikhandel. In *Die Römer zwischen Alpen und Nordmeer*, edited by L. Wamser, pp. 138–141. Philipp von Zabern, Mainz.

Biel, J.

1985 *Der Keltenfürst von Hochdorf*. Konrad Theiss, Stuttgart.

Bogatyrev, P.

1971 *The Functions of Folk Costume in Moravian Slovakia*. Translated by R. G. Crum. Mouton, The Hague.

Böhme-Schönberger, A.

1998 Das Grab eines vornehmen Kriegers der Spätlatènezeit aus Badenheim. *Germania* 76:217–256.

Brun, P., and B. Chaume (editors)

1997 *Vix et les éphémères principautés celtique: Les VIe–Ve siècles avant J.-C. en Europe centre-occidentale*. Éditions Errance, Paris.

Büchsenschütz, O.
2002 Die Entstehung von Wirtschaftszentren in Gallien. In *Dürrnberg und Manching: Wirtschaftsarchäologie im ostkeltischen Raum*, edited by C. Dobiat, S. Sievers, and T. Stöllner, pp. 63–76. Rudolf Habelt, Bonn.

Budd, P., and T. Taylor
1995 The Faerie Smith Meets the Bronze Industry: Magic versus Science in the Interpretation of Prehistoric Metal-Making. *World Archaeology* 27:133–143.

Burmeister, S.
1991 Ein Bleimodell einer Fibel aus dem Vicus von Seebruck. *Das archäologische Jahr in Bayern 1990*:92–93.

Cizmar, M.
1987 Laténské Sídliste ze Strachotína, Okr. Breclav (Eine latènezeitliche Siedlung aus Strachotín, Bez. Breclav). *Památky Archeologické* 78:205–230.

Collis, J.
1995 The First Towns. In *The Celtic World*, edited by M. Green, pp. 159–175. Routledge, London.

Csikszentmihalyi, M., and E. Rochberg-Halton
1981 *The Meaning of Things: Domestic Symbols and the Self*. Cambridge University Press, New York.

Czysz, W., K. Dietz, T. Fischer, and H.-J. Kellner (editors)
1995 *Die Römer in Bayern*. Konrad Theiss, Stuttgart.

Czysz, W., and G. Sorge
2000 Das römische Töpferdorf Rapis bei Schwabmünchen. In *Die Römer zwischen Alpen und Nordmeer*, edited by L. Wamser, pp. 142–144. Philipp von Zabern, Mainz.

Daim, F., and M. Mehofer
2003 Poysdorf. *Reallexikon der germanischen Altertumskunde* 23:327–331.

Dämmer, H.-W.
1978 *Die bemalte Keramik der Heuneburg*. Philipp von Zabern, Mainz.

Davies, R. W.
1989 *Service in the Roman Army*. Edinburgh University Press, Edinburgh.

De Navarro, J. M.
1972 *The Finds from the Site of La Tène, vol. 1: Scabbards and the Swords Found in Them*. Oxford University Press, London.

Dietler, M.
1990 Driven by Drink: The Role of Drinking in the Political Economy and the Case of Early Iron Age France. *Journal of Anthropological Archaeology* 9:352–406.

Drda, P., and A. Rybová
1995 Prostorové Rozlození Specializovaného Remesla v Zástavbe Keltského Oppida (Spatial Distribution of Specialized Crafts at a Celtic Oppidum). *Archeologické rozhledy* 47:596–613.

Drescher, H.
1955 Die Herstellung von Fibelspiralen. *Germania* 33:340–349.

Eicher, J.
1998 *Beads and Bead Makers: Gender, Material Culture, and Meaning*. Berg, Oxford.

Eicher, J. (editor)
1995 *Dress and Ethnicity*. Berg, Washington, D.C.

Flügel, C.
1996 Handgemachte Grobkeramik aus Arae Flaviae-Rottweil. *Fundberichte aus Baden-Württemberg* 21:315–400.
2000 Bronze- und Edelmetallverarbeitung. In *Die Römer zwischen Alpen und Nordmeer*, edited by L. Wamser, pp. 124–128. Philipp von Zabern, Mainz.

Frey, O.-H.
1995 Some Comments on Swords with Dragon-Pairs. In *Sites and Sights of the Iron Age*, edited by B. Raftery, pp. 163–176. Oxbow, Oxford.

Frey, O.-H., and H. Laumann
1978 Eine spätlatènezeitliche Töpferei bei Wehren, Gemeinde Fritzlar, Schwalm-Eder-Kreis. *Fundberichte aus Hessen* 17–18:137–150.

Furger-Gunti, A.
1977 Zur Herstellungstechnik der Nauheimer-Fibel. In *Festschrift Elisabeth Schmid*, edited by L. Berger, pp. 73–84. Geographisch-Ethnologische Gesellschaft, Basel.

Garbsch, J.
1982 *Terra Sigillata*. Prähistorische Staatssammlung, Munich.

Gebhard, R.
1991 *Die Fibeln aus dem Oppidum von Manching*. Franz Steiner, Stuttgart.

Gosden, C., and Y. Marshall
1999 The Cultural Biography of Objects. *World Archaeology* 31:169–178.

Gschwind, M.
1997 Bronzegiesser am raetischen Limes: Zur Versorgung mittelkaiserzeitlicher Auxiliareinheiten mit militärischer Ausrüstung. *Germania* 75:607–638.

2001 Zur kulturellen Eigenständigkeit der Provinz Raetien: Almgren 86, die raetische Form der sog. pannonischen Trompetenfibeln. *Germania* 79:401–416.

Haffner, A.

1989 *Zum Totenbrauchtum der Kelten und Römer am Beispiel des Treverer-Gräberfeldes Wederath-Belginum.* Philipp von Zabern, Mainz.

Harding, A. F.

2000 *European Societies in the Bronze Age.* Cambridge University Press, Cambridge.

Helms, M. W.

1993 *Craft and the Kingly Ideal.* University of Texas Press, Austin.

Hingley, R.

1997 Iron, Ironworking and Regeneration: A Study in the Symbolic Meaning of Metalworking in Iron Age Britain. In *Reconstructing Iron Age Societies*, edited by A. Gwilt and C. Haselgrove, pp. 9–18. Oxbow, Oxford.

Horn, H. G. (editor)

1987 *Die Römer in Nordrhein-Westfalen.* Konrad Theiss, Stuttgart.

Irlinger, W.

2002 Bemerkungen zur Verbreitung und wirtschaftlichen Struktur der offenen latènezeitlichen Siedlungen in Südbayern. In *Dürrnberg und Manching: Wirtschaftsarchäologie im ostkeltischen Raum*, edited by C. Dobiat, S. Sievers, and T. Stöllner, pp. 253–264. Rudolf Habelt, Bonn.

Jacobi, G.

1974 *Werkzeug und Gerät aus dem Oppidum von Manching.* Franz Steiner, Wiesbaden.

Jütting, I.

1995 Die Kleinfunde aus dem römischen Lager Eining-Unterfeld. *Bayerische Vorgeschichtsblätter* 60:143–230.

Keller, E.

1984 *Die frühkaiserzeitlichen Körpergräber von Heimstetten bei München und die verwandten Funde aus Südbayern.* C. H. Beck, Munich.

Kimmig, W.

1983a *Die Heuneburg an der oberen Donau.* 2nd ed. Konrad Theiss, Stuttgart.

1983b Zum Handwerk der späten Hallstattzeit. In *Das Handwerk in vor- und frühgeschichtlicher Zeit*, pt. 2, edited by H. Jankuhn, W. Janssen, R. Schmid-Wiegand, and H. Tiefenbach, pp. 13–33. Vandenhoeck and Ruprecht, Göttingen.

Kimmig, W. (editor)

2000 *Importe und mediterrane Einflüsse auf der Heuneburg.* Philipp von Zabern, Mainz.

Koch, H.

1997 Eine spätkeltische Siedlung mit Töpferofen von Landshut-Sallmannsberg (Berggrub). *Verhandlungen des Historischen Vereins für Niederbayern* 122–123:105–138.

Krämer, W.

1985 *Die Grabfunde von Manching und die latènezeitlichen Flachgräber in Südbayern.* Franz Steiner, Stuttgart.

Lasch, H.

1999 *Berching-Pollanten I: Die Keramik der spätkeltischen Siedlung Berching-Pollanten, Lkr. Neumarkt i.d. Oberpfalz.* Marie Leidorf, Rahden, Germany.

Maier, F.

1970 *Die bemalte Spätlatène-Keramik von Manching.* Franz Steiner, Wiesbaden.

1976 Ein Gefässdepot mit bemalter Keramik von Manching. *Germania* 54:63–74.

1991 The *Oppida* of the Second and First Centuries B.C. In *The Celts*, edited by S. Moscati, O.-H. Frey, V. Kruta, B. Raftery, and M. Szabó, pp. 411–425. Rizzoli, New York.

Maier, R.

1984 Römerzeitliche Töpferofen für "Norisches Bauerngeschirr" bei Kieling. *Das archäologische Jahr in Bayern 1983*: 113–116.

Mansfeld, G.

1973 *Die Fibeln der Heuneburg.* Walter de Gruyter, Berlin.

McKendrick, N., J. Brewer, and J. H. Plumb

1982 *The Birth of a Consumer Society: The Commercialization of Eighteenth-Century England.* Indiana University Press, Bloomington.

Meduna, J.

1970 Das keltische Oppidum Staré Hradisko in Mähren. *Germania* 48:34–59.

Müller-Karpe, H.

1955 Das urnenfelderzeitliche Wagengrab von Hart a.d. Alz, Oberbayern. *Bayerische Vorgeschichtsblätter* 21:46–75.

Nebelsick, L. D., and H.-G. Kohnke

1985 Eine hallstattzeitliche Siedlung mit Giesserei- und Schmiedeabfall von Niedererlbach (Niederbayern). *Archäologisches Korrespondenzblatt* 15:339–350.

Nierhaus, R.

1959 *Das römische Brand- und Körpergräberfeld "Auf der Stieg" in Stuttgart-Bad Cannstatt*. Silberburg, Stuttgart.

Oldenstein, J.

1976 Zur Ausrüstung römischer Auxiliareinheiten: Studien zu Beschlägen und Zierat an der Ausrüstung der römischen Auxiliareinheiten des obergermanisch-raetischen Limesgebietes aus dem zweiten und dritten Jahrhundert n. Chr. *Bericht der Römisch-Germanischen Kommission* 57:49–284.

1985 Manufacture and Supply of the Roman Army with Bronze Fittings. In *The Production and Distribution of Roman Military Equipment*, edited by M. C. Bishop, pp. 82–94. British Archaeological Reports, International Series 275. Oxford.

Pauli, L.

1972 *Untersuchungen zur Späthallstattzeit in Nordwürttemberg*. Helmut Buske, Hamburg.

1978 *Der Dürrnberg bei Hallein III*. C. H. Beck, Munich.

Pietsch, M.

2002 Eine neue spätkeltische Höhensiedlung bei Leonberg. *Das archäologische Jahr in Bayern* 2001:72–75.

Prammer, J.

1989 *Das römische Straubing*. Schnell und Steiner, Munich.

Rieckhoff, S.

1995 *Süddeutschland im Spannungsfeld von Kelten, Germanen und Römern*. Rheinisches Landesmuseum, Trier.

1998 Ein "keltisches Symposion": Spätrepublikanisches Bronzegeschirr vom Mont Beuvray als wirtschaftlicher und gesellschaftlicher Faktor. In *Studien zur Archäologie der Kelten, Römer und Germanen in Mittel- und Westeuropa*, edited by A. Müller-Karpe, H. Brandt, H. Jöns, D. Krausse, and A. Wigg, pp. 489–517. Dr. Faustus, Büchenbach, Germany.

Scarre, C., and B. M. Fagan

2003 *Ancient Civilizations*. 2nd ed. Prentice Hall, Upper Saddle River, N.J.

Schönfelder, M.

1998 Männer mit goldenen Ohrringen: Zu insignienhaften Gegenständen in der späten Hallstatt- und frühen Latènekulturen. *Archäologisches Korrespondenzblatt* 28:403–422.

Schwab, H.

1983 Châtillon-sur-Glâne. *Germania* 61:405–458.

Sievers, S.

1982 *Die mitteleuropäischen Hallstattdolche*. C. H. Beck, Munich.

1998 Vorbericht über die Ausgrabungen 1996–1997 im Oppidum von Manching. *Germania* 76:619–672.

2003 *Manching: Die Keltenstadt*. Konrad Theiss, Stuttgart.

Sørensen, M. L. S.

1997 Reading Dress: The Construction of Social Categories and Identities in Bronze Age Europe. *Journal of European Archaeology* 5:93–114.

Stöckli, W. E.

1979 *Die Grob- und Importkeramik von Manching*. Franz Steiner, Wiesbaden.

Striewe, K.

1996 *Studien zur Nauheimer Fibel und ähnlichen Formen der Spätlatènezeit*. Marie Leidorf, Espelkamp, Germany.

Struck, M.

1996 *Römische Grabfunde und Siedlungen im Isartal bei Ergolding, Landkreis Landshut*. Michael Lassleben, Kallmünz, Germany.

van Endert, D.

1991 *Die Bronzefunde aus dem Oppidum von Manching*. Franz Steiner, Stuttgart.

Völling, T.

1994 Studien zu Fibelformen der jüngeren vorrömischen Eisenzeit und ältesten römischen Kaiserzeit. *Bericht der Römisch-Germanischen Kommission* 75:147–282.

Volpert, H.-P.

2002 Neue Körpergräber der Heimstettener Gruppe. *Das archäologische Jahr in Bayern* 2001:79–82.

von Schnurbein, S.

1977 *Das römische Gräberfeld von Regensburg*. Michael Lassleben, Kallmünz, Germany.

1993 Nachleben in römischer Zeit? In *Das keltische Jahrtausend*, edited by H. Dannheimer and R. Gebhard, pp. 244–248. Philipp von Zabern, Mainz.

Waldhauser, J.

1987 Keltische Gräberfelder in Böhmen. *Bericht der Römisch-Germanischen Kommission* 68:25–179.

Webster, J.

1999 At the End of the World: Druidic and

Other Revitalization Movements in Post-Conquest Gaul and Britain. *Britannia* 30:1–20.

2001 Creolizing the Roman Provinces. *American Journal of Archaeology* 105:209–225.

Wells, P. S.

1983 *Rural Economy in the Early Iron Age: Excavations at Hascherkeller, 1978–1981*. Peabody Museum, Cambridge, Massachusetts.

1985 Material Symbols and the Interpretation of Cultural Change. *Oxford Journal of Archaeology* 4:9–17.

1993 *Settlement, Economy, and Cultural Change at the End of the European Iron Age: Excavations at Kelheim in Bavaria, 1987–1991*. International Monographs in Prehistory, Ann Arbor.

1995 Manufactured Objects and the Construction of Identities in Late La Tène Europe. *EIRENE: Studia graeca et latina* 31:129–150.

1996 Location, Organization, and Specialization of Craft Production in Late Prehistoric Central Europe. In *Craft Specialization and Social Evolution*, edited by B. Wailes, pp. 85–98. University Museum, Philadelphia.

1997 Zeugnisse einer wohlhabenden Familie vom Kelheimer Mitterfeld. In *Von Keltenkriegern und Kirchenmäusen: Archäologie im Landkreis Kelheim*, edited by M. M. Rind, pp. 147–150. Universitätsverlag, Regensburg.

1998 Identity and Material Culture in the Later Prehistory of Central Europe. *Journal of Archaeological Research* 6:239–298.

1999 *The Barbarians Speak: How the Conquered Peoples Shaped Roman Europe*. Princeton University Press, Princeton.

Werner, J.

1955 Die Nauheimer Fibel. *Jahrbuch des Römisch-Germanischen Zentralmuseums* 2:170–186.

1963 Aquileia-Velem-Hrazany: Palmettenförmige Gürtelschliessen aus pannonischen und boiischen Oppida. *Alt-Thüringen* 6: 428–435.

Wieland, G.

1993 Spätkeltische Traditionen in Form und Verzierung römischer Grobkeramik. *Fundberichte aus Baden-Württemberg* 18:61–70.

Willems, W. J. H.

1986 *Romans and Batavians: A Regional Study in the Dutch Eastern River Area*. University of Amsterdam, Amsterdam.

Wilson, S.

2000 *The Magical Universe: Everyday Ritual and Magic in Pre-Modern Europe*. Hambledon and London, London.

7

Multiple Crafts and Socioeconomic Associations in the Indus Civilization: New Perspectives from Harappa, Pakistan

JONATHAN MARK KENOYER AND HEATHER M.-L. MILLER

The study of crafts and changing patterns of technology and production has long been the focus of archaeological research in the greater Indus Valley region. During the initial excavations in the 1920s–1930s that first revealed the presence of the prehistoric Indus civilizations at sites such as Mohenjo Daro and Harappa, the interrelationships among many crafts were discussed in the excavation reports. Scholars such as Mackay (1938, 1943), Sanaullah and Beck (Marshall 1931), and many others provided invaluable analysis and interpretations of a wide range of crafts, from pottery making to faience production, shell working, stone working, metallurgy, and bead making. The groundbreaking studies of these and other early scholars established the foundation that has inspired subsequent generations of researchers working at large urban centers such as Mohenjo Daro and Harappa, as well as Dholavira. Other important sites for understanding the early crafts practiced in the Indus region include Mehrgarh (Barthélémy de Saizieu and Bouquillon 1994, 1997; Jarrige et al. 1995; Kenoyer 1995b; Lechevallier 1984; Marcon and Lechevallier 2000; Tosi and Vidale 1990; Vidale 1995) and the nearby site of Nausharo (Barthélémy de Saizieu and Bouquillon 1997; C. Jarrige 1997; J.-F. Jarrige 1994, 1997; Marcon and Lechevallier 2000; Quivron 1997; Samzun 1992; Vidale 1989), Rehmandheri (Durrani 1988; Durrani et al. 1991, 1995a; Hideaki 1996), and Ghazi Shah (Flam 1993a, 1993b). Smaller towns like Chanhudaro (Mackay 1943), Kalibangan (Lal 1979; Lal et al. 2003; Thapar 1975), Lothal (Rao 1979, 1985), and Kuntasi (Dhavalikar et al. 1996) have provided additional information on the range of crafts practiced in different regions of the Indus Valley. Excavations at specialized craft villages along the southern edge of the Rann of Kutch in Gujarat, such as Nageshwar (Bhan and Kenoyer 1980–1981), Nagwada (Bhan and Gowda 2003; Hegde et al. 1988), and most recently Gola Dhoro (also known as Bagasara [Sonawane et al. 2004]), have revealed some of the most exciting evidence for the association of specific crafts such as shell working, agate bead making, faience production, and copper working in small settlements.

The general picture that emerges from these sites is of a complex network of trade and craft relationships among settlements throughout the Indus Valley region and an equally complex set of associations among specific crafts within the sites themselves that can be examined both synchronically and diachronically. For example, at the site of Harappa during the Indus Integration Era (2600–1900 B.C.), raw materials and partially finished goods from distant resource areas were being sent to the urban center for processing and finishing. Some crafts, such as shell and ivory working, steatite and agate bead making, and stone tool production, appear to have been practiced together in distinct areas of the ancient settlement. These crafts can be grouped together as representing extractive or reductive technologies, employing processes such as chipping, grinding, carving, spinning and twisting, and so on to process raw materials into finished objects. Other crafts can be identified as transformative and involve the transformation of raw materials using various pyrotechnological or chemical processes, such as copper working, pottery firing, and the manufacture of faience and glazed steatite objects. At Harappa, some of these crafts, such as pottery firing and copper working, appear to have been undertaken in relative isolation, whereas others, chiefly faience and glazed steatite manufacturing, are found both alone and together with extractive-reductive crafts. Miller (2000, 2007) discusses these issues of the large-scale location and association of crafts at Harappa during the Integration Era (Harappa Period 3) using survey and excavation data and suggests a variety of reasons for the association and isolation of particular crafts. In this chapter, we focus on finer-scale associations of crafts as visible through excavation, which allows for greater insights into the associations of extractive-reductive crafts in particular. We suggest that for some crafts, these finer-scale locational associations may relate to aspects of the production process, especially the organization of labor, Harappan ideas of what constituted a single craft, and the relative value of crafts and their probable consumer base.

The most important contribution of this essay, however, is the presentation of new diachronic information, assembled primarily by Kenoyer, on the manufacture of a large variety of crafts from the Upper Paleolithic through the second millennium B.C. but focusing on the fourth and third millennia. This diachronic material shows how important it is to examine multiple crafts, as different crafts have shown different patterns of continuity and change in production organization, including specialization. The assessment of one craft provides excellent information about that craft, but it should not be assumed that similar sorts of changes took place in other crafts at the same time or at all. The data from the site of Harappa show a highly varied set of factors for the association and segregation of crafts. In other words, there is no simple pattern or rule, and each occurrence of a craft must be studied carefully to understand its relationship to other crafts at the site.

Ultimately the goal is to be able to compare the results of current research at Harappa with those from other regional settlements and follow changes in craft production and organization over time. Unfortunately, because of very different excavation techniques and the lack of systematic records from the early excavations, it has not been possible to utilize many of the previous excavation reports for comparative studies. Even the more recent excavations using modern scientific methods of excavation and recording and conducted by different excavation teams have used very different approaches, making it difficult to develop quantitative comparisons or to follow changes over time. However, some general qualitative patterns have been identified that reveal broad regional patterns of trade, technology, and craft associations, as well as some common patterns that link smaller sites to the larger urban centers. Specific aspects of these craft associations will be addressed below in the context of the analysis of crafts from the earlier periods at Mehrgarh and Naushharo, but the bulk of the data is from Harappa, where recent excavations have been able to collect information from a wide range of contexts from all phases of the site's history.

GEOGRAPHY AND CHRONOLOGY

The term *Indus Tradition* (also sometimes the *Indus Valley Tradition*) is used to refer to the long-term cultural trajectory that characterizes the large geographical region surrounding the ancient dual river system of the modern-day Indus River Valley (Figure 7.1; Kenoyer 1991, 1998a). This area extends from the highlands of Baluchistan, Pakistan, on the west across the vast Indus alluvial plains, to

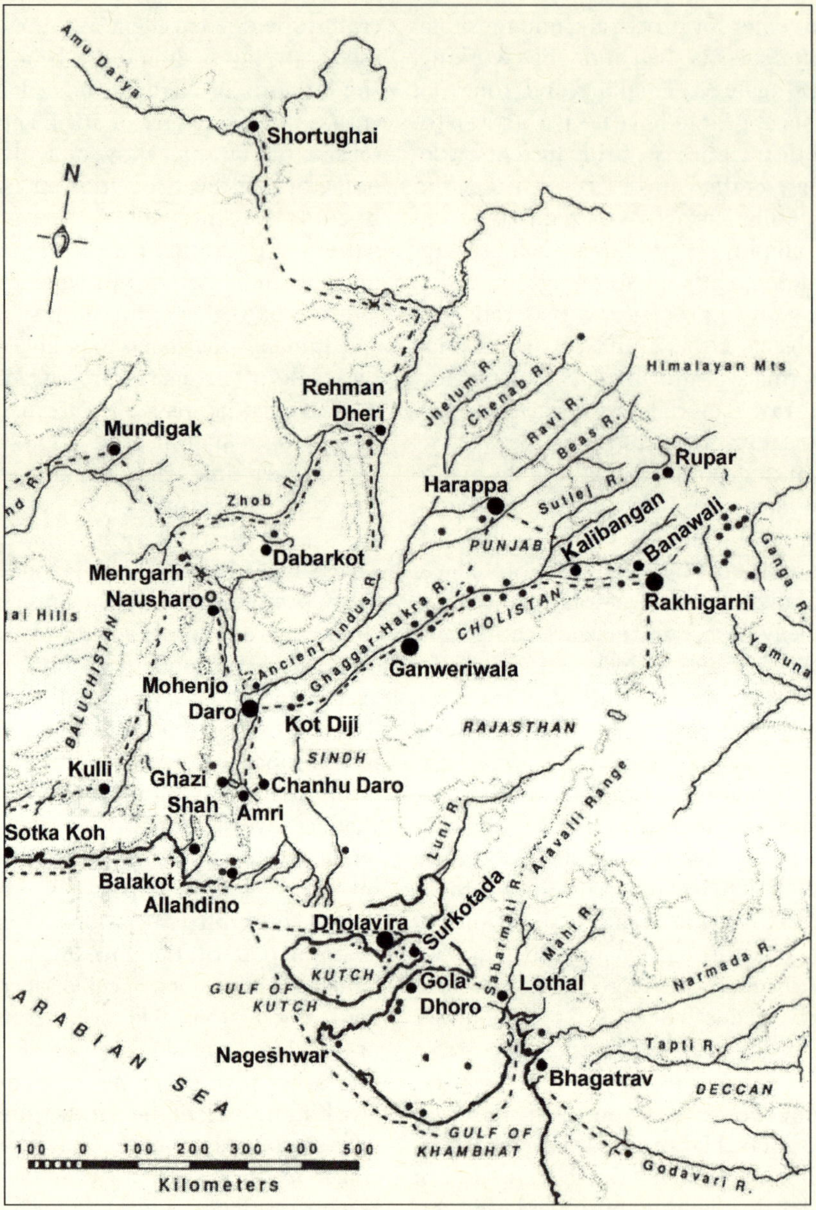

FIGURE 7.1. Major sites of the Indus Tradition.

the now-dried bed of the Ghaggar-Hakra River (also called the Sarasvati), to the deserts of Cholistan and Thar that form the border between Pakistan and India. From north to south the region stretches from the foothills of the Himalayas to the coastal regions of Makran, Pakistan, and the islands and mainland of Gujarat, India. This vast geographical area provides numerous diverse resource areas for raw materials as well as subsistence activities. The juxtaposition of these resources provided a stimulus for the establishment of exchange networks that were facilitated by numerous passes between the highlands and the alluvial plain, the presence of rivers, and relatively accessible coastal regions.

The chronology for the Indus Tradition has previously been divided into four eras (Shaf-

Table 7.1. Regional Chronology of Harappa, Mehrgarh, and Nausharo

Phase	Years B.C.
Localization Era	
Late Harappa Phase	ca. 1900–1300
Harappa: Periods 4 and 5	1900–1700
= Mehrgarh Period VIII	1800–1700
Integration Era	
Harappa Phase	2600–1900
Harappa: Period 3C, Final	2200–1900
= Nausharo, Period IV	2100–2000
Harappa: Period 3B, Middle	2450–2200
= Nausharo, Period III	
Harappa: Period 3A, Initial	2600–2450
= Nausharo, Period II	
Regionalization Era	
Early Harappa (Several Phases)	ca. 5500–2600
Harappa: Period 2, Kot Diji Phase	2800–2600
= Nausharo, Period I,	
= Mehrgarh, Period VII (Kot Diji Phase)	
Harappa: Period 1, A & B, Ravi/Hakra Phase (called the "Ravi Phase" in this article)	3300–2800
= Mehrgarh, Period IV–VI	3500–2800
Mehrgarh, Period III	4800–3500
Mehrgarh, Period II	5500–4800
Early Food Producing Era	
Neolithic	ca. 7000–5500
Mehrgarh, Period 1, Nonceramic	7000–5500
Foraging Era	>10,000–<2000
Mesolithic and Microlithic	

fer 1992), and now a fifth era has been incorporated into this framework to accommodate the parallel existence of hunting and foraging communities alongside the urban settlements (Table 7.1; Kenoyer 1998a). The Foraging Era (>10,000–<2000 B.C.) refers to the subsistence focus on wild plants and animals but also includes the early development of lithic and other craft technologies that provided a foundation for later craft developments. This era began in the Paleolithic and continues up until the present in some regions of the subcontinent. Mobile and sedentary foragers, including communities involved in hunting and fishing, continued to exist alongside sedentary agriculturalists and even interacted with larger urban centers. It is possible that many of the raw materials used in the cities, such as rocks, shells, antler, and other forest and desert products, were obtained through trade with such communities. The Early Food Producing Era (ca. 7000–5500 B.C.) is also commonly referred to as the Neolithic period and was a time when domestic plants and animals were first exploited in the Indus Valley. This era had an economy based on food production but lacking ceramics. During this era, the development of key trade networks began to link major resource areas and broadened the distribution of some raw materials and finished crafted objects. It is also the period during which more complex extractive and reductive technologies were developed and the foundations of transformative technologies were established. The Regionalization Era (5500–2600 B.C.) corresponds to a period of regional cultural development that is subdivided into various phases defined by specific artifact styles and regional cultural interaction. Distinct artifact styles, including ceramics, ornaments, seals, and architectural styles, clustered in time and space (without permanent fixed boundaries) and were connected by regional interaction networks. Recent excavations at the site of Harappa provide evidence for the emergence of an early Indus state around 2800 B.C. at the end of the Regionalization Era.

The Integration Era represents the major phase of state-level development and urbanism (2600–1900 B.C.). This era shows pronounced widespread homogeneity in some material culture as well as heterarchy and heterogeneity in craft traditions, reflecting intense interaction and fluctuating associations among social groups. Along with numerous regional variations of extractive and reductive technologies, the highest elaboration of transformative technologies is seen during this time period. The Localization Era represents a period during which there were significant changes in social and economic interaction, as well as changes in technology. At the same time there were continuities in some technologies depending on the accessibility of raw materials and the nature of the regional interaction networks. During this era at the site of Harappa, there is evidence for new applications of reductive technologies, such as microtubular drilling, and new transformative technologies, such as glass making. By the end of the Localization Era, the sociopolitical and ideological aspects of the Indus Tradition had been radically transformed and reflected the emergence of a new cultural tradition that incorporated a much wider geographical area extending from the Indus to the Ganga and Yamuna alluvial plains (Kenoyer 1995a).

HARAPPA EXCAVATIONS

The site of Harappa is one of the largest urban centers of the Indus Civilization, with a continuous sequence of occupational levels beginning with an early agricultural village, circa 3900 B.C., and continuing through the full cycle of Harappan urban development and decline, which ended around 1300 B.C. (Figure 7.2, Table 7.1). Since 1986, excavations at Harappa by the Harappa Archaeological Research Project (HARP) have focused on recovering quantitative as well as qualitative data on many different crafts to learn more about specific technologies and to better understand the relationships among crafts that were practiced in different parts of the site (Dales 1989; Meadow and Kenoyer 1997, 2001). A standardized system of excavation, recovery, and data processing has been used throughout the site during the course of 16 seasons of excavation. Fine screening (3 mm mesh) has been done for all occupation and craft activity areas plus many dump or debris deposits, as well as water screening (<1 mm to 4 mm mesh) of selected samples, resulting in the collection of microdebitage from a wide range of crafts. Primary craft production locations and changes in the nature of production over time can now be documented for many crafts during each major time period. The high-resolution analysis of manufacturing debris, tools, and finished objects has provided new insights into technological developments and changes over time (Kenoyer 1997). Most recently, provenience studies of raw materials and the reconstruction of trade patterns during each major period (Law 2006) are beginning to provide new information that may supply additional clues regarding the association of specific crafts because of common source areas for raw materials during different time periods. The experimental reconstruction of specific crafts combined with ethnoarchaeological studies has also been conducted for many of the crafts practiced in the ancient

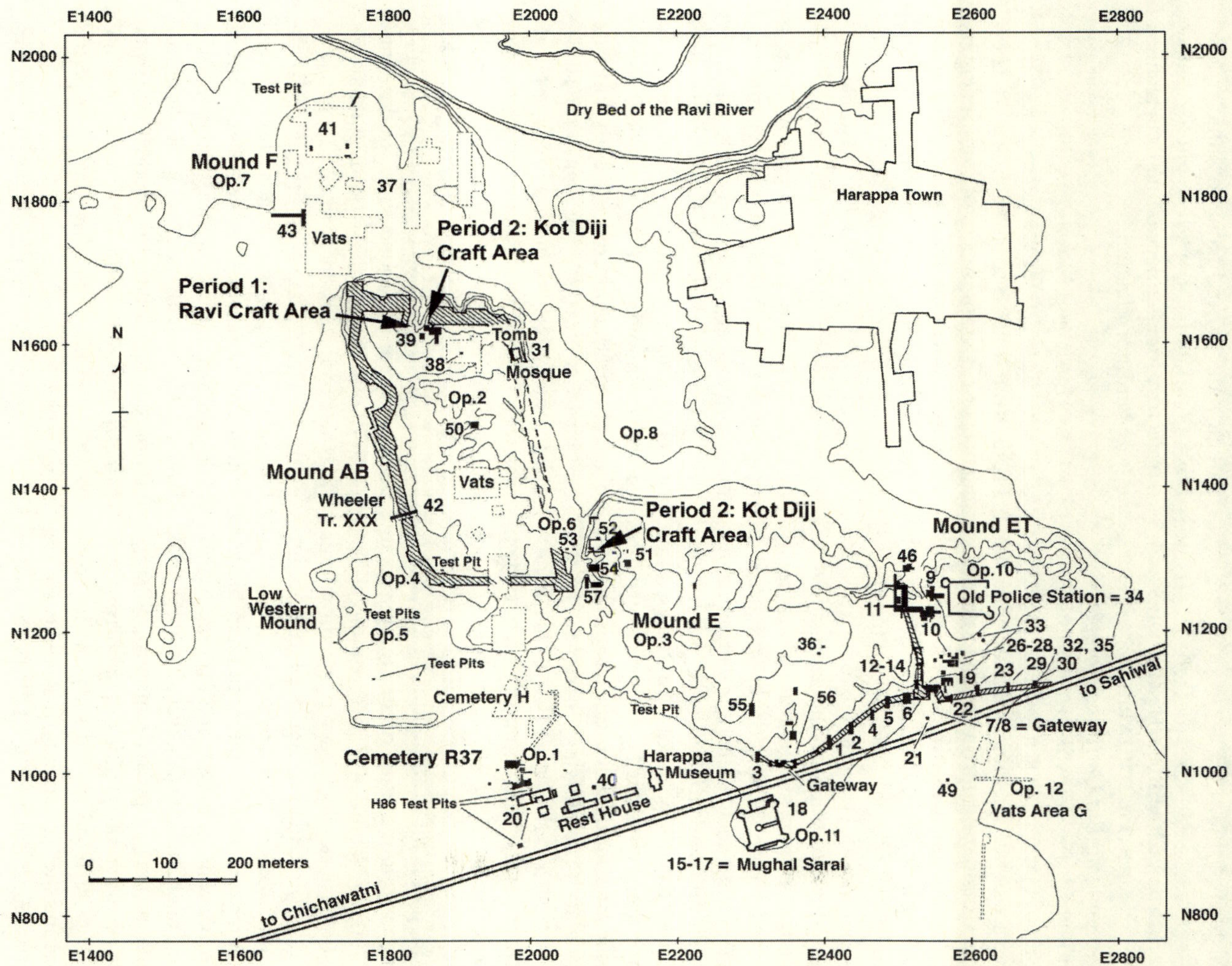

FIGURE 7.2. Harappa excavation areas, 1986–2001, showing the location of trenches.

settlement (Bhan et al. 2002; Kenoyer 1994a, 1998b; Kenoyer et al. 1991). The combination of these innovative and complementary studies has resulted in the collection of new types of data and new ways to examine crafts in the prehistoric period.

MAJOR CRAFT TECHNOLOGIES AND ASSOCIATIONS

Foraging Era Craft Technologies

Although this essay will focus primarily on the site of Harappa, it is important to provide some discussion of earlier craft traditions that set the foundation for later trade networks and technological associations. The earliest craft technologies dating to the Paleolithic period are not directly relevant in this discussion, but they did provide the foundation for basic reductive lithic technologies that involve chipping, pecking, grinding, and polishing. In the greater Indus region, the basic sources of raw materials such as chert, jasper, and agate for flake and blade tools were being exploited by Upper Paleolithic foragers, and these same source areas continued to be used throughout the later Indus Tradition. The use of heat treatment for modifying chert and jaspers can be traced to the Upper Paleolithic in various parts of the subcontinent (Allchin 1973; Kenoyer et al. 1983; Murty 1979) and continued to be used in the preparation of lithic raw materials for bead making in later times (Kenoyer 1986). The use of the highly specialized flaking technique of inverse indirect percussion has been documented in central India (Clark 1990) and also continued to be an important technique used during the Indus Tradition for producing some stone chert blades, the processing of beads, and eventually producing chert weights (Bhan et al. 1994, 2002). Different types of pecked- and ground-stone tools of different qualities of sandstone and quartzite were developed for grinding stone beads and shell ornaments. Flaked-stone tools such as microdrills made of chert and jasper were created to perforate shell and soft stone (Kenoyer et al. 1983). At sites such as Allahdino (Fairservis 1982; Shaffer 1974), the limited exposure of the earliest occupation levels makes it difficult to discuss associations of various crafts, but microlithic tool assemblages appear to be directly related to subsistence activities or tool maintenance activities. By around 10,000 B.C., foraging and hunting communities had developed a wide range of specific reductive techniques of lithic production and a familiarity with resources that set a trajectory for later developments in technology and craft production.

Early Food Producing Era Craft Technologies

From around 7000 to 5500 B.C. there is evidence for the presence of semisedentary agropastoral communities living along the edges of the Indus Valley and in the highland plateaus of Baluchistan at sites such as Mehrgarh, Kili Gul Mohammad, and Gumla (Figure 7.1; Kenoyer 1998a). Communities living at these sites had begun to exploit domestic plants such as wheat and barley and animals such as sheep, goat, and cattle (Meadow 1996). The earliest levels at Mehrgarh (Jarrige et al. 1995) predate the development of ceramic technology but reveal a rich assemblage of material culture from both domestic and burial contexts. At Mehrgarh the settlement consisted of simple mud-walled buildings with internal subdivisions as well as numerous burials, some with elaborate offerings (Jarrige et al. 1995). These offerings included animal sacrifices as well as bitumen-coated baskets, stone and bone tools, and a wide range of ornaments such as beads and bangles (Barthélémy de Saizieu 1990). Wide shell bangles were worn along with bracelets and anklets made from tabular beads of white shell or white limestone (Kenoyer 1995b). Many of these same materials continued to be used in later times, and with the development of more complex technologies for production, some came to be used as important symbolic and wealth objects during the Harappan period (Kenoyer 2000).

At Merhgarh, the production of utilitarian objects such as ground-stone celts and chipped-stone tools and microliths appears to have taken place throughout the settlement. Manufacturing material from both categories of objects has been found in most excavation areas. Stone beads made from locally available soft limestone may also have been produced at the site based on the presence of small stone drills, but no workshop areas have been reported. No evidence of the manufacture of exotic materials such as marine shell, carnelian, lapis lazuli, turquoise, or copper/malachite has been found from Mehrgarh during this early period, so it can be assumed that ornaments from these nonlocal materials were produced by craftspeople in distant resource areas and were traded to the settlement in finished form (Jarrige et al. 1995; Kenoyer 2000). This pattern suggests that some of the foraging or fishing communities along the coast may have begun to specialize in the production of local shell materials for trade, but no sites dating to this early preceramic Neolithic period have been discovered yet. Similarly, foraging communities and early agropastoral communities in the western Baluchistan highlands may have been active in the manufacture or trade of items such as lapis and turquoise beads and native copper beads, all of which have been found in early burials at Mehrgarh. The trade of exotic commodities to Mehrgarh indicates the presence of two major trade networks, one connecting the site to the highlands to the west and the other linking the site to the coastal areas of the Indus Valley, possibly along the piedmont or the river valley itself (Kenoyer 1995b).

The technologies used to create ornaments from shell and soft stone were not very complex during this period and involved relatively simple reductive processes of sawing with stone blades, chipping and grinding with hammer stones and abrasives, and drilling with chert microdrills (Vidale 1995). The initial production of shell and exotic stone beads may have been undertaken by different external communities, but by around 5500 B.C., both marine shell (Kenoyer 1995b) and exotic stone raw materials were being brought to Mehrgarh for local processing (Jarrige et al. 1995; Vidale 1995). This important change in production location indicates a major shift in economic strategies and corresponds to the beginnings of pottery making at the site. By bringing raw materials to the site, local producers began to control the production of materials that were used locally and could then be traded regionally. The association of shell and stone manufacturing debris in dumps along with domestic debris suggests that both types of reductive technology were being conducted in close proximity, but no primary workshop areas have been documented. Further research is needed to determine if both technologies were being undertaken by the same workshops or if they were segregated by households or perhaps by different craft specialists. The shift from trade for finished goods to the local production of specialized commodities is extremely important, and Mehrgarh is one of the only known sites where this can be properly studied.

Regionalization Era Craft Technologies

During the Regionalization Era, or Early Harappan period (ca. 5500–2600 B.C.), scattered agropastoral settlements became established on the extensive alluvial plains of the dual river system. As these settlements grew from small villages to large towns and market areas, they developed a high degree of regional and internal social differentiation. Most settlements were surrounded by massive mud-brick walls that may have served multiple purposes, including protection from external attacks by raiders, but there is no conclusive evidence for conflict. These walls also would have served economic purposes to control access into and out of the settlements and as revetments for protection from flooding. The need for economic control may have been linked to the virtual explosion in material culture and the emergence of numerous technologies for the production of utilitarian as well as prestige items.

At the site of Mehrgarh and the nearby site of Nausharo (Jarrige 1989), there is a dramatic increase in the amount of finished objects and manufacturing debris from all types of stone, shell, bone, and pottery objects during the Regionalization Era (Quivron 1994; Samzun 1992). The lithic technology is dominated by blade technology and the production of specialized microliths, drills, and scrapers (Lechevallier 1984, 2003; Marcon and Lechevallier 2000). The types of ornaments found in the burials include steatite bead necklaces and bracelets, along with pendants of lapis lazuli, carnelian, and other semiprecious stones (Barthélémy de Saizieu 2003; Tosi and Vidale 1990; Vanzetti and Vidale 1994; Vidale 1989). The varieties of stone beads increase, with new shapes and many different varieties of rock being used to produce attractive ornaments.

Significant remains of lapis lazuli, carnelian, calcite, garnet, and turquoise manufacturing waste and both finished and unfinished green jasper (pthanite) drills were found in one area of the site during the 1977–1978 excavations. Although the debris was not part of a primary workshop dump or working area, it was interpreted as waste from local workshops (Jarrige et al. 1995:214). Similar accumulations of shell manufacturing debris were found in different parts of the site from Periods II (fifth millennium B.C.) and III (fourth millennium B.C.) and suggest the presence of local shell working (Jarrige et al. 1995:214). Copper crucibles and fragments of copper also provide evidence for the local production of copper ornaments or tools during Period III (Jarrige et al. 1995:250).

One of the most important discoveries dating to Period II (fifth millennium B.C.) is a large rectangular mud-brick building subdivided internally with ten rectangular compartments that were probably used for storage. The surrounding open spaces revealed the concentration of numerous crafts, some directly linked to subsistence activities; others can be associated with the production of prestige items for local or regional exchange. The excavators feel that the scale of production around this building is "more complex than would be found in a village occupied by self-sufficient family units" (Jarrige et al. 1995:249). The building itself is thought to have been used as a granary because of the numerous grain impressions in the internal compartments and the discovery of two complete sickles made with stone blades with bitumen hafting in one of the internal compartments. To the south of the building was evidence for a steatite bead workshop, with finished and unfinished steatite beads, a few cut *Dentalium* shell beads, and some white talc beads colored with red ochre (Jarrige et al. 1995:248). The finished steatite beads were fired and glazed, but no firing containers or kilns were discovered. This could indicate that some of the production took place in a different location or that the firing structure was not preserved. To the east of the same building, large concentrations of bone tools, awls, and needles were interpreted as evidence for a leatherworking area, although the fact that these tools were found among "heaps of bones" raises the possibility of bone working instead of or in addition to leatherworking (Jarrige et al. 1995:248). A fireplace against the western wall contained numerous charred cottonseeds that were interpreted as evidence for nearby ginning of cotton. The oily cottonseeds may have been used as fuel or inadvertently charred along with cotton waste. Aside from these groups around the compartmented building, the manufacturing debris from shell and stone bead making as well as copper working are relatively distinct at Mehrgarh, so perhaps it can also be assumed that these distinctive objects were being produced by different groups of craftspeople. The only exception is the association of *Dentalium* shells with steatite bead-making debris in one working area.

A new technological development seen during the latest part of the Regionalization Era (3000–2800 B.C.) at Mehrgarh and Nausharo is the production of blue-green-glazed faience beads that combine reductive and transformative technologies. Two types of faience have been identified, one produced with ground steatite and a silica glaze and the other produced with ground quartz and a silica glaze (Barthélémy de Saizieu and Bouquillon 1997). The production of such ornaments required fairly high firing temperatures (ca. 940°C) as well as a specialized technology of frit and glaze preparation. Many new styles and shapes of pottery were also produced during this period, and specialized kilns were built for oxidizing and reduction firing processes (Audouze and Jarrige 1979; Samzun 1992; Santoni 1989; Wright 1995). At Mehrgarh and the nearby pottery-firing site of Lal Shah (Jarrige et al. 1995:461–462; Miller 1997; Pracchia 1985), distinct areas for pottery making have been identified on the basis of kilns, as well as stockpiles of unfired and fired pottery. These pottery production facilities do not appear to have been used for faience or fired steatite production. At present no specific production areas for faience or glazed steatite beads have been reported.

The sites of Mehrgarh and Nausharo thus provide evidence for the early development of both reductive and extractive crafts as well as the initial stages of transformative crafts. Generally speaking these crafts were being practiced in separate locations, presumably on a relatively small scale and at the household level, with no clear, consistent locational patterning or association between crafts, with two exceptions. The first is the close proximity of stone bead making, cotton processing, and bone tool manufacture and possibly leatherworking around a large granary structure. The second exception is the segregated pottery production at Mehrgarh, reaching its most extreme form in the separate pottery-firing site of Lal Shah. A wide range of pottery styles was clearly being produced on a large scale at the two sites, and pottery making may have been one of the most important production activities for local and regional trade. The segregation of pottery making from other crafts is an important pattern that is also documented later at Harappa and other sites. Although the site of Mehrgarh represents the best data we have for the early phase of the Regionalization Era along the periphery of the Indus Valley, the recent excavations at Harappa provide a unique opportunity to follow the same sort of development in the core area of the Indus region during the later part of the Regionalization Era.

Harappa: Ravi Phase Technologies and Associations

The earliest village at Harappa dates to around 3900 B.C., at the start of the Ravi Phase of the early Regionalization Era (Table 7.1). Excavations have revealed the local production of numerous types of finely crafted objects such as hard and soft stone beads, terra-cotta beads, pendants and bangles, shell bangles, bone button seals, painted pottery, and terra-cotta figurines. The versatility of the early craftspeople is reflected in both the styles of the objects created and the technologies developed to make unique objects (Kenoyer and Meadow 2000). The Ravi Phase occupation covers approximately 10 ha, but most of this area is buried beneath more than 15 m of later occupation levels. Two areas of the Ravi settlement have been excavated (approximately 82 m^2) at the north edge of Mound AB and the northwest corner of Mound E (Figure 7.2), but the most information for craft activities comes from excavations in two separate trenches in the northern portion of Mound AB: Trench 39S (4 × 4 m) and Trench 39N (6 × 11 m; Figure 7.3). Although the areas exposed are very small compared with the exposures for other time periods, a wide variety of craft production indicators and various workshop areas have been found, including shell working, soft and hard stone bead making, bone working, textile production (spindle whorls), and pottery making. All of these craft production activities appear to have been conducted in close proximity to domestic structures, and though there is a possible association between some crafts, others clearly took place independently or at different periods of time.

In Trench 39N the earliest evidence for craft production is seen in the high concentrations of agate, jasper, and some lapis lazuli and amazonite bead–making debris that accumulated on the natural sandy soil on which the first settlement was built (Figure 7.3). All of these relatively hard stone raw materials derive from distinct source areas located more than 300 km from Harappa and were probably obtained through different sets of trade networks. Agate and amazonite sources may be located in Gujarat and Rajasthan, the lapis lazuli probably came from northern Afghanistan, and the chert and jasper may have come from various localities in Baluchistan (Kenoyer 1997; Law 2006). The technology for processing hard stone combines reductive technologies of flaking and grinding with transformative techniques of heat treatment to make the agate and jasper easier to flake,

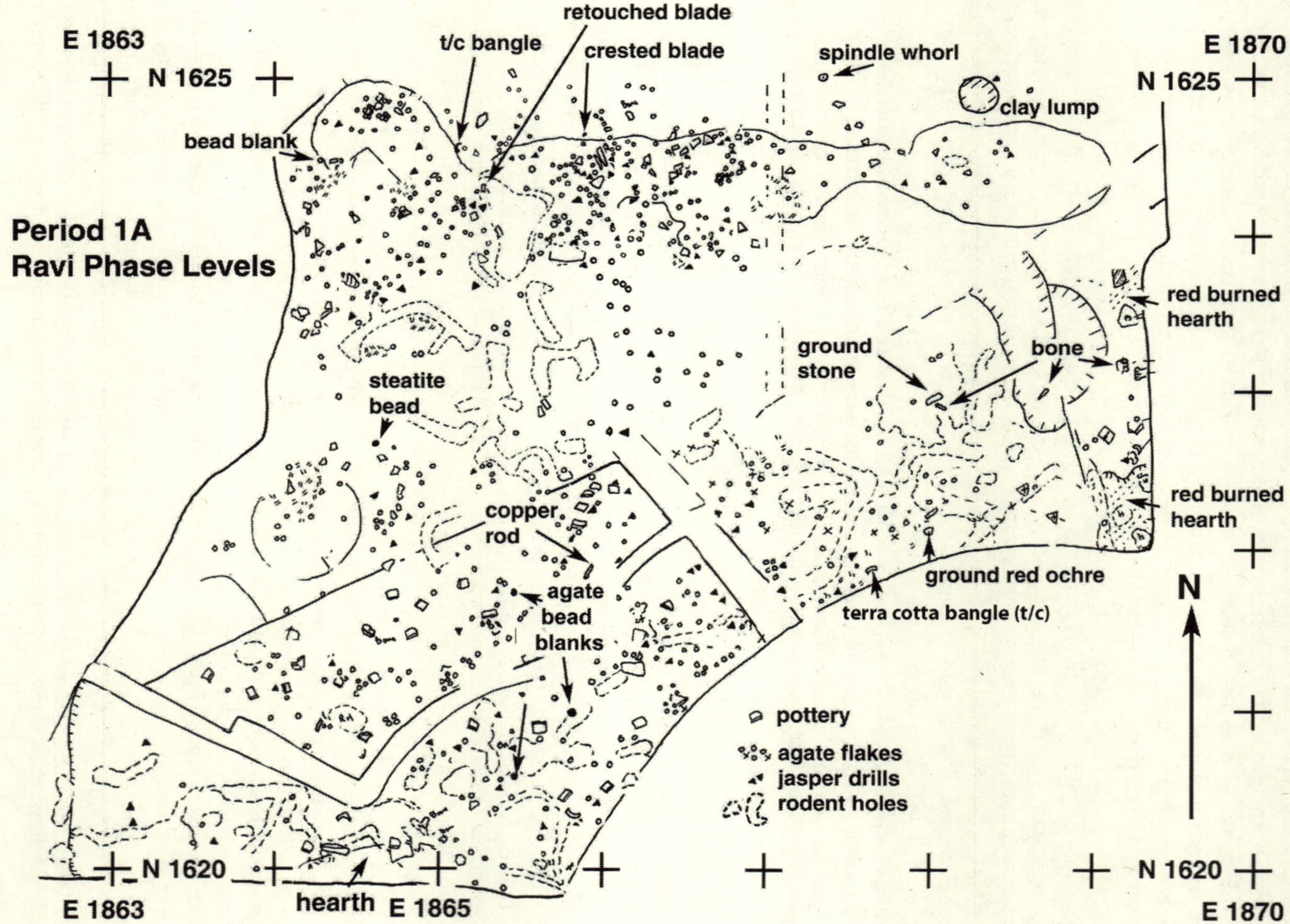

FIGURE 7.3. Ravi Phase agate-working area, Mound AB, Trench 39N, Harappa, 2000 season.

grind, and perforate. Lapis lazuli was probably processed using a combination of flaking, grooving with chert blades, and grinding, a process that is well documented at the site of Shahr-i-Sokhta (Piperno 1973). The production of drills involves the preparation of microflakes and then careful retouching to prepare the drill tip and tang for hafting. Microdebitage from the preparation of drill bits has been recovered along with the bead-making debris. The actual drilling process often results in the fracture or spalling of the drill tip. Again, microdebitage consisting of spalled drill tips indicates that actual drilling was also taking place in the same activity area as the drill preparation and agate bead blank preparation. Careful mapping of drills, flaking debitage, and unfinished beads reveals the concentration of manufacture in an open area with nearby hearths that may have been used for heat treatment. Microdebitage (<1 mm) obtained from fine screening and water screening confirms that primary manufacture took place in this location.

A careful analysis of the drills and beads themselves indicates the use of three different perforation techniques. The lapis beads were drilled with stone (chert or jasper) drills, whereas the carnelian beads were perforated by a pecking technique. Amazonite beads were drilled with a tapered cylindrical stone drill (Kenoyer and Vidale 1992). Different types of bead production and drilling techniques in the same area of the site indicate the diversity of bead technology during the Ravi Phase occupation. The differences among these technologies, from raw material acquisition to the final distribution of finished beads, are critical to understanding the relative value of beads and for defining the different ways in which production could have been controlled by elite groups (Kenoyer 2000; Vidale and Miller 2000).

Later storage pits and huts cut through the agate bead–making levels, and the open area in which agate working formerly took place was used for both domestic activities and what appears to be a pottery pit kiln. Large accumulations of pottery wasters were found in a series of superimposed hearthlike structures in one area of the site. The first of these irregular oval structures consists of a shallow depression excavated into earlier working floors that contained agate bead–making debris. A few meters to the north, on the other side of a possible hut, was a large dump of hand-formed bricks and pottery wasters that appear to be derived from a dismantled kiln that was probably located just outside of the excavated area.

In Trench 39S, evidence for glazed steatite bead production is seen in the large quantities of unfinished beads, some sawn fragments of manufacturing waste, and numerous finished beads that had been hardened and whitened by glazing and firing at high temperatures (around 940°C). Steatite bead making also uses a combination of reductive and transformative processes, including sawing, grinding, drilling, and glazing at high temperatures. Although most beads are white, some have a blue-green glaze that was probably made with powdered frit and copper oxide combined with a flux from plant ash (*sajji*), a process that is well documented for glazed faience (Kenoyer 1994b). Steatite sources for this time period are still being studied but may be located in areas in Baluchistan and the North-West Frontier Province of Pakistan (Law 2002). Copper could have been obtained as finished tools or ores from either Baluchistan or Rajasthan. Experimental studies of steatite and faience production have shown that the temperatures needed to fire and glaze steatite can be achieved by placing the beads in canisters made with straw and sand-tempered refractory clay and firing these in large bonfires (Kenoyer 2003, 2004). Pieces of vitrified, chaff-tempered clay were recovered in the debris on the floor areas of Trench 39S, and their presence suggests that the firing and glazing of the beads may have taken place in nearby areas. Some lumps of vitrified slag found in the northern area may derive from the firing and glazing of the steatite beads, but there is no direct evidence for steatite bead making in the northern area, and similar slags can be formed from vitrified dung. Saw marks on the steatite manufacturing waste indicate the use of a thin, irregularly denticulated copper saw, and drill marks also appear to have been made with a copper drill with a beveled tip. No stone drills were found in this area, and no agate flakes were recovered in association with the steatite bead–making areas. Although there is no stratigraphic profile that directly connects the two areas, the steatite bead making activities were probably slightly later in time than the hard stone bead production.

The total separation of steatite bead making, including glazed varieties, and all other types of bead making during the early Ravi Phase occupation is significant in terms of the organization of bead production. Clearly these two types of bead-making technologies were practiced by separate working groups, as also seems to be the case in later periods. No other primary production areas dating to the Ravi Phase have been documented, but domestic debris in both Trench 39N and Trench 39S included manufacturing debris from the production of chert tools, shell bangles, and bone tools, all of which can be defined as reductive technologies. The ubiquity of chert manufacturing debris indicates that stone tools were being made in many different localities and not in specialized workshops. Some microdebitage from chert tool production has been found in this area, but it may derive from the production of chert drills rather than from larger stone tool production. The discovery of terra-cotta spindle whorls and polished rib fragments in both trenches provides evidence for spinning and weaving, a reductive or extractive technology. Bone awls were also found in both areas and may have been used in basketry or leatherworking. These perishable materials could have been used for basic utilitarian and subsistence activities or as wealth items in the form of ornaments and clothing. Depending on their decoration and style, they could have been important indicators of status and ethnic identity. Shell ornaments also could have served similar functions and possibly also had ritual purposes. A few fragments of finished copper pins and arrow points have been found in both trenches, but no evidence for local copper production has been found during the Ravi Phase. Finally, two tiny flakes of gold foil were found in the Ravi levels. Gold foil was used in the later Harappa Phase to cover copper beads, a pattern that indicates a high value placed on gold. If gold-covered copper beads are found in future excavations of the Ravi Phase, it would indicate that the relative value of gold had a very long history at the site.

On the basis of current evidence, craft production at Harappa during the Ravi Phase involved the use of many local as well as exotic materials, brought to the site through different sets of trade networks. Local networks may have been maintained by individuals at the household level to supply materials for pottery making and textile and bone tool production. However, it is unlikely that the long-distance trade was carried out by each individual household, and exotic items may have been obtained from specialized traders or nomadic communities. The degree of specialization for shell bangle production is also unclear for this early period. The distinct production locations for both unglazed and glazed steatite beads and for hard stone beads, however, indicate the beginnings of occupational specialization, developing along lines that would continue in later periods. Another important pattern that may indicate the presence of hierarchical social stratification is the use of different types of raw materials to produce similar objects, such as terra-cotta and stone beads or terra-cotta and shell bangles (Kenoyer 2000). Further exposures of the Ravi Phase occupation levels are needed to confirm the presence of occupational specialization and social stratification, but these processes were undoubtedly taking place and would have contributed to the economic and political growth of the settlement.

Harappa: Kot Diji Phase Technologies and Associations

At Harappa, the Kot Diji Phase (2800–2600 B.C.) represents the final phase of the Regionalization Era and the transition to the Integration Era (Table 7.1). By around 2800 B.C., the settlement had grown into a small town of over 25 ha, with two distinct walled areas. Technologies developed during the earlier Ravi Phase became more elaborated, and glazed faience

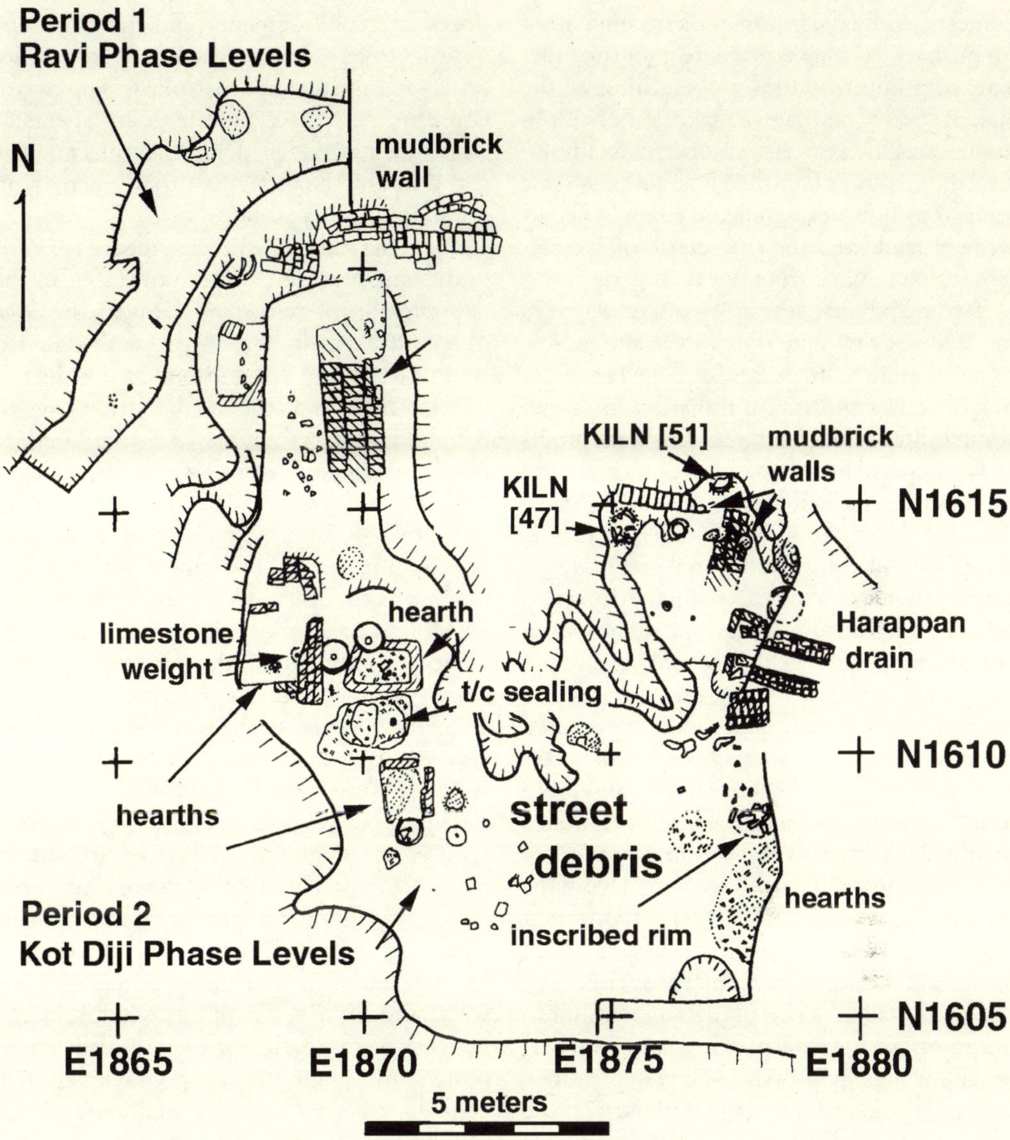

Figure 7.4. Kot Diji Phase kilns, Mound AB, Trench 39N, Harappa, 1996–2001.

bead making appears at about the same time that it developed at Mehrgarh to the southwest. Most of the Kot Diji Phase occupation levels at Harappa are buried beneath at least 8 m of later occupation deposits, and yet it has been possible to make some horizontal exposures in two areas, one on the northwestern edge of Mound E (approximately 15 × 3 m) and the other at the north edge of Mound AB, directly above the Ravi Phase exposures of Trench 39N (10 × 10 m; Figures 7.2, 7.4). Smaller exposures have been made in Trench 39S and in smaller trenches on Mound E and Mound ET. Although the total area exposed for this phase is much greater than that for the earlier Ravi Phase, only three small kilns have been discovered, one on Mound E and two on Mound AB (Figure 7.4). Various types of manufacturing debris have been recovered from domestic debris and dumps found in the middle of streets, but no primary bead-making area or other major craft production areas have been located. Chert tool manufacturing debris is also found in the street and domestic contexts, but there is no indication for primary flaking areas where intense or specialized production took place. Although the lack of specific craft areas makes it difficult to discuss specific associations of craft technologies, the analysis of finished artifacts and craft indicators indicates significant changes in patterns of trade and the emergence of specialized technologies during this time period.

Two major changes can be documented in the trade of raw materials to the site, access to new resource areas, and the development of more efficient forms of transport for heavy commodities. Initial studies of raw materials at Harappa indicated new links to chert resources from the Rohri Hills, some 400 km to the south. New studies by Randall Law (2002, 2006) have identified additional raw material sources that began to be exploited during this period, such as lead, steatite, and possibly copper. The major change in transport can be associated with the increasing use of boats and wheeled carts. This change is demonstrated indirectly by the presence of numerous terra-cotta toy cart and wheel fragments and more directly by the presence of specific artifacts themselves. Although one possible terra-cotta toy cart fragment has been recovered from the Ravi levels, the utilization of several different types of carts appears to have become more prominent during the Kot Diji Phase (Kenoyer 2004). The introduction of carts has major implications for the transport of heavy raw materials such as mud brick, stone, and timber, along with other types of materials. The use of boats would have allowed the movement of heavy goods up and down the major rivers, and carts facilitated the transport of these commodities from the river docks to the growing urban centers and outlying towns and villages. The construction of massive mud-brick walls around the settlement at Harappa was probably made possible through the use of carts. Another important indicator of this exchange is the presence of large discarded fragments of grinding stones as well as increasing amounts of chert debitage from gray, tan, and banded chert obtained from the Rohri Hills. During the earlier Ravi Phase all groundstone objects were reused until they were reduced to small fragments, and chert debitage appears to have been reused to make smaller tools such as drills and microliths. The lack of concern about discarding large fragments of stone during the Kot Diji Phase at Harappa can probably be attributed to an increase in the supply of raw stone.

As noted, the only direct evidence for craft workshops during this period is seen in the construction of two small kilns at the edge of a courtyard along a north–south-oriented street on Mound AB and one at the edge of a domestic area at the northwestern edge of Mound E (Figures 7.2, 7.4). All three kilns are highly vitrified on the interior walls, which is evidence for repeated use at relatively high temperatures (800–1,000°C). Each kiln is between 30 and 60 cm in diameter and between 40 and 50 cm deep. Two of the kilns, the one on Mound E and one of those on AB, show evidence for reduction firing, in that the pottery wasters used to cover the kilns and to serve as a flexible floor above the fuel in the lower levels are heavily blackened. No finished objects were recovered from sealed deposits within the kilns, but fragments of gray- or red-fired terra-cotta bangles as well as anthropomorphic and animal figurine fragments have been recovered from the upper fill layers or in surrounding deposits along with pottery wasters. These objects may indicate that the kilns were used in firing small terra-cotta objects, but they were clearly not used for firing large pottery. Although it is also possible that they could have been used to fire small canisters for glazing steatite and faience beads, no canister fragments, unfinished steatite beads, or faience manufacturing debris have been found in close proximity to the kilns.

The production of terra-cotta animal and human figurines, sometimes painted with black and red designs, became more common during the Kot Diji Phase. Some of these figurines could have been fired in small pit kilns on an ad hoc basis, but the more finely made and painted varieties may have been fired in small kilns similar to those described above or in larger pottery-firing kilns. The decorations on both animal and human figurines provide important clues regarding the increase in ornament styles and decorative textiles during this period. Many bull figurines have painted bands and stripes on their legs and humps that may represent decorative blankets, as are commonly used on draft animals throughout the region even today. Ram figurines have holes to attach wheels so that they can be pulled as toys. Although most of the human figurine fragments from Harappa have no preserved painted decorations, one female figurine has a unique hairstyle painted with black paint and a necklace with pendants painted around her neck. She has painted bangles on both arms and is depicted wearing a skirt made with a plaid or checkered design. Similar shaped figurines but without preserved painted designs have been found at the sites of Gumla (Dani 1970–1971) and Rehmandheri (Durrani 1988; Durrani et al. 1995b). These figurines from the northern Indus Valley are quite different from those found at the sites of Mehrgarh and Nausharo and in central Baluchistan during this same period (C. Jarrige 1997). The detailed study of terra-cotta figurines at Harappa indicates that this craft was itself quite complex and highly varied (Clark 2003; Dales 1991).

The variety of terra-cotta bangles increased dramatically during this phase, and the small kilns may have been used by household specialists to produce the many different styles. Though most of the bangles were made with red-fired terra-cotta, sometimes with painted or pinched designs, in the later Kot Dijian levels there are gray-fired bangles with highly polished and incised surfaces. As was first noted in the Ravi Phase, both wide and narrow bangles were made from the marine shell *Turbinella pyrum* using simple techniques of incising with stone tools, grinding, and polishing. However, during the Kot Diji Phase many more shell bangles were being produced, and it is not unlikely that a copper saw was being used to separate the shell discs. Finished copper objects are much more common during this phase and include bangles, pins, knives, and arrow points. Although some copper working was probably going on at the site, no production areas have been located yet.

No large pottery production areas or kilns have been found at Harappa during this period, but the initial study of the pottery itself indicates the predominant use of the fast wheel by around 2800 B.C., though some types of small vessels continued to be made by hand forming. Most of the pottery was probably fired in updraft kilns similar to those found at Mehrgarh and Lal Shah (Miller 1997; Pracchia 1985). A gradual shift from polychrome painted designs to red slip with black painting or plain unslipped wares in relatively standardized shapes suggests mass production of pottery by craft specialists for local and regional trade rather than household production.

A considerable variety of manufacturing debris and finished objects has been found in secondary street deposits dating to the Kot Diji period. These craft indicators are evidence for primary workshops that must be located somewhere outside the excavated areas. Although only a few domestic areas have been excavated, the absence of any manufacturing debris in the houses suggests that unlike the case during the earlier Ravi Phase, craft production was not conducted in most household areas but, rather, in relatively distinct areas. Debris from such areas was thrown out onto the streets and washed down the open drains. The wide street on Mound AB contained fragments of unfinished and finished glazed steatite seals and sawn steatite fragments from bead making. The finished artifacts found here include pottery, worked bone and worked shell fragments, shell bangle fragments, gray- and red-fired terra-cotta bangles, copper bangle fragments, gold beads and sequins, and highly refined compact faience beads. These types of artifacts and manufacturing waste provide evidence for increased technological complexity during this time period and also probably reflect an increase in craft specialization. In other words, most of these objects were probably being made by specialists for use by others rather than for personal or household consumption, and once the actual workshops have been located we will have a much more complete understanding of this shift toward increasing specialization in craft production (Kenoyer 1989, 1992).

In summary, all of the basic technologies used at Mehrgarh and Nausharo (see above) appear to have been present at Harappa during the same general time period. Both sites show a gradual increase in the diversity of both reductive and transformative crafts. Even though actual production areas have not been located for the Kot Diji Phase at Harappa, the dominant use of the wheel for pottery making indicates an increase in the overall scale of production, similar to that seen at Mehrgarh and Lal Shah. The major difference between these sites is the fact that at Harappa there is long-term occupational continuity and continuity in general site layout between the different periods of the Regionalization Era (Early Harappan) and the subsequent Harappan period of urban integration. At Harappa, the layouts of the north–south-oriented streets and of the city walls in both these periods represent new developments in site organization and the emergence of a more highly differentiated society. Specialized crafts would have been important for providing new forms of status markers and social elaboration (Vidale and Miller 2000). These crafts would have contributed to the development and maintenance of socioeconomic hierarchies, as well as the inclusion of new occupational specialists. The emergence of writing, seals, and standardized weights is also documented during this final transitional stage (Kenoyer and Meadow 1997, 1999) and serves to confirm the development of a more complex socioeconomic organization that would benefit from these techniques of communication and administration. All of these developments set the stage for the elaboration of craft technologies and the intensification and expansion of craft production during the subsequent Integration Era.

Indus Craft Technologies: Integration Era

The Integration Era, Harappa Phase (2600–1900 B.C.), was a period of urban growth and expansion of interaction networks throughout the greater Indus Valley region. During this time period, Harappa grew to over 150 ha in area by the addition of several walled suburbs and additional habitation areas outside of the two original walled sectors. The site of Mohenjo Daro was even larger, with several walled mounds covering over 250 ha (Kenoyer 1998a). Recent excavations at Harappa have been able to subdivide the 700 years of the Harappa Phase (called Period 3 at Harappa) into three subphases, designated Periods 3A, 3B, and 3C (Table 7.1). Because of the heavy overburden of later occupations, the earliest 3A levels (2600–2450 B.C.) have only been accessible in a few areas of the site, and only one area along the western part of Mound E has been exposed horizontally. Most of our information on Indus craft technologies is based on surface surveys and excavations of the 3B (2450–2200 B.C.) and 3C levels (2200–1900 B.C.).

In 1986 through 1988, surface surveys of the entire site were conducted to identify potential craft areas for excavation. Generally speaking, all of the uppermost levels of the site have been badly disturbed by the removal of brick walls by brick robbers. Subsequent test trenches revealed that some areas with surface concentrations of craft indicators were the result of erosion from brick robber debris, and no actual workshops were found in the undisturbed underlying areas of the excavation. In other locations, surface concentrations did in fact reflect the presence of workshops or primary dumps located inside rooms, in courtyard areas, or along the streets. Through careful stratigraphic excavation after the removal of brick robber debris, it has been possible to separate out specific craft areas and isolate production according to specific crafts. Subsequent full-coverage surface surveys of the entire mounded site were carried out by Heather M.-L. Miller (1999, 2000) in 1994–1995, along with geomagnetic surveys of selected areas. These surveys provided additional information on the locations of craft areas, and some of these areas were subsequently excavated. In the following section we will focus on the evidence of craft activities from excavated areas, which provide more fine-scale information on craft associations, especially for reductive crafts, and compare this information with what was observed on the basis of surface surveys.

Harappa: Period 3A Craft Technologies and Associations

Craft activities dating to the 3A period have been located along the western edge of Mound AB (Trench 42) and the western edge of Mound E (Trench 54; Figures 7.2, 7.5). Both areas have discrete accumulations of steatite bead–making debris that could be the result of actual working areas or dumps from a nearby workshop. The concentrations are located at the edges of streets and include large quantities of beads and bead blanks in all of the initial stages of manufacture, with a few fired beads mixed in with the unfired beads. The concentration on Mound AB also includes a copper saw that was used for sawing the initial bead blanks. Although the initial processing of steatite beads is a reductive process, the final heating to harden and whiten the steatite is a transformative process. These concentrations of bead-making debris reflect primarily the reductive process of manufacture and some evidence for the transformative stage of production, but the actual firing areas have not been located. The absence of other craft debris from these concentrations suggests that steatite bead makers may have been working as semi-independent craftspeople and were not practicing their trade in conjunction with other reductive or transformative crafts. This pattern would reflect continuity in the pattern seen during the earlier Ravi Phase, when hard stone bead making and steatite bead making were practiced in distinct activity areas even though they were in the same general area of the site.

Ethnographic observations of modern steatite bead makers in Pakistan indicate that they are usually itinerant craftsmen, often carrying their entire stock and tools in a small cloth bag. They move from one town to the next, following major markets or festivals, and work along the edge of the street or in any available open area (Vidale 1992; Vidale and Shar 1991). The craftsman sells his finished product directly to consumers or to an established ornament merchant who has a more permanent shop. It is possible that the early steatite bead makers at Harappa were independent or itinerant craftsmen moving about the city and working along the main streets in a similar fashion. This pattern did change during the subsequent period, as will be noted below.

Adjacent to the steatite concentration on the western side of Mound E was a large pit filled with pottery wasters and kiln debris that had been deposited after the accumulation of the steatite concentration. The wasters include kiln slag, vitrified kiln walls and broken pottery, and flat pottery discs used for covering the pottery in the kiln prior to the plastering over of the kiln top. A wide range of painted and plain pottery styles is represented in the broken pottery fragments, along with molds and terra-cotta bats for use on the potter's wheel. A few fragments of terra-cotta bangles and some terra-cotta toy cart and wheel fragments were recovered in the upper layers of the pit along with some copper fragments and a few pieces of chert manufacturing waste. The presence of a few unfinished steatite beads in these same layers suggests that most of the nonpottery artifacts are present because they mixed with other street debris and do not represent actual kiln waste. In the lower layers of the pit, a few more terra-cotta bangles and some additional steatite bead fragments were recovered, but there was a significant decrease in other types of nonpottery waste. The absence of figurines or terra-cotta toys in the lower layers is important to note because it suggests that the potters were not engaged in other types of ceramic production. The general impression from the analysis of this pottery dump is that the specialized production of painted and utilitarian pottery was relatively segregated from other terra-cotta object manufacturing during the 3A period. This pattern is seen again in Period 3B, as described below and by Wright (1991). Dumping the pottery firing debris into a pit dug into the street or an open space resulted in the association of the pottery debris with other craft debris, such as the steatite bead–making concentration, but this was secondary to the actual pottery production activity.

Most of the corresponding deposits to the south of the pit had eroded away, and because no kiln was discovered in the large areas

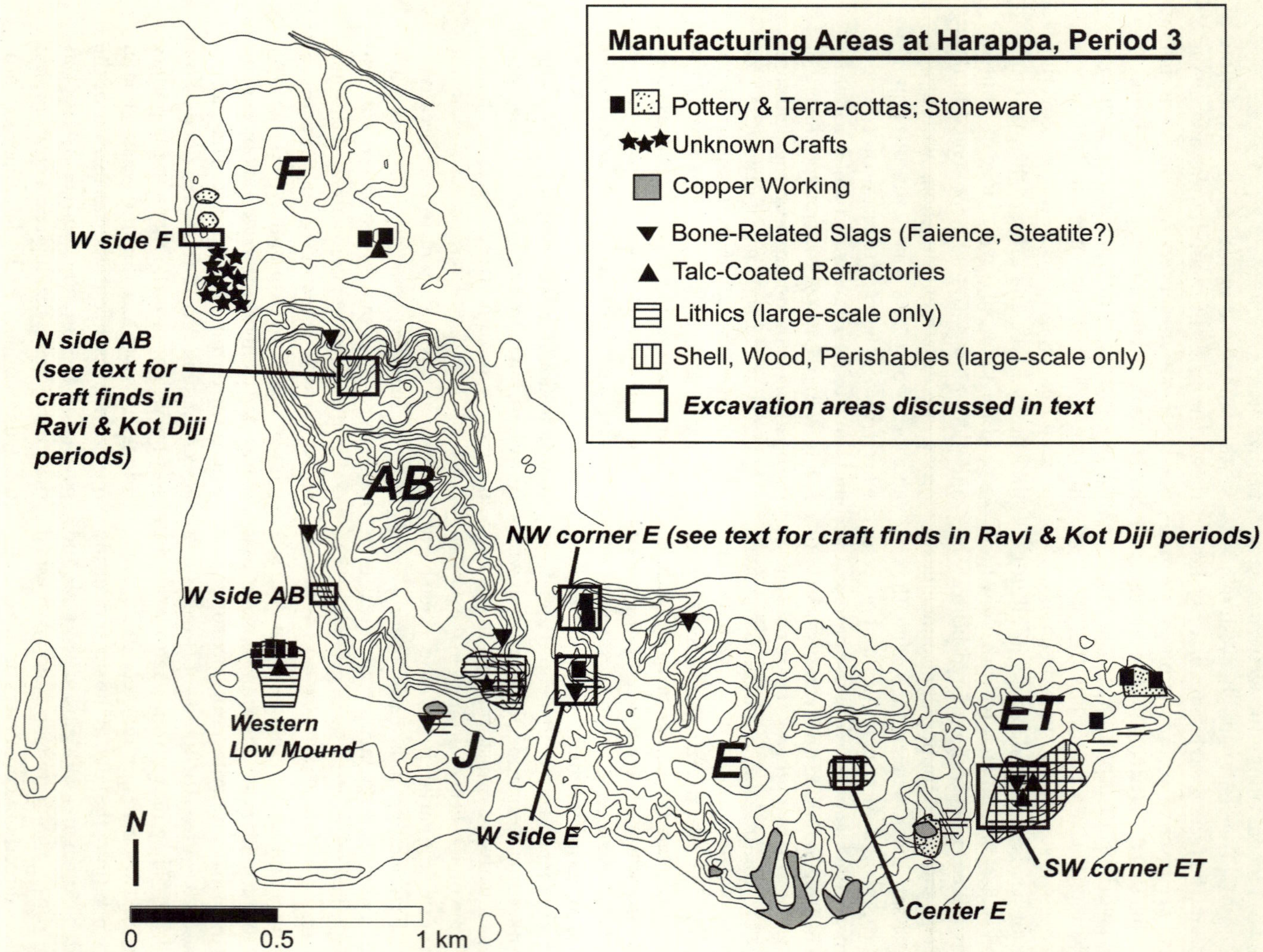

FIGURE 7.5. Harappa Phase (Period 3) craft activity areas, Harappa. Data compiled from H. M.-L. Miller survey and M. S. Vats, Berkeley/Harappa Archaeological Research Project excavations.

exposed on all sides of this pit, it is possible that the actual kiln structure has been lost through erosion. Nevertheless, it seems reasonable to assume that a kiln was somewhere in the vicinity, inside the city wall but at the western edge of the habitation area. As described below, in the 1988 through 1990 seasons, three kilns were excavated at the northwest corner of Mound E, some 75 m to the north of the Period 3A pottery dump and inside the city wall but situated at the edge of the habitation area. One of these kilns is from the Kot Diji Phase and was likely used for bangle or figurine production, as discussed above, and the other two are from the later Harappa Phase (3B) and were primarily used for pottery production, as discussed in the next section.

Harappa: Period 3B Craft Technologies and Associations

Although Period 3B occupation levels have been found in most of the excavation areas at Harappa, only the two excavated areas with craft production debris from Period 3B will be discussed in this essay. One area represents pottery production, with little evidence for other craft activities (Miller 1997; Wright 1991), and the other area has evidence for the manufacture of faience molded tablets, steatite tablets, and beads from both steatite and faience (Kenoyer 2005).

In the course of surface surveys in 1986, a large pottery kiln was identified eroding from the northwest corner of Mound E (Trench 52). Subsequent excavations between 1987 and 1989 revealed a sequence of firing structures. The earliest kiln dates to the Kot Diji Phase (discussed above), and just above and to the east was a small pit kiln dating to the Harappa Phase, Period 3B. This small pit kiln was filled with ash and a layer of broken pottery sherds. Large hand-formed clay nodules were placed at the mouth of the kiln to aid in heat retention once the fuel had burned away. Although no complete vessels were found in this kiln, because of its size it is probable that the kiln was used for firing only small vessels, figurines, or terra-cotta bangles. A much larger, double-chambered updraft kiln was located just to the north but stratigraphically later than the small pit kiln. The larger kiln measures 2 m east–west and 3 m north–south and had a large central column that would have supported a floor. The walls of the upper chamber would have been built up and plastered over once all the pottery had been stacked into the firing chamber. The lower firebox was heavily vitrified and had been replastered several times, indicating long-term reuse of this structure. Large quantities of vitrified pottery were found on floor areas around the kiln and on the eroding slope. The kiln was eventually abandoned and gradually was covered with more layers of ceramic production debris. Numerous red burned floor layers above the kiln indicate continuity in pyrotechnological activities in this area of the site on into Period 3C.

The types of pottery fired in this structure include dish-on-stand vessels, medium to large painted pots and jars, perforated vessels, and medium to large plain wares, but there is no evidence for the production of pointed-base goblets or smaller plain wares. (See Wright 1991 for additional details.) Molds for making the bases of large storage jars, pieces of hematite for making pigments, and bone spatulas and chert blades used for scraping and burnishing pottery were found in floor levels associated with the use of the kiln. Patches of fine clay that may represent areas where the potters mixed or wedged the clay were also discovered near the kiln. A few pieces of carved bone and antler were also discovered in these layers, but no other specialized craft debris was found.

Basically this craft area appears to have been used exclusively by potters and ceramic workers during Period 3B (2450–2200 B.C.) and probably even later, as is suggested by the pottery wasters and red burned floor levels found stratigraphically above the large kiln. When combined with the evidence from the Kot Diji Phase (2800–2600 B.C.) kiln, this part of the site was used for firing clay objects for more than 600 years. This is a very significant pattern for site use and for the segregation of craft activities.

The other major craft area that is exclusively associated with Period 3B is located to the south of these kilns, in Trench 54 on the western side of Mound E. This area was selected for excavation because of surface concentrations of craft debris thought to be associated with faience manufacturing (Miller 2000, 2007, in press). After screening all of the eroded surface deposits there was clear evidence for chert tool production that may have been associated with woodworking and the manufacture of drills for various other types of crafts, as well as agate and steatite bead making. Shell manufacturing debris from bangle making and the production of other types of shell objects was also noted. Debris from faience production and gold working was also found in the eroded deposits. After initial surface clearing, wall voids from brick robbing were identified and cleared to reveal stratified occupation levels associated with mud-brick walls or baked-brick wall voids. Excavations in the central part of this trench revealed the steatite debris and pottery dump dating to Period 3A. Most of the craft indicators found in these eroded deposits appear to have derived from already destroyed portions of the site, and only one primary workshop was discovered. This working area was located in a restricted area of the southern eroded slope, where large accumulations of "faience slag," a frothy vitreous slag containing bone with drops of blue-green glaze, had originally been noted in surface surveys by Miller (in press). Subsequent surface surveys in this area also produced several molded faience tablets and faience bead fragments.

The faience slag area turned out to be a small room that had partly eroded, but the preserved section (2.5 × 2.5 m) was filled with debris from faience inscribed tablet and bead production (Figure 7.6; Kenoyer 1997, 2005; Meadow et al. 2001). In addition to complete as well as broken beads and faience tablets, the debris from this room contained a few chert blades, lumps of sawn steatite, fired steatite beads, bone tools, terra-cotta cakes, and large masses of vitrified faience slag with whitened bone, some of which had drops of greenish to light blue glaze. Inscribed and uninscribed steatite tablet blanks were found in the ashy debris within the workshop along with a broken fan-shaped steatite tablet with deeply incised script that was used as a mold to make faience tablets. The techniques used to make faience tablets have long been a mystery, as no clay or steatite molds had ever been recovered. This new discovery suggests that some of the unfired and apparently unfinished steatite "seals" may in fact be molds for faience tablets. Future searches to match inscriptions and designs incised on known steatite objects of this type with designs on known faience tablets should resolve this point. Large fragments of vegetable-tempered canisters, conical setters with splayed ends, and thick disc-shaped lumps that may have been used as kiln setters were also recovered. The surfaces of many of the canisters and setters are partially vitrified, indicating that they were used in the firing of faience and steatite objects. A careful analysis of the canisters and vitrified setters along with experimental reconstructions suggests that these were used as a mini-kiln that reached sufficiently high temperatures (ca. 940°C) to transform steatite and glaze faience. Experimental reconstructions of stoneware bangle production have shown that canister kilns can in fact achieve such high temperatures (Kenoyer 1994a, 1997).

Kenoyer suggests that this workshop was highly restricted and probably directly controlled by elites, as the debris from faience object production was not dumped in the streets or nearby houses but, rather, was concentrated in a single room, suggesting that even the debris from production was carefully controlled. Kenoyer (1998a) has also suggested that the faience tablets may have been used as a form of credit or communication, for economic and possibly even for ritual purposes. It is clear that the workshop produced tablets that were then redistributed throughout the site, for some of the tablets made in this workshop are identical to ones found in other areas of the city, such as Mounds D and F. With this excavation, the period of such interactions can now be dated to Harappa Period 3B (ca. 2450–2200 B.C.).

Along with the inscribed faience tablets, the craftspeople produced faience and steatite beads. The discovery in the erosion gully ad-

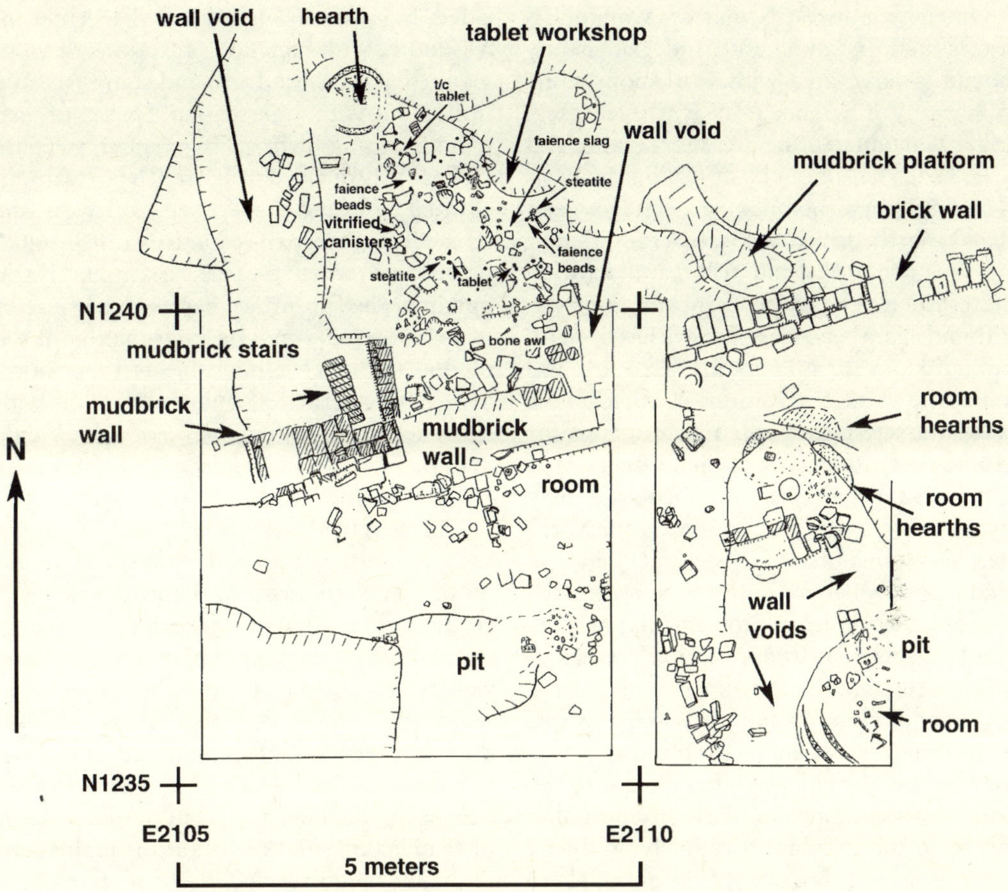

Figure 7.6. Harappa Phase (Period 3B) inscribed steatite and faience tablet workshop, Mound E, Trench 54 South Extension, Harappa, 2001 season.

jacent to the workshop of a tiny faience bull figurine leg made of two different colors of faience suggests that they may have also been involved in the production of other types of objects for elite consumption or trade. Several of the faience tablets found in the workshop are made from the same two colors of faience and further support the idea that the bull figurine leg was made in the same workshop as the tablets, possibly even by the same craftsperson. Because the firing area was probably quite smoky and dusty, the actual preparation of the faience beads and tablets may have taken place in one of the nearby rooms. A grinding stone and pestle found in a room to the east may have been used for the preparation of faience paste or for grinding the edges of steatite tablets, but firing waste or other manufacturing debris was found in surrounding rooms or nearby structures. The lower levels of the workshop and adjacent structures remain to be excavated and will undoubtedly provide invaluable information on the origin and development of this production facility.

Harappa: Period 3C Craft Technologies and Associations

As noted above, most of our evidence for craft activities comes from the Harappan Period 3C (2600–1900 B.C.) occupations. The most important excavated craft areas will be presented here to illustrate some finer-scale patterns of craft associations. General large-scale patterns, and exceptions to these large-scale patterns provided by excavated areas, are discussed in more detail by Miller (2007).

On the southwest corner of Mound ET, near a main gateway into the walled city mound, a large area with workshop debris from many different crafts was excavated. Primary manufacturing areas for steatite and agate beads have been identified on the basis of nodule fragments, flaking debitage, bead blanks, partly drilled beads, several different types of spent drills, and drill manufacturing waste. Microdebitage from all stages of steatite and agate bead manufacture has been recovered from some floors inside rooms and courtyard areas. Excavations in adjacent areas of the site have yielded no evidence for steatite or agate bead working. Other crafts in this area include shell ornament and inlay manufacture, as well as bone and ivory working, chert weight manufacture, and chert tool production, which may have been associated with woodworking or furniture inlay setting. All of this debris is from reductive crafts and appears to have come from manufacturing locations in close enough proximity to one other that their manufacturing waste was combined in the streets and washed into the major drains leading out of the city, although they were not necessarily originally produced in the same workshops or activity areas. There is also a small amount of evidence for faience ornament production and gold working, both of which are transformative crafts but which also produce products of high value.

The overall pattern in this part of the site indicates that craft production took place in a small area inside the gateway, as is seen in the craft bazaars of traditional cities of the subcontinent today. It is noteworthy that most of the activities represented produced objects of apparently high value. Although the location of these crafts in a restricted area inside the gateway cannot be interpreted as direct control by urban elites, it does indicate a degree of indirect control through the regulation of movement into and out of the craft area. The presence of chert weights and weight manufacturing debris in the area just inside the gateway may indicate the taxation of goods coming into or leaving this walled sector of the city (Kenoyer 1997).

A second lithic production area was located by surface surveys in the center of Mound E, with high concentrations of vesuvianite flakes, chert drills, and some ernestite drills along with some scattered pieces of shell manufacturing debris. Two small trenches (collectively referred to as Trench 36) were excavated to locate the workshop. This is one of the few extractive-reductive craft production areas in which the surface scatters clearly mirrored the subsurface deposits; because of its location in a small basin at the top of the mound, the area was not subject to deposition and only experienced minimal erosion with some deflation (Miller 1999, 2000; cf. Vidale and Balista 1990; Vidale and Lechevallier 1994). Actual working areas for the manufacture of chert blades and drills as well as flaking areas for vesuvianite bead blanks were found inside a small room in one of the trenches. Microdebitage analysis of floor levels revealed tiny flakes (<1 mm to 4 mm) of vesuvianite as well as agate, carnelian, and chert. Some tiny fragments of shell were also recovered from these same levels. Unfortunately, the presence of historical burials made it impossible to fully expose the working areas, but the association of bead making and shell working in this area of the site is consistent with the pattern of associated lithic and shell manufacturing noted above for Mound ET and elsewhere at the site (Figure 7.5).

A third excavation area with unique evidence for the association of multiple crafts within one room is on the west side of Mound F (Trench 43). Although all of the larger buildings had been robbed of their bricks, some brick walls that had apparently collapsed shortly after this part of the site was abandoned completely sealed and preserved the contents of several rooms. The floors of the rooms were littered with manufacturing waste from a range of very different crafts. In addition to the presence of chert manufacturing debris, which is found over much of the site, room 202 contained a partially worked shell used for making shell inlay and other ornaments, a flat stone piece used as a saw, several hammer stones, angular pestles, and a perforated ring stone. The finished objects include 17 terra-cotta toy tops, 18 faience beads, 22 steatite disc beads, and 11 stone beads. Three lumps of grayish blue clay that appears to have been unfired faience paste were recovered from one area of the room, although testing has not yet been concluded. Three important inscribed objects were also found on the floor mixed in with all of the other types of artifacts. These include one inscribed and fired steatite fragment with red pigment that may have been a test piece, a faience swastika seal, and a broken and worn, but unfired, rectangular steatite seal. The discovery of all of these objects together in one room and on one sealed floor indicates that people associated with one house were undertaking several different types of activities. These activities include shell working and crafts that would have required the use of stone pestles and grinding stones, possibly faience manufacture. Although some of the ground-stone objects also may have been used for domestic activities and food processing, the ground-stone piece used as a saw is clearly not a tool for food processing. Microdebitage from soil samples is being analyzed from this and the other rooms in order to obtain further substantiation of the types of materials worked in this part of the site.

Finally, the results of earlier as well as recent excavations and surveys on Mound F have revealed a large number of kilns, most probably for pottery firing and dating to Period 3C, although we cannot be sure for the earlier excavations (Miller 1997, 1999, 2000; Vats 1940). Some of these kilns from the early excavations were apparently found in streets or open areas right in the center of a major habitation area. The construction of a kiln in the middle of a street or next to a house would normally be interpreted as the reuse of an area after abandonment. However, we should not assume that this was necessarily the case. This pattern could be attributed to urban congestion and the lack of access to firing areas at the edge of the settlement. During Period 3C Harappa grew to its largest size, and there may have been incredible pressures on space throughout the city. The location of some kilns in Mound F in what may have been less than desirable locations could reflect congestion and some degree of breakdown in civic order. Because of the lack of detailed stratigraphic recording by the earlier excavators, it is not possible to resolve this question, but later excavations by HARP give us no reason to assume that this area represents an abandonment and reoccupation of Mound F.

CONCLUSION

The study of Indus craft traditions and their associations is still a work in progress, and more information is needed from other sites to compare with the preliminary findings from the site of Harappa. Final analyses of microdebitage and detailed quantification of craft indicators from Harappa by Kenoyer may result in some minor modifications to the summaries provided above. Nevertheless, through the study of both single crafts and multiple crafts and their relationships, it has been possible to gain a more balanced perspective on the complex nature of the relationships among Indus craft technologies, particularly in an urban setting.

One of the earliest craft associations seen in the greater Indus region is the separation of pottery production from other crafts, as seen in the most extreme case in the specialized pottery production site of Lal Shah near Mehrgarh during the Regionalization Era. Such specialized pottery production sites are also found during the Integration Era, as found by Mughal's (1997) surveys in Cholistan along the dry Ghaggar-Hakra riverbed. Within the sites of Mehrgarh and Harappa themselves, whether or not pottery production was relatively isolated is not entirely clear for the early Regionalization Era (the Ravi Phase at Harappa). There are no clear pottery kilns yet found at Harappa from the Kot Diji Phase (later Regionalization Era), only small kilns likely used for firing terra-cotta figurines or bangles, as is further discussed below. The large-scale patterns seen for the Harappa Phase (Integration Era) are quite striking, however. Except for Mound F, always an anomaly, the pottery kilns at Harappa during Period 3 are all on the northern halves of mounds, always on the edges of the mound, and no other craft activities are associated with them (Figure 7.5). This northern location is in contrast to the pattern for almost all other crafts, as is discussed for metalworking below. Likely reasons for this unusual large-scale pattern for pottery firing are assessed by Miller (2007), with the most likely being the functional explanation of avoiding wind during firing, as uncontrolled flare-ups of the fire caused by wind gusts are a disaster for potters. The location of pottery firing on the edges of the habitation areas also makes sense in terms of the requirements of production, as sufficient space would be needed to spread out pottery for drying (perhaps perched on the city walls themselves?), and it would make it easier to dump wasters off the edge of the settlement, avoiding the extra work of removing debris. In addition, locations on the edges of the mounds might help to avoid sending excessive smoke and cinders into habitation areas.

Interestingly enough, there are hints of some differences in production organization between pottery and small terra-cotta objects during the Kot Diji period. There are three small kilns from two different mounds (north side of AB and northwest corner of E) that were used for terra-cotta bangle or figurine production but not for pottery, indicating that the production of pottery was already separated from the production of small terra-cotta objects, at least for the firing stage. These small kilns are associated with domestic structures and so perhaps represent continued household-level production of small terra-cotta objects, while pottery was produced by specialists. In the Harappa Phase, however, the one small kiln likely used for terra-cotta object production is found next to the large pottery kiln on the northwest corner of Mound E, so it may be that the specialist potters had now taken over terra-cotta small object production as well. These two types of products were still fired separately, though, as is borne out by the apparent use of the nearby kiln on the western side of Mound E for pottery alone. This is somewhat surprising, for all of these objects were oxidation fired; by the Harappa Phase, only the highly specialized stoneware bangles were reduction fired. In the earlier Kot Diji period some bangles were reduced (fired black), whereas pottery was oxidized (fired red), so that different kinds of kilns would have been required; note that two of the three Kot Diji Phase small kilns were definitely used for reduction firing. Perhaps the continued separate firing was part of this legacy, or perhaps the two sets of objects were still produced by separate groups, who shared the firing areas. In either case, the continued use of the northwestern corner of Mound E for terra-cotta object production in both the Kot Diji and Harappan phases is evidence for some very long-term continuities in the uses of production space at Harappa.

Pottery was not the only segregated craft. In fact, during the Harappa Phase (Period 3), most transformative crafts were segregated, not only from reductive crafts but also from each other (Miller 2000, 2007). Different transformative crafts may have been segregated for different reasons. For instance, we have hardly mentioned copper production in this chapter, in part because the vast majority of the evidence for copper melting at the site of Harappa is found during Period 3 on the south side of Mound E; it is the only craft so localized in distribution (Figure 7.5). There are a few fragments of copper-melting debris on the southern side of Mound AB, although it is not clear if these are from furnaces in these areas or debris transported from the area on the south side of Mound E. The actual activity areas on Mound E were all destroyed during nineteenth-century A.D. brick robbing, but it is clear from the scale of the spread of debris across the southern edge of Mound E that multiple activity areas or workshops were engaged in copper melting. Such a strong aggregation of all the copper working in one region of the site sounds like elite or civic control of this craft. Elite control of metal production is a classic pattern of wealth monopolization. However, excavations in the southern area turned up very little evidence for control in the form of seals, weights, or other possible bureaucratic devices in the deposits containing the copper-melting debris (Miller 2005). It is possible, even likely, that such a noxious production stage was simply restricted to this region by some civic authority because of the smoke and fumes. If this is the case, then it is surprising that pottery firing, equally if not more smoky, is found in many different locations. One explanation is functional; if copper-melting furnaces used wind draft to raise their temperatures, then locations perched along the high southern side of Mound E (and perhaps even Mound AB) might have taken advantage of the wind as well as the updraft from the slope of the mound. Another explanation is symbolic; most of the production areas at Harappa are on the southern half of each of the mounds, with the striking exception of pottery production, and this could relate to Harappan civic patterns of order (Miller 2007).

Another very early spatial pattern is the separation of glazed and unglazed steatite bead production from the production of other types of stone beads, especially agates, even when both were produced in the same general region of the site. A particularly clear example of this is already evident during the early Regionalization Era (Ravi Phase) at Harappa, in the specialized activity areas for these crafts on the north side of Mound AB. In the Kot Diji period, even in mixed street deposits in this same area, only steatite bead–making debris is found, not hard stone beads. In the Period 3A levels on the western edges of Mounds AB and E, steatite bead production is again found alone, but in Period 3B levels on the western edge of E, steatite bead production is found in association with both faience and steatite objects of various kinds, including inscribed tablets. In the Period 3C levels, the large area on the southwest corner of Mound ET containing debris from hard stone bead production, steatite production, other lithic crafts, and shell working is too mixed for us to sort out specific activity areas. The less disturbed area at the center of Mound E, however, contains only hard stone bead production debris and some shell working, as noted above.

We would like to know much more about the practice of bone working, leatherworking, and fiber or textile production, as these are glimpsed early on, during the Regionalization Era at Mehrgarh and Harappa (Ravi Phase), but then become invisible in later periods. Perhaps bone working and leatherworking were moved outside of the walled city because of their strong smell. Textile production at Harappa is very mysterious; spindle whorls essentially disappear after the Ravi Phase, and even if some have been mistakenly counted as large beads, there are still far fewer than we would expect, especially for an early center of domesticated cotton. Kenoyer has suggested that the disappearance of the spindle whorls corresponds to the appearance of the spinning wheel, made completely of perishable wooden materials. Future fiber studies may be able to support this, but it seems the most parsimonious explanation at this time.

Overall, by the Harappa Phase (Period 3) at Harappa, extractive-reductive crafts, particularly hard stone bead manufacturing and shell working, were generally aggregated in the same regions of the site, although not in the same exact activity area or workshop. This may relate to factors such as the sharing of menial labor, raw materials, and the need to combine many of the different materials into more complex objects. The scheduling of working processes such as the time-consuming grinding, polishing, and drilling of various materials would have made it possible for workers to participate in more than one workshop. No specialized knowledge is required to grind many types of stone beads, and this work could have been done by a shell worker or a part-time or itinerant laborer. Even though the actual workshops may have been distinct, the same workers may have participated in more than one craft, moving from one workshop to the next to make the best of their specific skills. The close proximity of different crafts would have allowed for this type of easy movement. During the terminal phase of the Harappan period, intensified production may also have led to a need for the concentration of many crafts into the crowded space within the city walls. However, these patterns of aggregation seem to parallel patterns seen in earlier periods, as noted above, so that the concentration of crafts because of crowding may have followed along preestablished lines of craft association.

Several exceptions to this general pattern

point out some interesting alternatives or additions to any production-based explanations. The eroded surface debris from Trench 54 on the western side of Mound E and the craft areas inside the gateway on the southwestern corner of Mound ET have combinations of both extractive-reductive and transformative crafts. In both of these cases, the transformative crafts are from the steatite–faience glazing complex, what Miller (2007) refers to as "bridging" crafts, together with some evidence for small-scale gold working. Again, this pattern may reflect an intensification of production and the congestion of craft working areas during Harappa Period 3C. These exceptions suggest that the overall patterns of aggregation and segregation might also be about the value of end products. All three of the areas mentioned above appear to represent production areas for objects of high value, either economically or symbolically.

Alternatively, this pattern might relate to Harappan perceptions of what constituted a group of associated crafts, in addition to the sorts of favorable organization of production arrangements noted above for extractive bead and shell working—or even because of such arrangements. All workers in stone and shell may have been seen as belonging to the same craft, regardless of the material type or the end object, or perhaps all ornament makers were perceived to be engaged in one craft, regardless of the type of raw materials employed (Miller 2007). This suggestion is particularly attractive given the chronological continuity of stone bead and shell production associations through time.

The gradual development and specialization of craft production at sites such as Mehrgarh and Harappa are undoubtedly linked to more fundamental social processes of crafting power (Helms 1993) that were going on during the initial phase of urban development. The creation of powerful symbols was something that could be done through special technologies or by using specific raw materials that were not easily accessible for most people (Kenoyer 2000). Although the knowledge of specific craft technologies was probably passed on from one generation to the next through kin networks and various forms of ritual practice (Kenoyer 1989), the access to specific materials could have been carefully regulated by controlling trade. At both Harappa and Nausharo, the building of massive mud-brick walls around the settlements would have been the most effective way to control access to raw materials. The walls and gateways would also have allowed for control of the export trade in finished commodities. Although there is little evidence for direct control of most crafts at Harappa, all of the crafts that were taking place inside the city wall could thus have been indirectly controlled by the city elites through taxation. Taxation at the city gates would have been the most effective method for gaining benefit for the city. The only crafts whose production may have been closely monitored for the control of trade and wealth appear to be chert weight manufacture, the manufacture of steatite seals, and the production of faience molded tablets or steatite inscribed tablets, based on current hypotheses about the use of these objects (Kenoyer 2000). Future research at Harappa and other settlements needs to focus on refining our understanding of the nature of these possible tools of bureaucratic or elite control.

Acknowledgments

Most of the data presented in this essay derives from recent excavations at Harappa, and we would especially like to thank the Department of Archaeology, Government of Pakistan, for facilitating our continued work at Harappa. Our thanks go to all the colleagues who have participated in the research at Harappa and have helped to collect and analyze data, with special thanks to Dr. Richard Meadow, Dr. Rita Wright, Mrs. Barbara Dales, Dr. Massimo Vidale, Dr. William Belcher, Sharri Clark, Randy Law, Brad Chase, Peter Eltsov, Aasim Dogar, Nadeem Ghouri, Saeed Haderi, and all the excavators at Harappa. Our ongoing research on craft technologies at Harappa and the Indus Valley Civilization in general have been supported by numerous organizations: the National Science Foundation, the National Endowment for the Humanities, the National Geographic Society, the Smithsonian Institution, the American School of Prehistoric Research (Peabody Museum of Archaeology and Ethnology, Harvard University), the University of Wisconsin, the George Dales Foundation, and private donors.

References Cited

Allchin, B.

1973 Blade and Burin Industries of West Pakistan and Western India. In *South Asian Archaeology 1971*, edited by N. Hammond, pp. 39–50. Duckworth, London.

Audouze, F., and C. Jarrige

1979 A Third Millennium Pottery-Firing Structure at Mehrgarh and Its Economic Implications. In *South Asian Archaeology 1977*, edited by M. Taddei, pp. 213–221. Istituto Universitario Orientale, Naples.

Barthélémy de Saizieu, B.

1990 Le Cimetiere Neolithique de Mehrgarh (Balouchistan Pakistanais) Apport de L'analyse Factorielle. *Paléorient* 16(1): 23–43.

2003 *Les Parures de Mehrgarh: Perles et Pendentifs du Néolithique Préceramique à la Période pré-Indus: Fouilles 1974–1985.* Editions Recherche sur les Civilisations, Paris.

Barthélémy de Saizieu, B., and A. Bouquillon

1994 Steatite Working at Mehrgarh during the Neolithic and Chalcolithic Periods: Quantitative Distribution, Characterization of Material and Manufacturing Process. In *South Asian Archaeology, 1993*, edited by A. Parpola and P. Koskikallio, pp. 47–60. Suomalainen Tiedeakatemia, Helsinki.

1997 Evolution of Glazed Materials from the Chalcolithic to the Indus Period Based on the Data of Mehrgarh and Nausharo. In *South Asian Archaeology 1995*, edited by B. Allchin and R. Allchin, pp. 63–76. Oxford and IBH, New Delhi.

Bhan, K. K., and D. Gowda

2003 Shell Working at Nagwada (North Gujarat) with Special Reference to Shell Industries of the Harappan Tradition in Gujarat. *Man and Environment* 18(2): 51–80.

Bhan, K. K., and J. M. Kenoyer

1980–1981 Nageshwar, an Industrial Centre of the Chalcolithic Period. *Puratattva* 12:115–120.

Bhan, K. K., M. Vidale, and J. M. Kenoyer

1994 Harappan Technology: Methodological and Theoretical Issues. *Man and Environment* 19(1–2):141–157.

2002 Some Important Aspects of the Harappan Technological Tradition. In *Indian Archaeology in Retrospect, vol. 2: Protohistory: Archaeology of the Harappan Civilization*, edited by S. Settar and R. Korisettar, pp. 223–272. Indian Council of Historical Research, New Delhi.

Clark, J. D.

1990 Why Change? An Example of the Technology from India's Enduring Past. *Bulletin of the Deccan College* 49:83–98.

Clark, S.

2003 Representing the Indus Body: Sex, Gender, Sexuality, and the Anthropomorphic Terracotta Figurines from Harappa. *Asian Perspectives* 42(2):304–328.

Dales, G. F.

1989 Harappa: A New Look at the Type Site of the Indus Civilization. In *Old Problems and New Perspectives in the Archaeology of South Asia*, edited by J. M. Kenoyer, pp. 127–132. Wisconsin Archaeological Reports, Madison.

1991 Some Specialized Ceramic Studies at Harappa. In *Harappa Excavations 1986–1990*, edited by R. H. Meadow, pp. 61–70. Prehistory Press, Madison, Wisconsin.

Dani, A. H.

1970–1971 Excavations in the Gomal Valley. *Ancient Pakistan* 5:1–177.

Dhavalikar, M. K., M. R. Raval, and Y. M. Chitalwala

1996 *Kuntasi: A Harappan Emporium on West Coast.* Deccan College Post-Graduate Research Institute, Pune.

Durrani, F. A.

1988 Excavations in the Gomal Valley: Rehman Dheri Excavation Report No. 1. *Ancient Pakistan* 6:1–232.

Durrani, F. A., I. Ali, and G. Erdosy

1991 Further Excavations at Rehmandheri, 1991. *Ancient Pakistan* 7:61–151.

1995a The Beads of Rehman Dheri. *Ancient Pakistan* 10:15–81.

1995b Miscellaneous Terracotta Objects. *Ancient Pakistan* 10:164–197.

Fairservis, W. A. J.

1982 Allahdino: An Excavation of a Small Harappan Site. In *Harappan Civilization: A Contemporary Perspective*, edited by G. L. Possehl, pp. 107–112. Oxford and IBH Publishing, New Delhi.

Flam, L.

1993a Excavations at Ghazi Shah 1985–1987: An Interim Report. *Pakistan Archaeology* 28:131–158.

1993b Excavation at Ghazi Shah, Sindh, Pakistan. In *Harappan Civilization: A Recent Perspective*, edited by G. L. Possehl, pp. 457–467. Oxford and IBH Co. Pvt. Ltd., New Delhi.

Hegde, K. T. M., V. H. Sonawane, D. R. Shah, K. K. Bhan, P. Ajitprasad, K. Krishnan, and S. P. Chardnan

1988 Excavation at Nagwada—1986 and 1987: A Preliminary Report. *Man and Environment* 12:55–65.

Helms, M. W.

1993 *Craft and the Kingly Ideal: Art, Trade, and Power*. University of Texas Press, Austin.

Hideaki, S.

1996 Rehman Dheri and Kot Diji: Gomal Plain Culture of the Pre-Harappan Period. *Indo-Koko-Kenkyu: Indian Archaeological Studies* 18:112.

Jarrige, C.

1997 The Figurines from Nausharo Period I and Their Further Developments. In *South Asian Archaeology 1995*, edited by B. Allchin and R. Allchin, pp. 33–43. Oxford and IBH, New Delhi.

Jarrige, C., J.-F. Jarrige, R. H. Meadow, and G. Quivron (editors)

1995 *Mehrgarh Field Reports 1975 to 1985—From the Neolithic to the Indus Civilization*. Department of Culture and Tourism, Government of Sindh, and the French Foreign Ministry, Karachi.

Jarrige, J.-F.

1989 Excavation at Mehrgarh-Nausharo 1987–88. *Pakistan Archaeology* 24:21–67.

1994 The Final Phase of the Indus Occupation at Nausharo and Its Connection with the Following Cultural Complex of Mehrgarh VIII. In *South Asian Archaeology, 1993*, edited by A. Parpola and P. Koskikallio, pp. 295–314. Suomalainen Tiedeakatemia, Helsinki.

1997 From Nausharo to Pirak: Continuity and Change in the Kachi/Bolan Region from the 3rd to the 2nd Millennium B.C. In *South Asian Archaeology 1995*, edited by B. Allchin, pp. 11–32. Oxford and IBH, Bombay.

Kenoyer, J. M.

1986 The Indus Bead Industry: Contributions to Bead Technology. *Ornament* 10(1): 18–23.

1989 Socio-Economic Structures of the Indus Civilization as Reflected in Specialized Crafts and the Question of Ritual Segregation. In *Old Problems and New Perspectives in the Archaeology of South Asia*, edited by J. M. Kenoyer, pp. 183–192. Department of Anthropology, University of Wisconsin, Madison.

1991 The Indus Valley Tradition of Pakistan and Western India. *Journal of World Prehistory* 5(4):331–385.

1992 Harappan Craft Specialization and the Question of Urban Segregation and Stratification. *Eastern Anthropologist* 45(1–2):39–54.

1994a Experimental Studies of Indus Valley Technology at Harappa. In *South Asian Archaeology, 1993*, edited by A. Parpola and P. Koskikallio, pp. 245–262. Suomalainen Tiedeakatemia, Helsinki.

1994b Faience from the Indus Valley Civilization. *Ornament* 17(3):36–39, 95.

1995a Interaction Systems, Specialized Crafts and Culture Change: The Indus Valley Tradition and the Indo-Gangetic Tradition in South Asia. In *The Indo-Aryans of Ancient South Asia: Language, Material Culture and Ethnicity*, edited by G. Erdosy, pp. 213–257. W. DeGruyter, Berlin.

1995b Shell Trade and Shell Working during the Neolithic and Early Chalcolithic at Mehrgarh. In *Mehrgarh Field Reports 1975 to 1985—From the Neolithic to the Indus Civilization*, edited by C. Jarrige, J.-F. Jarrige, R. H. Meadow, and G. Quivron, pp. 566–581. Department of Culture and Tourism, Government of Sindh, and the French Foreign Ministry, Karachi.

1997 Trade and Technology of the Indus Valley: New Insights from Harappa, Pakistan. *World Archaeology* 29(2):262–280.

1998a *Ancient Cities of the Indus Valley Civilization*. Oxford University Press, Karachi.

1998b Craft Traditions of the Indus Civilization and Their Legacy in Modern Pakistan. *Lahore Museum Bulletin* 9(2[1996]): 1–8.

2000 Wealth and Socio-Economic Hierarchies of the Indus Valley Civilization. In

Order, Legitimacy and Wealth in Early States, edited by J. Richards and M. Van Buren, pp. 90–112. Cambridge University Press, Cambridge.

2003 Uncovering the Keys to the Lost Indus Cities. *Scientific American*, July: 67–75.

2004 Die Karren der Induskultur Pakistans und Indiens (Wheeled Vehicles of the Indus Valley Civilization of Pakistan and India). In *Rad und Wagen—Der Ursprung einer Innovation: Wagen im Vorderen Orient und Europa (Wheel and Wagon—The Origin of an Innovation: Wagon in the Near East and Europe)*, edited by M. Fansa and S. Burmeister, pp. 87–106. Verlagg Philipp von Zabern, Mainz.

2005 Steatite and Faience Manufacturing at Harappa: New Evidence from Mound E Excavations 2000–2001. *Museum Journal (National Museum of Pakistan)* 3–4 (January–December 2002):43–56.

Kenoyer, J. M., D. Mandal, V. D. Misra, and J. N. Pal

1983 Preliminary Report on Excavations at the Late Palaeolithic Occupation Site at Baghor I Locality. In *Palaeoenvironments and Prehistory in the Middle Son Valley*, edited by G. R. Sharma and J. D. Clark, pp. 117–142. Abhinash Prakashan, Allahabad.

Kenoyer, J. M., and R. H. Meadow

1997 New Inscribed Objects from Harappa. *Lahore Museum Bulletin* 9(1[1996]):1–20.

1999 Harappa: New Discoveries on Its Origins and Growth. *Lahore Museum Bulletin* 12(1):1–12.

2000 The Ravi Phase: A New Cultural Manifestation at Harappa. In *South Asian Archaeology 1997*, edited by M. Taddei and G. De Marco, pp. 55–76. Istituto Italiano per l'Africa e l'Oriente, Rome; and Istituto Universitario Orientale, Naples.

Kenoyer, J. M., and M. Vidale

1992 A New Look at Stone Drills of the Indus Valley Tradition. In *Materials Issues in Art and Archaeology, III*, edited by P. Vandiver, J. R. Druzick, G. S. Wheeler, and I. Freestone, pp. 495–518. Materials Research Society, Pittsburgh.

Kenoyer, J. M., M. Vidale, and K. K. Bhan

1991 Contemporary Stone Bead Making in Khambhat India: Patterns of Craft Specialization and Organization of Production as Reflected in the Archaeological Record. *World Archaeology* 23(1):44–63.

Lal, B. B.

1979 Kalibangan and the Indus Civilization. In *Essays in Indian Protohistory*, edited by D. P. Agrawal and D. K. Chakrabarti, pp. 65–97. B. R. Pub. Corp., Delhi.

Lal, B. B., B. K. Thapar, S. P. Joshi, and M. Bala

2003 *Excavations at Kalibangan: The Early Harappans (1960–1969)*. Memoirs of the Archaeological Survey of India, No. 98. New Delhi.

Law, R.

2002 Potential Steatite Source Areas of the Indus Valley Civilization. In *Proceedings of the International Colloquium on Indus Civilization, April 6th–8th, 2001*, edited by M. A. Halim, pp. 158–169. UNESCO, Islamabad.

2006 Regional Interaction in the Prehistoric Indus Valley: Initial Results of Rock and Mineral Sourcing Studies at Harappa. In *South Asian Archaeology 2001*, edited by C. Jarrige and V. Lefèvre, pp. 179–190. Éditions Recherche sur les Civilisations, Paris.

Lechevallier, M.

1984 The Flint Industry of Mehrgarh. In *South Asian Archaeology 1981*, edited by B. Allchin, pp. 41–51. Cambridge University Press, Cambridge.

2003 *L'Industrie Lithique de Mehrgarh: Fouilles 1974–1985*. Editions Recherche sur les Civilisations, Paris.

Mackay, E. J. H.

1938 *Further Excavations at Mohenjodaro*. Government of India, New Delhi.

1943 *Chanhu-Daro Excavations 1935–36*. American Oriental Society, New Haven.

Marcon, V., and M. Lechevallier

2000 Lithic Industries of the Indo-Iranian Borderlands: Technological Approach of Blade Debitage in the Assemblages of Mehrgarh, Nausharo and Miri Qalat in Balochistan, Pakistan. In *South Asian Archaeology 1997*, edited by M. Taddei and G. De Marco, pp. 215–235. Istituto Italiano per l'Africa e l'Oriente, Rome.

Marshall, S. J.

1931 *Mohenjo-daro and the Indus Civilization*. A. Probsthain, London.

Meadow, R. H.

1996 The Origins and Spread of Agriculture and Pastoralism in South Asia. In *The*

Origins and Spread of Agriculture and Pastoralism in Eurasia, edited by D. R. Harris, pp. 390–412. Smithsonian Institution Press, Washington, D.C.

Meadow, R. H., and J. M. Kenoyer

1997 Excavations at Harappa 1994–1995: New Perspectives on the Indus Script, Craft Activities and City Organization. In *South Asian Archaeology 1995*, edited by B. Allchin and R. Allchin, pp. 139–172. Oxford and IBH, New Delhi.

2001 Recent Discoveries and Highlights from Excavations at Harappa: 1998–2000. *Indo-Koko-Kenkyu: Indian Archaeological Studies* 22:19–36.

Meadow, R. H., J. M. Kenoyer, and R. P. Wright

2001 Harappa Archaeological Research Project: Harappa Excavations 2000–2001. Report submitted to the Director General of Archaeology and Museums, Government of Pakistan, Karachi.

Miller, H. M.-L.

1997 Pottery Firing Structures (Kilns) of the Indus Civilization during the Third Millennium B.C. In *Prehistory and History of Ceramic Kilns*, edited by P. Rice and W. D. Kingery, pp. 41–71. American Ceramic Society, Columbus, Ohio.

1999 *Pyrotechnology and Society in the Cities of the Indus Valley.* Ph.D. dissertation, University of Wisconsin–Madison. University Microfilms, Ann Arbor.

2000 Reassessing the Urban Structure of Harappa: Evidence from Craft Production Distribution. In *South Asian Archaeology 1997*, edited by M. Taddei and G. De Marco, pp. 77–100. Istituto Italiano per l'Africa e l'Oriente, Rome.

2005 Investigating Copper Production Technology at Harappa: Surveys, Excavations, and Finds. In *South Asian Archaeology 2001, vol. 1: Prehistory*, edited by C. Jarrige and V. Lefèvre, pp. 245–252. Éditions Recherche sur les Civilisations, Paris.

2007 Associations and Ideologies in the Locations of Urban Craft Production at Harappa, Pakistan (Indus Civilization). In *Rethinking Specialization in Complex Societies: Archaeological Analysis of the Social Meaning of Production*, edited by Z. X. Hruby and R. Flad, pp. 37–51. Archaeological Papers of the American Anthropological Association, Vol. 15. University of California Press and the American Anthropological Association, Berkeley.

In press The Indus Talc–Faience Complex: Types of Material, Clues to Production. In *South Asian Archaeology 1999*, edited by K. R. van Kooij, E. M. Raven, and G. L. Possehl. International Institute of Asian Studies, Leiden.

Mughal, M. R.

1997 *Ancient Cholistan: Archaeology and Architecture.* Ferozsons, Lahore.

Murty, M. L. K.

1979 Recent Research on the Upper Paleolithic Phase in India. *Journal of Field Archaeology* 6(3):301–320.

Piperno, M.

1973 Micro-Drilling at Shahr-i-Sokhta; the Making and Use of the Lithic Drill-Heads. In *South Asian Archaeology 1971*, edited by N. Hammond, pp. 119–130. Duckworth, London.

Pracchia, S.

1985 Excavations of a Bronze-Age Ceramic Manufacturing Area at Lal Shah, Mehrgarh. *East and West (New Series)* 35(4): 458–468.

Quivron, G.

1994 The Pottery Sequence from 2700–2400 BC at Nausharo, Baluchistan. In *South Asian Archaeology 1993*, edited by A. Parpola and P. Koskikallio, pp. 629–644. Suomalainen Tiedeakatemia, Helsinki.

1997 Incised and Painted Marks on the Pottery of Mehrgarh and Nausharo-Baluchistan. In *South Asian Archaeology 1995*, edited by B. Allchin and R. Allchin, pp. 45–62. Oxford and IBH, New Delhi.

Rao, S. R.

1979 *Lothal: A Harappan Port Town (1955–62)*, Vol. 1. Archaeological Survey of India, New Delhi.

1985 *Lothal: A Harappan Port Town (1955–62)*, Vol. 2. Archaeological Survey of India, New Delhi.

Samzun, A.

1992 Observations on the Characteristics of the Pre-Harappan Remains, Pottery, and Artifacts at Nausharo, Pakistan (2700–2500 B.C.). In *South Asian Archaeology 1989*, edited by C. Jarrige, pp. 245–252. Prehistory Press, Madison, Wisconsin.

Santoni, M.

1989 Potters and Pottery at Mehrgarh during the Third Millennium B.C. (Periods

VI and VII). In *South Asian Archaeology 1985*, edited by K. Frifelt and P. Sørensen, pp. 176–185. Curzon Press, London.

Shaffer, J. G.

1974 *Allahdino and the Mature Harappan: A Preliminary Report on the Cultural Stratigraphy*. Department of Anthropology, Case Western Reserve University, Cleveland.

1992 The Indus Valley, Baluchistan and Helmand Traditions: Neolithic through Bronze Age. In *Chronologies in Old World Archaeology*, 3rd ed., edited by R. Ehrich, pp. 441–464. University of Chicago Press, Chicago.

Sonawane, V. H., P. Ajitprasad, K. K. Bhan, K. Krishnan, S. Pratapachandran, A. Majumdar, A. K. Patel, and J. Menon

2004 Excavations at Bagasara—1996–2003: A Preliminary Report. *Man and Environment* 28(2[2003]):21–50.

Thapar, B. K.

1975 Kalibangan: A Harappan Metropolis beyond the Indus Valley. *Expedition*, winter: 19–32.

Tosi, M., and M. Vidale

1990 4th Millennium B.C. Lapis Lazuli Working at Mehrgarh, Pakistan. *Paléorient* 16(2):89–101.

Vanzetti, A., and M. Vidale

1994 Formation Processes of Beads: Defining Different Levels of Craft Skill among the Early Beadmakers of Mehrgarh. In *South Asian Archaeology 1993*, edited by A. Parpola and P. Koskikallio, pp. 763–776. Suomalainen Tiedeakatemia, Helsinki.

Vats, Madho Sarup

1940 *Excavations at Harappa: Being an Account of Archaeological Excavations at Harappa Carried Out between the Years 1920–21 and 1933–34*. Government of India Press, Delhi.

Vidale, M.

1989 Early Harappan Steatite, Faience and Paste Beads in a Necklace from Mehrgarh-Nausharo (Pakistan). *East and West* 39(1–4):291–300.

1992 *Produzione Artigianale Protostorica: Etnoarcheologia e Archeologia*. Universita degli Studi di Padova, Padua.

1995 Early Beadmakers of the Indus Tradition: The Manufacturing Sequence of Talc Beads at Mehrgarh in the 5th Millennium B.C. *East and West* 45(1–4):45–80.

Vidale, M., and C. Balista

1990 A Geoarchaeological Approach to the Formation Processes of Craft Activity Areas of Moenjodaro. *Prospezioni Archeologiche Quaderni* 1:51–62.

Vidale, M., and M. Lechevallier

1994 Example of Recent Transformation Processes of a Chalcolithic "Activity" Area at Mehrgarh (Pakistan). In *South Asian Archaeology 1993*, edited by A. Parpola and P. Koskikallio, pp. 789–802. Suomalainen Tiedeakatemia, Helsinki.

Vidale, M., and H. Miller

2000 On the Development of Indus Technical Virtuosity and Its Relation to Social Structure. In *South Asian Archaeology 1997*, edited by M. Taddei and G. De Marco, pp. 115–132. Istituto Italiano per l'Africa e l'Oriente, Rome.

Vidale, M., and G. M. Shar

1991 Zahr Muhra: Soapstone-Cutting in Contemporary Baluchistan and Sind. *Annali (Estratto da Annali dell'Istituto Universitaro Orientale, Napoli)* 50(1):61–78.

Wright, R. P.

1991 Patterns of Technology and the Organization of Production at Harappa. In *Harappa Excavations 1986–1990*, edited by R. H. Meadow, pp. 71–88. Prehistory Press, Madison, Wisconsin.

1995 Fine Ware Traditions at Mehrgarh. In *Mehrgarh Field Reports 1975 to 1985—From the Neolithic to the Indus Civilization*, edited by C. Jarrige, J.-F. Jarrige, R. H. Meadow, and G. Quivron, pp. 662–671. Department of Culture and Tourism, Government of Sindh, and the French Foreign Ministry, Karachi.

8

Co-Craft and Multicraft: Section-Mold Casting and the Organization of Craft Production at the Shang Capital of Anyang

YUNG-TI LI

Bronze metallurgy in ancient China followed a very different path from that of its Western counterparts. When metal production appeared in the Central Plains region during the first half of the second millennium B.C., bronze vessels, rather than implements or sculptures, were being cast in large-scale operations with a sophisticated use of clay models and molds (Bagley 1999; Chen 2006; Franklin 1983b; Liu and Chen 2003).[1] Research on both the finished products and the manufacturing remains over the past few decades has clearly indicated that, even though the end product of ancient Chinese bronze production was metal, the operation was also based on the use and extensive knowledge of clay.

It is the interplay between the use of metal and clay that led Ursula Franklin (1983a) to propose the term *prescriptive technology* for the production organization of the Shang bronze foundry. Franklin argues that the different expertise required of the clay workers and the metal casters meant that the operation of a Shang bronze foundry would have been carefully divided into different sections, each designed for workers with different skills. To Franklin, the Shang bronze foundry foreshadowed the assembly-line production in a modern factory. Bronze production in ancient China, therefore, was not merely the production of metal. It was a co-craft based on the mastery of two very different materials, metal and clay. *Co-craft* is defined here as a craft-producing activity that depends on the close collaboration of craftspersons specializing in two or more different crafts in order to produce the final products.

Excavations in the past seven decades have uncovered several craft production sites across the Anyang region, the last capital of the Shang dynasty, including bronze foundries and bone and lithic workshops. Although current data do not permit a full understanding of craft-producing activities at the Shang capital, they do suggest that there were several craft production precincts located throughout the Anyang region, each with multiple workshops specializing in different crafts.

This chapter examines two aspects of craft production at Anyang: the co-crafting nature of the bronze industry and the presence of multicraft production at the Shang capital. It first reviews the history of research on the section-mold casting technology of Bronze Age China and summarizes the current reconstruction for the manufacturing sequences of ancient Chinese bronze casting. The summary is intended to demonstrate how clay was employed in the production of bronzes and how a significant part of the Shang bronze workshop must have been devoted to clay working. Franklin's model of holistic and prescrip-

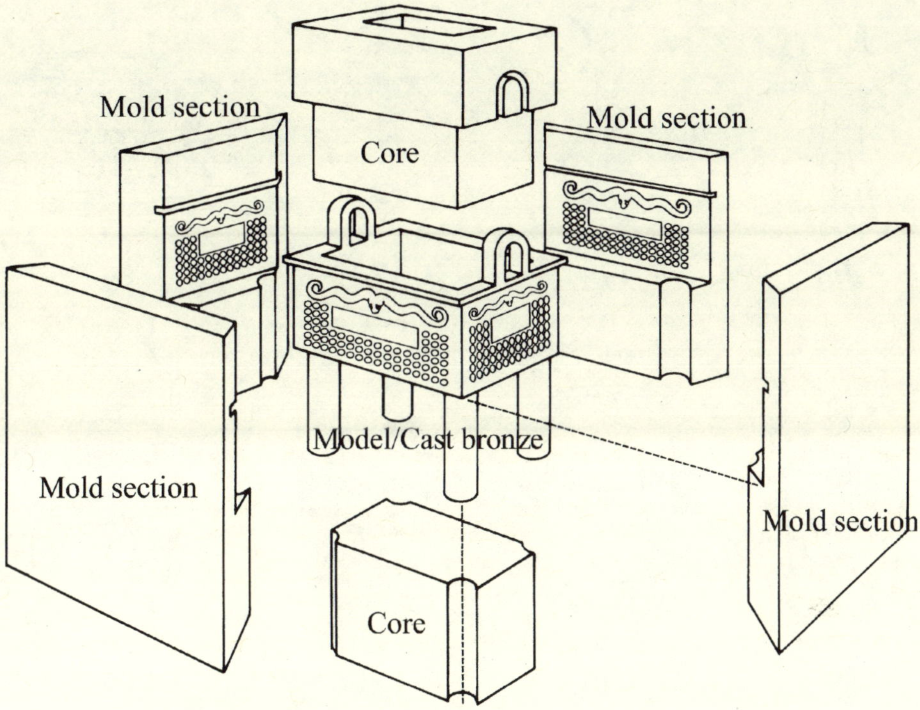

FIGURE 8.1. Parts of a mold assemblage for a *fangding* vessel. After Chase 1991:24, fig. 1.

tive production is then briefly examined and modified by using parameters I propose for identifying the two forms of production in the archaeological records. These parameters are then used to examine the production organization of the Anyang bronze foundries. The final section of the chapter examines the archaeological evidence from the Anyang craft production precincts and attempts to look at multicraft production in the sociopolitical context of the Shang capital. Preliminary interpretations for the control over and the patronage of craft production during the Late Shang period are also discussed.

SECTION-MOLD CASTING IN BRONZE AGE CHINA

The intricate design and shape of ancient Chinese bronzes led early Western scholars to argue that Shang bronzes were cast using the same "cire perdue" method known for Greek and Roman bronze sculptures (e.g., Aitchison 1960; Maryon and Plenderleith 1954; Yetts 1929; see Gettens 1969:21–23 for a summary of such early views). The fact that ancient Chinese metalworkers used section-mold rather than lost-wax casting was only firmly established in the 1960s, a surprisingly late date considering how widely and readily this view is accepted today (Bagley 1990:7). The following section summarizes the history of research on section-mold casting in two areas: (1) reconstruction of the technology through foundry remains and finished bronzes and (2) replication experiments. Figures 8.1 and 8.2 introduce some of the basic terminologies and vessel types discussed in this chapter. Readers are directed to the existing literature, such as work by Bagley (1987, 1990) and Chase (1983, 1991), for a more comprehensive overview of the section-mold technology.

History of Research on Section-Mold Casting

Karlbeck's 1935 study of Anyang mold fragments was the first to argue for a section-mold casting technique with regard to bronze making at Anyang. Karlbeck obtained a number

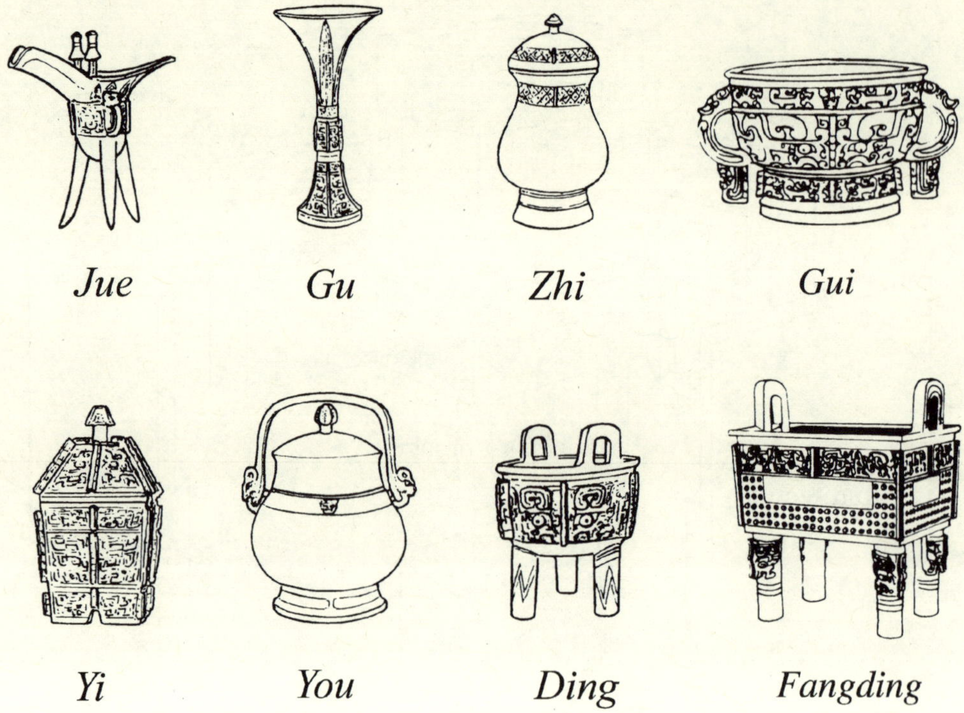

FIGURE 8.2. Common bronze types of the Late Shang and Early Western Zhou periods. Vessels are not drawn to the same scale. Adapted from Fong 1980: f, fig. 1 (*gui* and *ding*); and Loewe and Shaughnessy 1999:xxii–xxiii.

of decorated clay fragments from dealers that were said to have been found at Xiaotun and its vicinity in the early twentieth century. Karlbeck noticed that, upon close inspection, the inner side of the clay pieces showed signs of having been scorched, whereas the backside was often left unfinished. He was also able to identify the types of bronze vessels represented by these clay fragments. Karlbeck reached the conclusion that these clay pieces were in fact clay molds for casting bronzes and the scorched inside surface meant that they came into direct contact with molten metal. Karlbeck's groundbreaking and thorough study of Anyang molds provided valuable insights into the mold-making methods, although the importance of his work was not fully acknowledged in the West until further studies demonstrated that Shang bronzes were indeed cast by using section molds (Bagley 1990).

Chinese archaeologists, especially the original excavators of Anyang, soon recognized the importance of Karlbeck's discovery. Thirteen consecutive seasons of excavations by the Institute of History and Philology, Academia Sinica (hereafter IHP), in the 1930s revealed the first bronze foundry site to be found at Anyang, Xiaotun Northeast (Kuo 1933; Li 1930:240–241; Shih 1947, 1955, 1959:329–332). Shih Chang-ju (1955), one of the excavators of Xiaotun Northeast, cites Karlbeck's work and identifies fragments of clay molds, and especially clay models, for various vessels, weapons, and chariot fittings in his pioneering study of Anyang foundry remains and section-mold casting. Shih also attempts to reconstruct section-mold assemblies for the major known bronze types, although, in hindsight, the reconstructions are hypothetical rather than factual.

Although some Western scholars still wrote about the lost-wax casting of early Chinese bronzes in the 1960s (e.g., Aitchison 1960), researchers paying close attention to archaeological finds in the newly founded People's Republic of China had noticed the clay mold and model fragments unearthed from ancient foundries of Shang and Zhou dates at sites such as Zhengzhou, Anyang, and Houma (Barnard 1961; Fairbank 1972; Gettens 1969). With the new discoveries of bronze foundry remains and the preliminary publications that followed (e.g., Liao 1957; Zhang 1962), it became clear that ancient Chinese bronzes were cast differently from bronzes of the Near East.

Barnard's (1961) early effort at reconstructing the section-mold casting technique helped to win acceptance among Western scholars of the view that ancient Chinese bronzes were cast without the use of a wax model (Fairbank 1972; Gettens 1969; Smith 1981:260). Barnard systematically discusses how clay models and molds could have been used by the Shang casters, basing his arguments on observations of excavated foundry remains, mostly on close inspections of the mold marks on bronzes. He (1961: app. A) provides extensive illustrations of reconstructed mold assemblies for various vessels from the Shang to the Eastern Zhou periods and demonstrates in detail the manufacturing procedures he reconstructed.

Barnard's study, by representing the alternative means through which Shang and Zhou bronzes could have been made, is a detailed counterargument against the lost-wax casting argument for Chinese bronzes. Some of the specific details in Barnard's reconstructions have been refuted by more recent technological studies, especially those carried out by Chinese scholars on newly excavated pieces (Hua et al. 1981). Barnard's reconstruction of mold assemblies with a multitude of mold sections has also been deemed unfeasible from the viewpoint of modern foundry practices (Maryon 1963; Smith 1962–1963). But for the most part, the systematic survey by Barnard laid the groundwork for later scholars to improve on, and his emphasis on horizontal mold division is to some degree supported by recent archaeological finds (Li et al. in press).

An important contribution to the study of Shang and Zhou bronze-casting technology comes from technical studies conducted by Gettens (Bagley 1990). Combining archaeological data, metallurgical studies, and technological observations, Gettens (1967, 1969) has painstakingly examined the Shang and Zhou bronzes in the Freer Gallery of Art to reconstruct their manufacturing techniques. Gettens's original work shows how the ancient Chinese craftspersons were able to surmount the technological constraints of section molds and produce bronzes of intricate shapes and designs by using a number of joining techniques such as precasting, casting on, and soldering. His meticulous investigation and technological expertise prove beyond doubt that the Shang and Zhou bronze makers used section molds to cast bronzes of complicated forms. Gettens's work has since set the standard for technological studies of Chinese bronzes, his effort continued by the equally thorough work of Chinese scholars such as Wan Chia-pao (1972), Hua Jueming (Hua et al. 1981), and Su Rongyu (Su et al. 1988). It is now standard practice to include a section on technological studies of bronzes in Chinese archaeological site reports.

In his study of Eastern Zhou bronzes of the sixth and fifth century B.C., Robert Bagley (1993, 1995, 1996) explores the behavioral and social aspects behind the Chinese bronze-casting technology, especially those related to the foundry of the State of Jin at Houma. Through examination of bronzes of Houma origin and comparison with the model and mold fragments unearthed at the same site, Bagley argues that the Houma craftspersons produced decorated bronzes by replicating decorations with pattern blocks, namely, pieces of decorated and fired clay tablets upon which wet slabs of clay were applied to make negative impressions of the decoration. The plasticity of the negatives and their ability to be trimmed meant that the same decoration could be applied to multiple vessels of very different shapes and sizes. Bagley points out that the pattern-block replication method separated decoration carving, which was probably the most time-consuming task, from the rest of the production process and allowed skillfully executed work by a small number of master carvers to be used by a larger number of less skilled workers. The pattern-block technique therefore indicates a more efficient use of labor at the Houma workshop, which may have led to or allowed further changes in the division of labor and workshop organization aimed at increasing productivity (Bagley 1995). Bagley's study addresses issues beyond the mere reconstruction of ancient technology and has hence revealed aspects of the production organization and underlying sociopolitical systems of bronze production in ancient China, an approach that has yet to be widely adopted in the study of Chinese bronzes.

Replication Experiments

A different but equally important line of approach in the study of Shang bronze casting is replication experiments, which employ the same raw material used by the Shang casters in order to reconstruct ancient foundry practices. Such experiments combine information obtained from archaeological remains and from technical and metallurgical studies of the bronzes themselves. Results thus gained do not necessarily reflect the actual Shang foundry practices but nevertheless provide invaluable insights into the ancient technology.

For scholars of Chinese bronzes, the most immediate question to be answered was if and how clay models and molds were used in the Shang bronze technology. Replication experiments conducted by Wan Chia-pao (1972) of the IHP, Feng Fugen and associates (Feng et al. 1980; Feng et al. 1982) of the Institute of Archaeology, Chinese Academy of Social Sciences (hereafter IA), and Tan Derui (1999) of the Shanghai Museum, for instance, have thus understandably focused mainly on the making of clay models and molds. Of the various replication experiments, two will be discussed briefly here: the one by Wan Chia-pao and the one by Tan Derui.[2]

Early Attempts by the Institute of History and Philology, Academia Sinica. The first systematic and controlled replication experiment was carried out by Wan Chia-pao (1972), a metallurgist by training, at the Academia Sinica in the 1960s. Wan compared the chemical compositions of Anyang clay molds with clay samples taken from brick factories in northern Taiwan. He (1972:20–21) notes that the clay used by the Shang mold makers was not too different from clay used for modern-day bricks. The replication experiment was published with detailed photographs demonstrating the casting of a *jue* (Wan 1972:22–23, pl. 9). Wan also succeeded in replicating bronze *ding*, *jia*, and *gu* with clay models and molds. In his brief account of the replication experiment, Wan (1972:21, see also photographs in pl. 9) describes the following steps in bronze casting: (1) model carving, (2) firing the model, (3) preparing clay slabs for mold sections, (4) applying clay slabs onto the model, (5) trimming mold sections and preparing mortises and tenons, (6) withdrawing mold sections, (7) firing mold sections, (8) scraping down the model to be used as the core, (9) assembling mold sections and the core, (10) casting, (11) retrieving the bronze by breaking open the mold, and (12) retouching and finishing the bronze.

Wan Chia-pao's pioneering replication studies demonstrated for the first time that the ancient casters in China employed a bronze-casting technology that was based on extensive use of clay. His work laid the foundation for later replication experiments, inspiring other scholars to carry out similar projects. Wan's study, however, was mainly to demonstrate the feasibility of casting bronzes with clay models and molds. Although the outcome was successful and convincing, the documentation is concise rather than detailed. For instance, Wan mentions adding certain temper to the clay to prevent uneven shrinkage but does not specify the actual material used. Also, without attempting to address further issues concerning foundry practices or workshop organization, the success of Wan's project lies mainly in reconstructing the most basic technological aspects of the Shang bronze industry. Wan's project nonetheless marked the beginning of the use of replication experiments in the study of Shang bronze technology.

Replication Experiments at the Shanghai Museum. The Shanghai Museum project, headed by Tan Derui, combined lab research, field investigation of traditional clay mold casting, and replication experiments for its study of ancient clay mold casting (Huang 1993; Tan 1999; Tan et al. 1993). As traditional metal craftspersons in southern China still employ clay molds for metal casting, the manufacturing procedures at these modern foundries served as ethnotechnological guidelines for the replication experiments, and the results of lab analyses of the ancient mold pieces provided the basic parameters against which the replication experiments were measured.

The technology of section-mold casting with clay models was lost to the Chinese sometime after the Zhou period. However, metal casting with clay molds from a wax model is still carried out in modern China, both by traditional craftspersons (Ruan and Huang 1959, cited in Tan 1999) and by casters making fake bronzes (Shih 1955:108). Traditional technological treatises on casting, such as *Tiangong kaiwu* (*The Exploitation of the Works of Nature*, dated 1637), also describe a process that involved the use of wax or resin for making models and the use of clay for making molds (see Hua 1986 for a discussion of these texts). Scholars of ancient Chinese metallurgy did not hesitate to explore traditional clay-mold casting techniques for the purpose of studying Shang and Zhou bronze casting, and the results were fruitful.

In light of the long history of the use of clay molds in traditional metal production, the Shanghai Museum team began by conducting field research on traditional metal casting of bronze vessels, religious paraphernalia, and iron objects such as ladles, pots, and other utilitarian pieces (Huang 1993:31). Observations of the manufacturing procedures at iron foundries in Suzhou show that local metalworkers use several types of clays, prepared according to different formulas, for the inner layer of the mold, the outer layer or backing material, and the slip for shaping and retouching the mold. Different ratios of clay, water, plant remains, and various grades of ashes and coal powder are mixed together to form the different clays, and the mixed clay is further kneaded, aged, and soured before it can be used for making the mold (Huang 1993: 31–32).

The addition of plant remains and ashes to the clay by modern traditional casters prompted the Shanghai team to look further into the function and purpose of such a practice, especially as molds thus manufactured display physical attributes similar to those observed in ancient molds. These attributes include resistance to high temperature, clean and smooth mold surface, and randomly distributed fine pores of various sizes that can be seen in a freshly cut section of the mold (Huang 1993:33). Test samples of pure clay, clay mixed with wheat stalks, and clay mixed with the ashes of burned wheat stalks were prepared and tested for shrinkage rate and ease of carving. The results showed that samples with plant ashes had the lowest shrinkage rate and were best for carving decoration, whereas samples made of pure clay, although easy to carve, had the highest shrinkage rate (Huang 1993:33, tab. 1). As a high shrinkage rate of the clay resulted in a greater tendency of the mold to warp and lose the intended shape during drying or firing, rendering the molds unusable for casting, the researchers have concluded that the function of adding plant remains and ashes in the clay in traditional clay-mold casting is mainly aimed at preventing uneven shrinkage of the mold during manufacture (Huang 1993).

To determine if casters in Bronze Age China also followed similar practices, phytolith analysis was conducted on the Bronze Age clay mold fragments (Tan 1999; Tan and Huang 1996; Tan et al. 1993). The findings indicate that various forms of phytolith were indeed present in the ancient molds. When compared with local soil and pottery collected from the foundry sites, most of the ancient clay molds proved to have a higher SiO_2 (silicon dioxide) content than the pottery, and some have a higher SiO_2 content than the soil sampled from contemporary strata. Tan and his colleagues therefore argue, basing their hypothesis partly on the field research on traditional clay-mold casting, that the additional SiO_2 would have been added in the form of plant ashes.

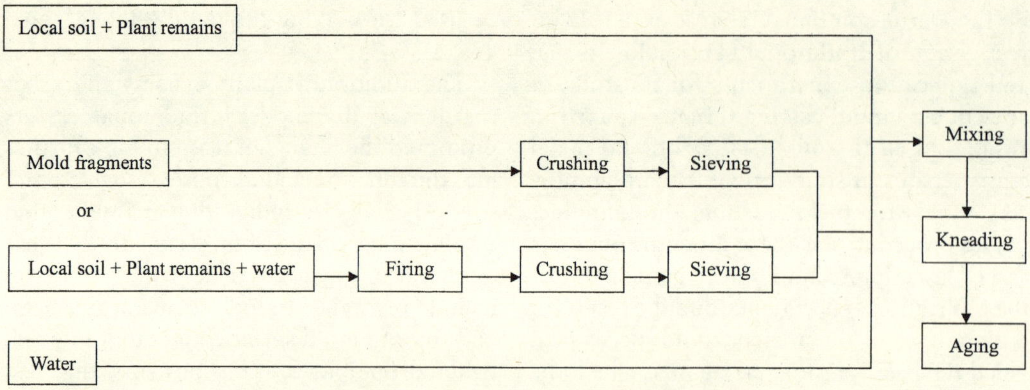

FIGURE 8.3. Process of preparing clay for the back layer of the mold. Based on Tan 1999:227.

Further analysis for density, specific heat, and heat conductivity was performed on two sets of samples: ancient molds and local soil from Shanghai and the Houma foundry site. The results show that with the addition of plant ashes, the density of the mold decreases, the specific heat increases, and the overall heat conductivity drops significantly (Tan 1999: fig. 7). Because the lower the heat conductivity, the better the mold is able to sustain the temperature and fluidity of the molten metal, it follows that the addition of plant ashes is crucial for successful casting (Tan 1999: fig. 7). These results demonstrate that the ability of the ancient Chinese to cast bronzes with fine detail and intricate shapes was achieved by lowering the heat conductivity with the addition of plant ashes into the mold, rather than by producing molds with good air permeability, as in the case of modern sand casting.

In addition to plant ashes, fired clay—which can easily be obtained by recycling used mold fragments—is also crushed and mixed into the clay in traditional clay-mold casting (Ruan and Huang 1959, cited in Tan 1999). Although the presence of such fired clay in the mold cannot be detected by lab analysis, replication experiments have shown that it is an important ingredient for reducing the shrinkage rate and preventing uneven shrinkage (Tan 1999:226). After the organic material and the crushed fired clay are mixed in, the clay then has to be repeatedly kneaded and aged (Huang 1993; Tan 1999). Figures 8.3 and 8.4 illustrate the reconstructed operational sequence for preparing clay for the inner and outer layers of a mold by the Shanghai Museum project.

As uneven shrinkage and warping can cause irreparable defects in the mold, the Shanghai Museum team adopted a few more procedures used in traditional clay-mold casting to prevent deformation of the mold during the mold-making process (Tan 1999). When the mold is being formed on the model, the clay has to be patted repeatedly. Once finished and removed from the model, the mold sections are reassembled, tightly bound, and allowed to dry in a slow and controlled process. During firing of the mold, kiln temperature also has to be raised in slow increments to prevent cracking or warping. Both the field investigation of traditional metal crafts and the replication experiments indicate that considerable time must be invested in the making of clay molds to prevent warping and uneven shrinkage.

The Shanghai Museum team then proceeded to replicate bronzes by using the reconstructed clay recipe and by following procedures for making clay molds in traditional Chinese metal casting. The Shanghai team chose several common types of bronze vessels, including a jue (Erligang period, 1500–1300 B.C.), a gu, a *zhi*, and a double-owl *you* with a rope-shaped swing handle (all three of

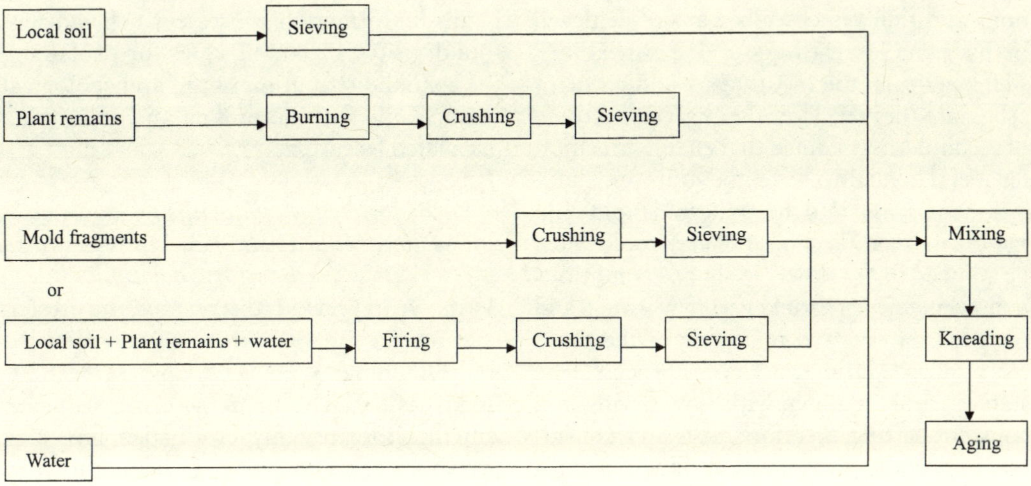

Figure 8.4. Process of preparing clay for the inner layer of the mold. Based on Tan 1999:228.

Table 8.1. Workdays Invested in the Main Production Steps for Making *Jue*, *Zhi*, *Gu*, and *You*

Bronze Type	Model	Mold Sections/ Parts	Core	Inscription	Casting	Retouching	Appendages	Total
Jue	11	34.5	2	0	1	1	0	49.5
Zhi	10	37	6	0	1	1	0	55
Gu	12	25	9	5	1	1	0	53
You	35	33	13	2	1	11	15	110

Source: Adapted from Tan 1999.

the Late Shang period), to be replicated (Tan 1999). Technical studies were first conducted on these vessels to determine how mold divisions, decoration, and inscriptions, if present, were prepared. The composition of the metal was also analyzed to ensure that the alloy for the replicated bronze would match the original. The amount of time invested in each main manufacturing step was recorded, and the total workdays spent on each vessel were calculated. Among the four vessels, the *you* is the most elaborate and complicated in shape; thus the time and labor spent on its replication were significantly greater than those for the other vessels.

The manufacturing processes used in the Shanghai replication experiments (Tan 1999) are summarized in Table 8.1.[3] It shows that the most time-consuming parts of bronze casting with clay molds are the preparation of the clay model and the mold and above all the making and firing of mold sections. For vessels of simple shape and decoration, such as the Erligang-type jue replicated by the Shanghai Museum, the time invested in model making was 11 workdays, whereas the time needed to complete the mold assembly before firing was 35.5 workdays. The actual casting took up only one workday.

Through its extended investigations, and after numerous failed replication experiments, the Shanghai Museum team came to the following conclusions regarding the making of clay molds in Bronze Age China: First, the ability of the ancient Chinese metalworkers to cast bronzes with finely detailed decorations and thin vessel walls was not achieved by using molds with good air permeability. The key lies in the fact that by adding plant ashes into the clay, a low thermal conductivity of the molds is obtained that enables the molten metal to retain high temperature and fluidity when poured into the preheated mold (Tan 1999:239–241). Second, the clay mold must be allowed to dry slowly in shaded conditions to prevent uneven shrinkage and warping, and repeated patting and pounding of the mold are required during the drying process. The stage of mold making, with slow drying and constant patting, therefore takes up a considerable portion of the entire production time, about 60–70 percent of the total invested work hours (Tan 1999:240). Third, although the clay molds used by ancient Chinese casters were fired, they were not fired to the sintering point and are hence technically "terra-cotta" (Rice 1987:5, tab. 1.2). The relatively low firing temperature, between 850 and 900°C, allowed gases to escape from the clay without causing the mold to change shape or become too hard. Molds fired too hard are more difficult to remove from the cast bronze and may cause the bronze to crack when the metal solidifies, because as the bronze cools, it shrinks more than the mold does.

The Shanghai Museum project is a landmark in the study of Shang and Zhou bronze-casting technology. Observations and conclusions were supported by data obtained from controlled lab procedures, field investigations, and actual replication experiments, all of which have been published in good detail, enabling more rigorous research on the subject.

However, almost all replication experiments to the present day have focused on bronze casting with clay models and molds. Metal melting with reconstructed furnaces has yet to be attempted (cf. Xu et al. 2001). Such experiments may further our understanding, which is at this point fairly insufficient, of how the ancient furnaces were built, operated, and maintained and may shed light on issues concerning the organization of labor and the scale of operation in ancient foundries. Replication experiments might also help to explain site formation processes at the ancient foundry sites, providing clues for explaining the spatial distribution of the archaeological remains and for interpreting and identifying excavated features.

Reconstructed Manufacturing Sequence of Section-Mold Casting and Implications for Workshop Organization

As shown in previous discussions, the present research has provided a reasonably detailed understanding of ancient Chinese section-mold casting. The manufacturing sequence can therefore roughly be divided into clay preparation, model making, mold making, core making, furnace building, metal melting, casting, and finishing. Because studies have yet to provide clear details on furnace building and furnace operation, the procedures related to metalworking cannot be described in as much detail as those for clay working:

- Clay preparation: Mining clay; levigation; mixing in plant remains, ashes, and crushed fired clay as temper; aging and souring. Different clays are prepared for the model, the core, and the inner and outer layers of the mold.
- Model making: Shaping; carving decoration; firing below sintering temperature.
- Making mold sections: Building clay around the model according to predefined sections; repeated patting during slow drying; preparing mortises and tenons; releasing mold sections from the model; reassembling mold sections; binding mold sections tightly together and allowing them to dry slowly; firing below sintering temperature.
- Making cores: Either scraping down the model or adding a sheet of clay on the inside of the mold and then building the core within the assembled mold sections. Cores are also fired below sintering temperature.
- Casting: Reassembling mold sections and the core; securely positioning the mold assembly; preheating the mold assembly; melting metal in furnaces; casting.
- Finishing: Breaking the mold off the cast bronze after the mold and bronze have cooled; removing excess metal, for example, pouring inlets and vents, now filled

with metal; polishing and retouching the bronze.

The early works of Fairbank (1972) and Barnard (1961) already recognized the close relationship between ancient Chinese bronze casting and pottery technology. They point out that the advanced pottery kilns capable of reaching high firing temperatures served as the precondition for metal technology, and the long tradition of pottery making by using molds must also have foreshadowed the making of clay models and molds for bronze casting. The connection between bronze casting with clay models and molds and pottery making, however, extends even further. Preparing clay for bronze casting molds is not too different from preparing clay for pottery. Both have to go through steps such as levigation, kneading, aging, souring, and tempering. The procedures to prevent uneven shrinkage and warping also mean that the ancient foundry people had at their command the extensive skills and knowledge in handling clay so that the specific physical constraints required in bronze casting, namely, low heat conductivity, resistance to high temperature, and good physical strength, could be met.

Because working with clay and working with metal may require very different skills, as Franklin (1983a) has already pointed out, a logical point for the division of labor in the workshop organization of a Shang bronze foundry might be between the stages that precede casting and the stages of casting and finishing. The Shanghai Museum and other reconstruction projects have demonstrated that considerable labor and resources are invested in the preparation of clay and the making of models, cores, and mold sections. The proportion of time and labor invested in clay processing should also be reflected in the physical features of the workshop, along with other more common finds of foundry debris such as mold fragments, slag, and furnace fragments. For instance, at a Shang bronze foundry, we may expect to find facilities related to clay working, such as levigation tanks, storage for aging and souring clay, low-firing kilns for prefiring molds, or shaded storing facilities for drying molds. These facilities may well parallel those found at pottery workshops such as Jingdezhen. Evidence for clay processing facilities was found at Miaopu North (Li 2003), and a recent salvage project may have located features for preparing and storing clay at Xiaomintun (Yue and Liu 2006). Overall, however, there is relatively little information for workshop features at Anyang bronze foundry sites (see below).

Franklin and the Holistic and Prescriptive Models

Franklin (1983a) was among the first to recognize that the different skills required in working with clay and metal may have shaped the workshop organization of the Shang bronze foundry. She argues that the Shang bronze workshop was probably subdivided into independent but related production units, with craftspersons specializing in only the tasks of the particular units they belonged to. With Shang bronze production in mind, Franklin (1983a) proposes a classification of technologies according to the way the production procedure is organized. A technology is prescriptive if the production procedure is broken down into predetermined production units and carried out by individuals with independent skills in each unit, such as the section-mold casting of the Shang bronze technology or, to use a more familiar example, the assembly-line or flow-line production (Wild 1975:5–7) in modern factories. It is "specialization by process" (Franklin 1992:20). A technology is holistic if the manufacturing process involves "a single, stepwise approximation toward the final object" (Franklin 1983a:96) and is performed by the same individuals through the entire manufacturing process. It is "specialization by product" (Franklin 1992:19). Stone knapping and smithing are examples. A holistic process is therefore a "sequential and linear development" (Franklin 1983a:96), whereas a prescriptive process depends on coordination among different groups of workers in subdivided production units. The emphasis of Franklin's model is on the party who controls the production of the object. In a holistic technology, the craftsperson is in charge of the manufactured object and the entire manufacturing procedure. Individual workers in a prescriptive technology, however, do not necessarily possess sufficient knowledge of the whole production process; instead, they are confined to a limited range of tasks in one or a few steps of the process. The overall coordination and control of production, and hence of the finished product, go to a different party, usually an administrative body (Franklin 1983a, 1992).

In essence, Franklin's thesis parallels works in anthropological archaeology under the theme of *chaîne opératoire* (e.g., Dobres 1999, 2000; Lemonnier 1992; Roux and Matarasso 1999). However, with many of her ideas not further developed or given concrete examples, Franklin's work remains chiefly a set of illuminating definitions. In particular, although the concepts of holistic and prescriptive technologies are instructive in describing forms of production organization, Franklin does not provide interpretive models for archaeological data. In other words, how can a holistic or prescriptive technology be identified archaeologically? Can a workshop layout in the archaeological record reflect the type of production organization and hence the social and political milieu of the particular workshop? The notion of holistic and prescriptive technologies is also not sufficient to discuss the various forms of craft production systems that may have existed in traditional Chinese craft industries.

The long tradition of handicrafts such as pottery and porcelain making in premodern China offers valuable insights into how large-scale craft production can be organized and controlled. Porcelain production at Jingdezhen and stoneware production at Yixing, for instance, are classic examples of what Franklin would characterize as prescriptive (Jingdezhen) and holistic (Yixing) technologies. Based on the ethnohistorical data of Yixing and Jingdezhen, I (2003, 2007) have argued elsewhere that certain parameters can be isolated to identify Franklin's two forms of production organization in the archaeological record. These parameters include the distribution of workshop features, the range of production procedures represented in the workshop debris, and the diversity of products manufactured by the workshop. They are used to examine the archaeological remains of Anyang bronze foundries in the section on the bronze foundry sites. What follows is a discussion of the distribution of workshops and how they reflect the organization of multicraft production at Anyang in relation to the settlement pattern and the political context of the Shang capital.

ANYANG AS THE CENTER OF CRAFT PRODUCTION

As of today, craft production remains an understudied topic in Anyang archaeology. The emphasis has traditionally been on the historicity and cultural history of the site, and the fact that the Shang court was also an important consumer of various resources, as evident from the splendor of its material culture, is often ignored. In fact, Anyang, like any other centers in ancient civilizations, was also the capital of craft and, indeed, "multicraft" production (Shimada, this volume). The second half of this chapter reviews current data on craft production at Anyang and explores its role in the Shang state within the context of settlement patterns and the social and political landscape.

The Late Shang Capital at Anyang

Anyang is located in northern Henan Province, central China, approximately 450 km south of Beijing (Figure 8.5). It has been a focus of attention for historians, epigraphers, art historians, and archaeologists since the late nineteenth century when oracle bone inscriptions were first recognized and their source was traced back to Anyang. Since 1928, Anyang has been continuously investigated—interrupted only by the Sino-Japanese war and the subsequent civil war—by field archaeologists, first from the Institute of History and Philology, Academia Sinica, now relocated to Taipei, and later from the Institute of Archaeology, Chinese Academy of Social Sciences, Beijing. It remains one of the most important sites of Bronze Age China. The dates of the oracle bone inscriptions found at Anyang place

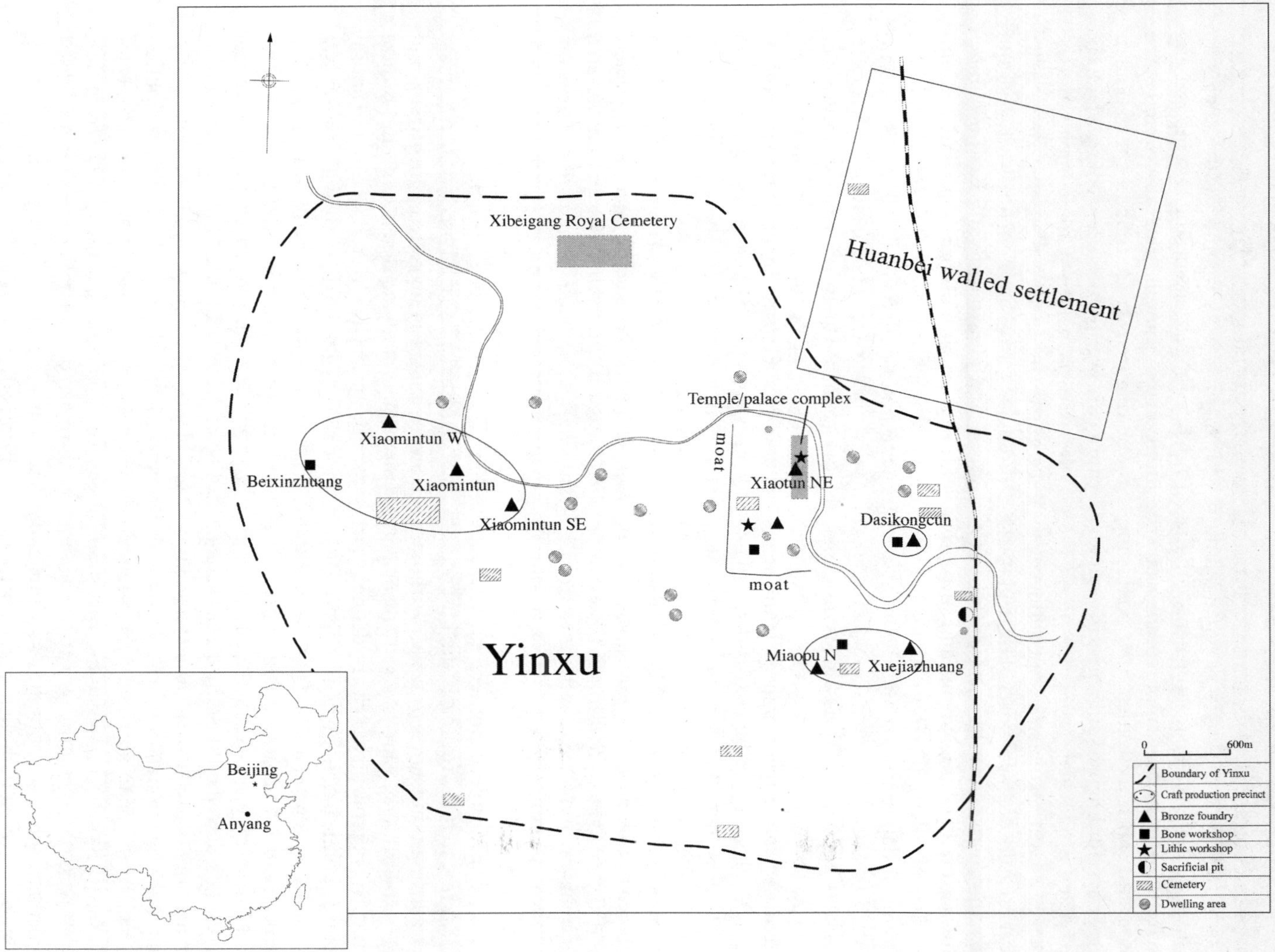

FIGURE 8.5. Map of archaeological sites in the Anyang region.

the time of the Late Shang settlement, known as Yinxu, between 1200 and 1000 B.C.[4]

Settlement Pattern at Yinxu

The pioneering archaeological research at Anyang by the IHP in the 1930s sketched out the basic outline of the Yinxu settlement with the discovery of the palace/temple complex at Xiaotun and the royal cemetery at Xibeigang (Figure 8.5; for English summaries of the finds, see Bagley 1999; Chang 1980; Li 1977).[5] Extensive excavations have been conducted in other areas by archaeologists of the IA (2001) since the 1950s, unearthing a wealth of new information that complements and surpasses the early work by the IHP in terms of the amount of data retrieved and the areas covered. In the 1960s, IA archaeologists defined an area of approximately 24 km^2 as the extent of the Late Shang occupation in the Anyang region (Anyang Archaeological Team, Institute of Archaeology, Chinese Academy of Social Sciences [hereafter Anyang] 1961); more recent archaeological finds have extended the area first to over 30 km^2 (Zheng 1988:933) and then to over 36 km^2 (Figure 8.5; Yue et al. 2006:335). Anyang archaeologists also note an expansion of the Yinxu settlement from 12 km^2 during Yinxu Phase I to 30 km^2 during Phases III and IV (IA 2003:295–296).

The long unsolved mystery for archaeological research at Anyang is the absence of an enclosing rammed-earth wall.[6] Without a rammed-earth enclosure, the settlement pattern of Anyang appears less clearly defined as an urban or political center when compared with the earlier settlement at Erligang, Zhengzhou (see An 1986; Chang 1980; Thorp 2006). The late Professor Kwang-chih Chang (1980:130, fig. 38) has described the Anyang settlement as a network of archaeological sites, each representing a functional node in the larger capital region: the Xiaotun royal palace–temple block, with its service accessories, workshops, and graves; the royal cemetery at Xibeigang; residential hamlets of both the lesser elite and commoners, often with their own cemeteries; and workshops or workshop quarters. Chang has also noted that "these nodules were not spatially contiguous so as to form an urban area of dense houses and streets" (1980:130). The sparse distribution of archaeological sites, however, may have more to do with the state of archaeological research in the region than with the actual ancient layout.

If Anyang lacks a rammed-earth enclosure to define the settlement, the two intersecting moats in the Xiaotun village discovered in the 1960s (Figure 8.5; IA 1987) seem to demarcate a core area for the Shang court, which encompasses the palace/temple complex and the accompanying sacrificial pits; the three caches of oracle bones, Xiaotun Northeast, Xiaotun South, and Huayuanzhuang East; a number of bronze and lithic workshops; and medium-sized but richly furnished elite burials, such as the tomb of Fuhao and Tomb No. 54 at Huayuanzhuang. Despite the fact that new fieldwork calls into question the exact location of the moats (see below), concentrations of activities related to the Shang ruling elite still indicate that the area bounded by the moats and the Huan River may have been the locus of the Shang court.

Anyang archaeologists have argued that a lineage-based cemetery system and perhaps a lineage-based settlement pattern existed during the Late Shang occupation at Anyang (Anyang 1979; IA 2003:336–338; Tang 1998; Yang 1983, 1986; Zheng 1995). The notion that Shang and Zhou societies were structured around a lineage-based social organization has long been suggested by historians and epigraphers studying oracle bone and bronze inscriptions.[7] Guo Moruo's (1961) proposal that the glyphs on Late Shang bronzes are "clan signs" provided the means to look for lineage organization in the archaeological record (see also Barnard 1986). Anyang archaeologists have examined the burial goods assemblages, the spatial segmentation of the cemeteries, and, most of all, the distribution of clan signs inscribed on bronzes in the cemeteries and argue that the clusters of burials seen in Yinxu cemeteries were lineage based (Anyang 1979:114; IA 2003; Yang 1986; cf. Li in press). It has been further inferred that the settlements nearby could also have been lineage based (Zheng 1995).

The precise location and the defensive

function of the moats have recently come under reexamination by Anyang archaeologists, as have the date and layout, although not the function, of the palace/temple complex. Excavations revealed that sections of the western moat may not have been connected and that various parts of the moat near the Anyang field station may have served as the loci for human sacrifices and as water ditches during different periods of Yinxu (Yue et al. 2006:334–335). The results of extensive coring over Xiaotun Northeast also show that the layout of the palace/temple complex may have consisted of a series of rectangular buildings with long corridors that resemble the palatial structures at other pre-Anyang walled settlements, such as Yanshi and Huanbei (Yue Hongbin, personal communication regarding the preliminary results of systematic coring of the palace/temple complex at Xiaotun Northeast, Anyang, Henan, November 2005; Tang 2004). Additional features are also located in the palace/temple complex, such as a large reservoir and a number of accompanying rammed-earth structures (Yue et al. 2006: 337–340). The reservoir is situated west of Architecture Groups A and B and connected to the Huan River at its northern end. It is likely that the water ditches unearthed by the IHP archaeologists in the 1930s (Chang 1980:77, fig. 13) were connected to the reservoir to channel water away from the palace/temple complex. Anyang archaeologists now believe that with an area of only 0.7 km^2, the size of the core area, as previously defined by the two moats, is relatively small when compared with earlier walled centers at Erlitou, Yanshi, and Huanbei; the palace/temple complex may have extended beyond the western moat, reaching as far west as the modern village of Sipanmo (Yue et al. 2006:335–336).

I look forward to discoveries and breakthroughs in the field of Anyang archaeology that will dramatically reshape our understanding of Yinxu. The following discussion, however, will be conducted within the framework of the published archaeological data, with the understanding that my conclusion will need to be revised once new studies and new discoveries become available.

Multicraft Production at Yinxu

Archaeologists working at Yinxu in the past seven decades have located a number of craft production sites, including pottery kilns, bronze foundries, and bone and lithic workshops (Figure 8.5).[8] Craft activities are found both in the core area and across the Anyang regions. As will be shown in the following discussion, the location, the products, and the scale of production of these workshops indicate that they were most likely "specialized craft production" (Sinopoli 2003:1) controlled by the elite.[9] Bronze and bone workshops outside of the core area also form into individual craft production areas or precincts, which were probably run by local elites residing in the nearby settlements. The ceramic dates for the workshops indicate that multiple workshops operated simultaneously during the Yinxu period (Figure 8.6).

The production of pottery, the most ubiquitous artifact, is beguilingly the least represented in the Anyang archaeological data. Only four poorly preserved kilns have been found, with no other related workshop features or remains that may indicate large-scale production (IA 2003:300). These kiln sites have yet to be fully documented and therefore are not included in this discussion.

Bone and Lithic Workshops. Bone working was a common type of craft activity at habitation sites at Yinxu, suggesting local household production (IA 1987:283). Anyang archaeologists, however, have discovered three large bone workshops at Dasikongcun South, Beixinzhuang (IA 1987:70–89), and Miaopu North (He Yuling, personal communication regarding the salvage excavation north of Miaopu North that located a large bone workshop, Anyang, Henan, December 2003). A bone refuse pit was also found in the core area at Huayuanzhuang (Anyang 1992). The refuse pit, H27 at Huayuanzhuang, is located outside of the palace/temple complex, about 600 m south of the Xiaotun South oracle bone cache (Figure 8.5). The excavators estimate that the pit contained over 300,000 fragments of animal bones, mostly cattle. Bone fragments were mainly from the axial skeleton,

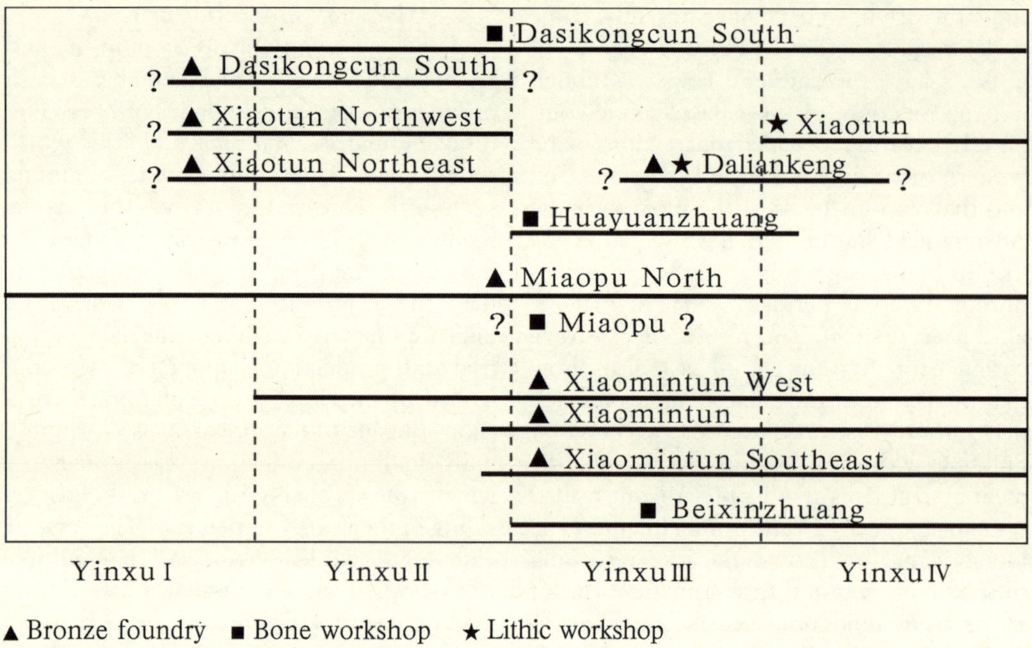

FIGURE 8.6. Duration of Anyang craft production sites in terms of Yinxu period pottery chronology.

with very few limb bones. Only some 80 bone fragments show signs of processing, such as saw marks, with no finished or semifinished bone artifacts present. The excavators suggest that the pit was used mainly as a dumping ground for refuse bones from a nearby butchering site or bone workshop (Anyang 1992). The pit was in use from Yinxu Phase III to Early Yinxu Phase IV.

The three bone workshops are scattered outside of the core area and often near the known bronze foundries (Figure 8.5). The bone workshop at Dasikongcun South was in operation during the Yinxu Phase II (Miaopu Phase II) and continued production into the second half of the Yinxu period (Miaopu Phase III). The bone workshop at Beixinzhuang was in operation during the later half of the Yinxu period (Miaopu Phase III). The recent find of a bone workshop at Miaopu, not yet published, is located just north of the Miaopu foundry (He Yuling, personal communication, 2003; see also IA 2003:300). All three yielded large volumes of debris, semiprocessed bones, and bone artifacts that indicate unusually high levels of bone-making activity. The Beixinzhuang and Dasikongcun South workshops produced bone hairpins in large quantities (Figure 8.7). Hairpins appear to have been part of the elite costume; although they may not have been important status markers, they are often found in large numbers in richly furnished elite tombs. For instance, over 400 hairpins of different styles were found in a wooden box in the tomb of Fuhao (IA 1980:208–213). Other bone artifacts found include arrowheads, awls, and plain spatulas, all of utilitarian function (IA 1987). None of the bone workshops, however, yielded any high-status objects. The exquisitely executed bone artifacts seen in the royal tombs, such as the bone spades (Figure 8.8) and spatulas from Royal Tombs HPKM1001 and 1003, and the ivory cup from the tomb of Fuhao were made elsewhere. The lack of sumptuous products for royal use at the bone workshops forms an interesting contrast with the lithic and bronze workshops.

Although jade, marble, and other types of lithic objects are often encountered in burial and residential contexts at Anyang (IA 2001: 323–383), only two lithic workshops have

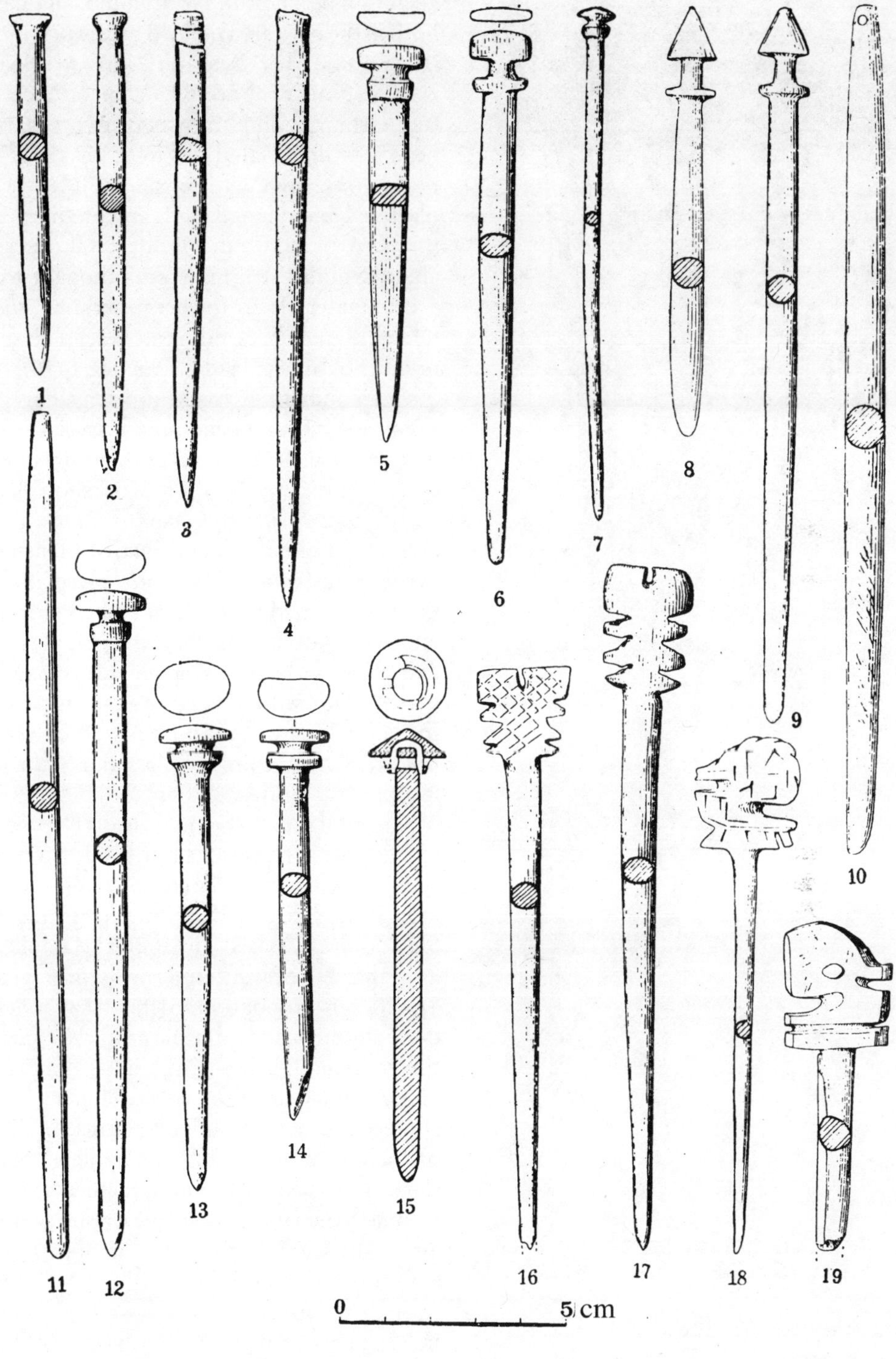

FIGURE 8.7. Various types of bone hairpins found at Anyang. Institute of Archaeology, Chinese Academy of Social Sciences 1987:190, fig. 144.

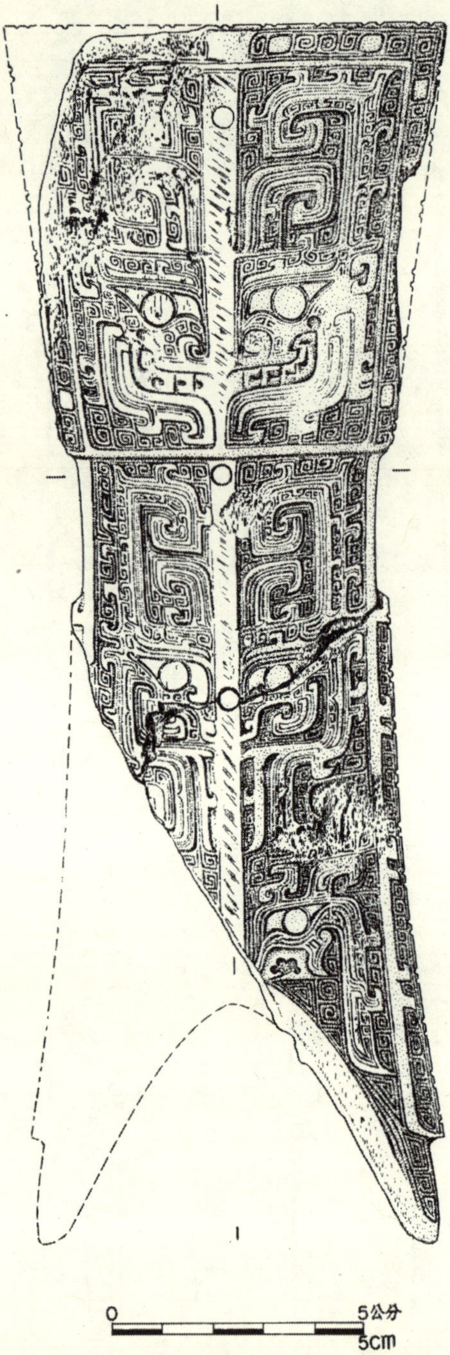

Figure 8.8. A bone spade with elaborate incised decoration from Royal Tomb M1001, Xibeigang, Anyang. Liang and Kao 1962: pl. CXCIII.

been found, Xiaotun North and Daliankeng, both in the core area defined by the two moats (Figure 8.5). The Xiaotun North workshop is located near the western edge of Architecture Group C and consists of two subterranean dwellings dated to Yinxu Phase IV. The houses were built with considerable care, with plastered and painted walls and fire-hardened floors. Artifacts found inside the houses include cylindrical lithic pieces thought to be precut materials for further processing, *zhang* tablet fragments, grinding stones, and a small number of finished jades (Anyang 1976).

A second lithic workshop may have existed in the Daliankeng area between Architecture Groups A and B in Xiaotun Northeast (Figure 8.9; Li n.d.). Workshop debris and artifacts from the Daliankeng area suggest that craft-producing activities using a range of different media took place, including bronze casting, lithic and shell working, and most notably marble working. The evidence for marble working includes elongated marble slabs with cut marks left untreated on the edge (Figure 8.10a); fragments of vessels (Figure 8.10b), chime stones, and anthropomorphic sculptures; and finished plaques and ornamental pieces (Figure 8.10c). The coexistence of bronze casting, lithic production, and shell working at Daliankeng would seem to indicate a pattern of both "multicrafting," defined as crafts based on similar technologies as in the case of lithic and shell working, and "multicraft production," defined as the "concurrent practice of multiple crafts" (Shimada, this volume). However, the date, the extent, and the exact nature of the Daliankeng workshop cannot be determined because of a lack of documentation and the incomplete excavations of the 1930s. The mold fragments mainly belong to the Type II mold, indicating a date as late as the second half of the Yinxu period for the bronze-casting activities.[10]

Artifacts made of marble were among the most exclusive high-status objects in Shang civilization. Bronze vessels and jade objects were clearly prestige items, but their presence in mortuary contexts is still extensive rather than exclusive, as bronzes and jades can be found in burials of modest size and

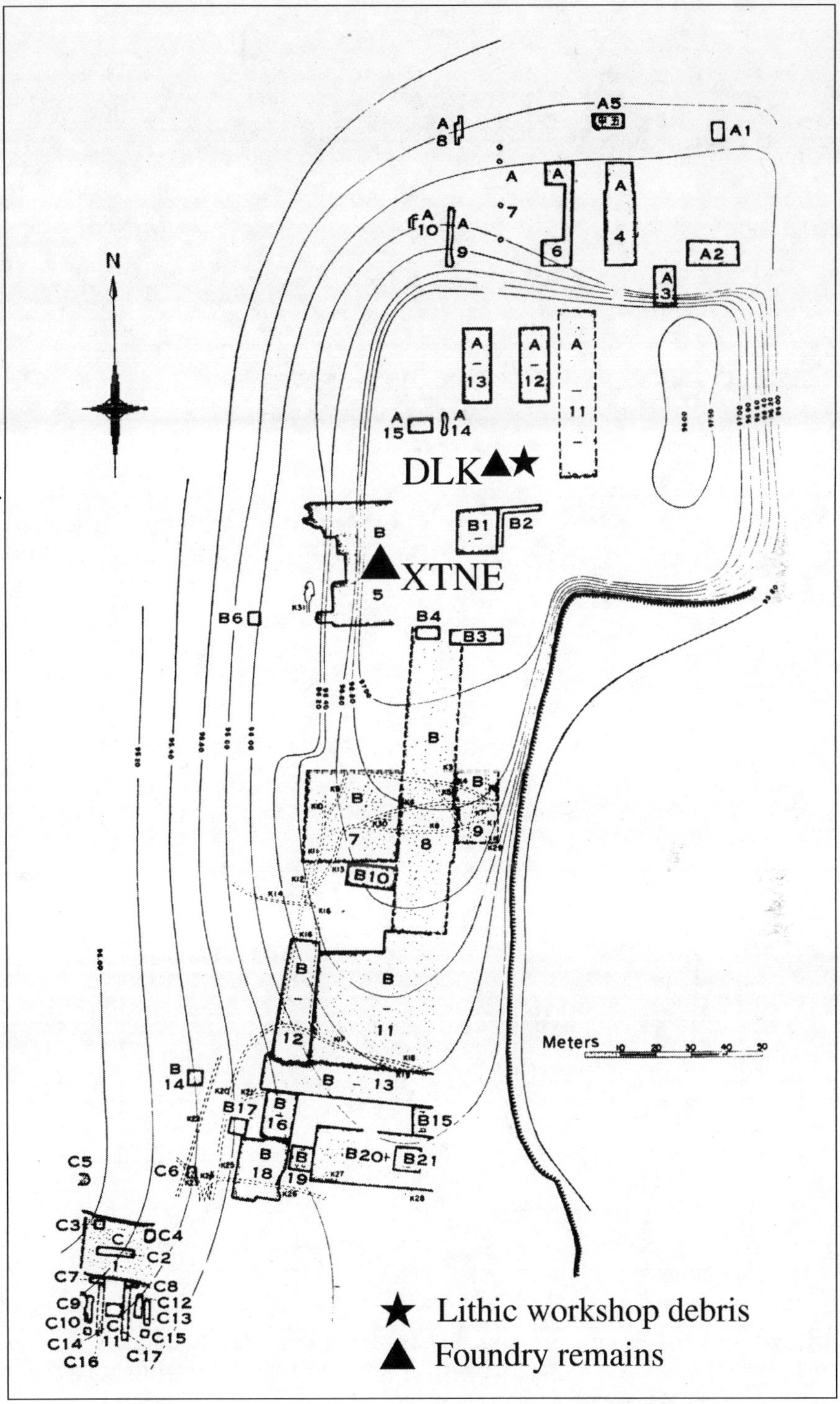

FIGURE 8.9. Locations of the lithic and bronze workshops at Daliankeng (DKL) and Xiaotun Northeast (XTNE) in the palace/temple complex, Xiaotun. Adapted from Chang 1980:94, fig. 23.

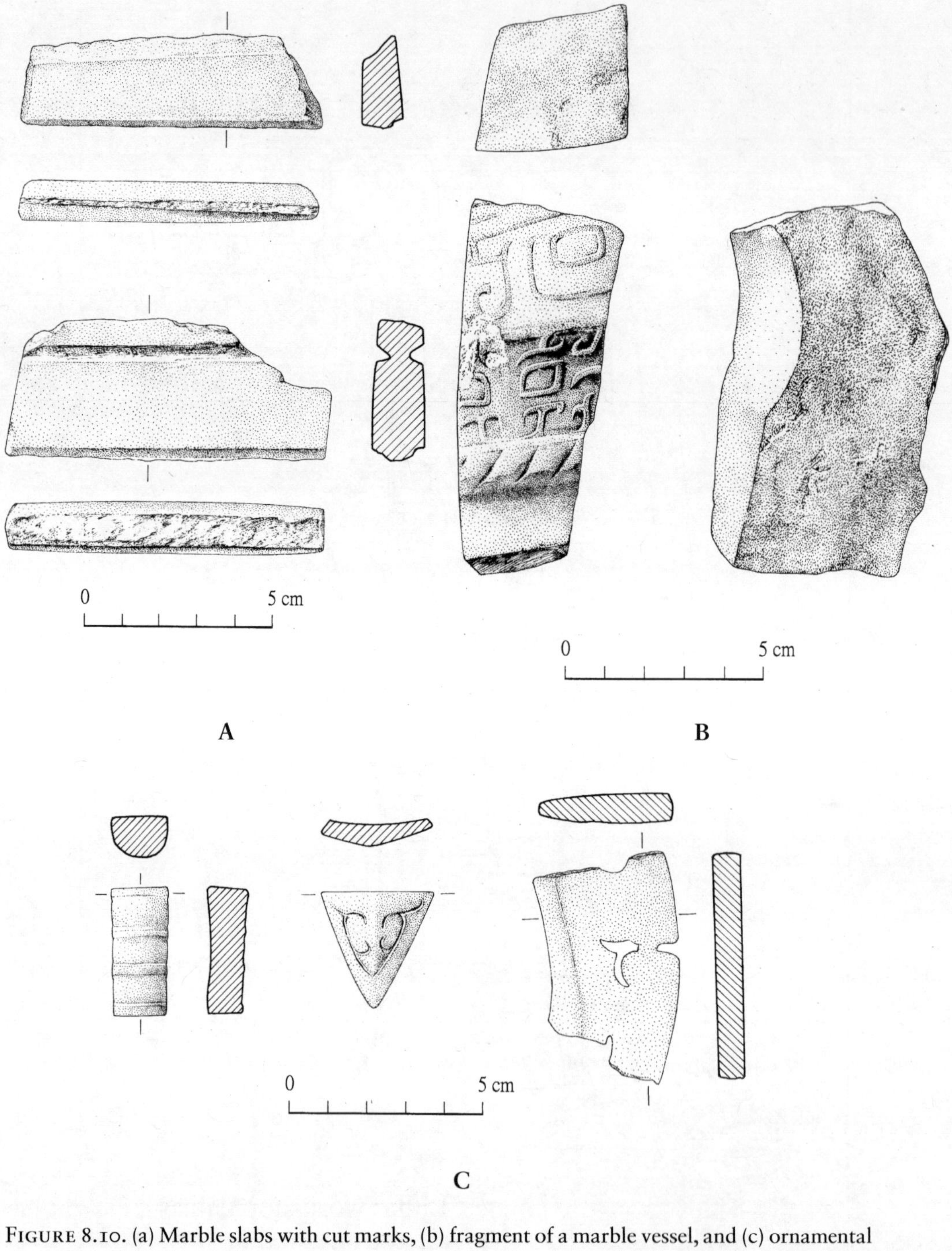

FIGURE 8.10. (a) Marble slabs with cut marks, (b) fragment of a marble vessel, and (c) ornamental pieces from Daliankeng, Xiaotun, Anyang. Original drawings by Lai Shu-li; copyright © the Institute of History and Philology, Academia Sinica.

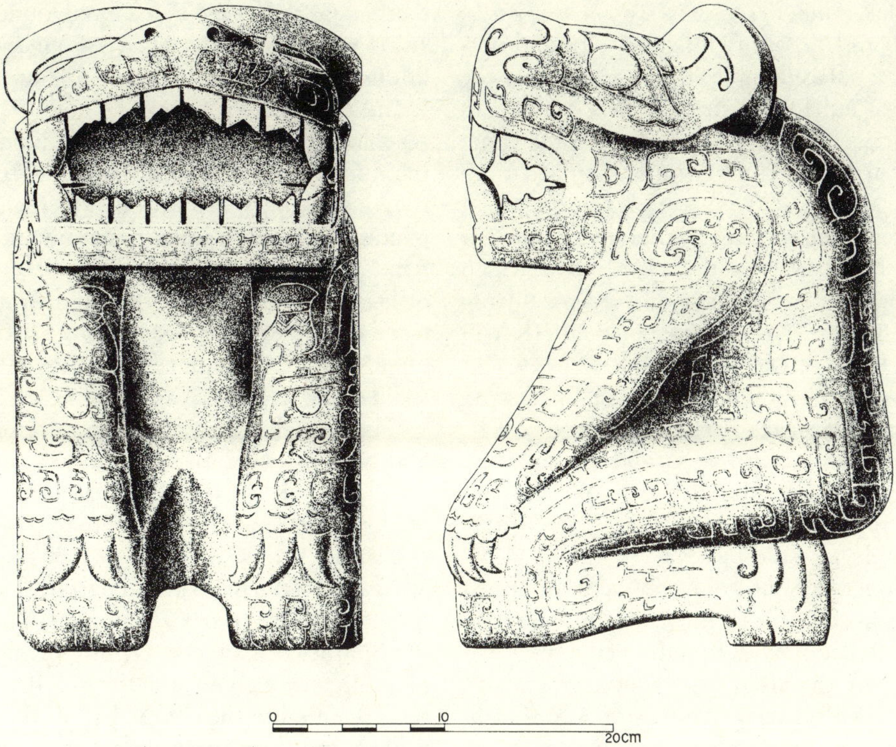

Figure 8.11. A marble sculpture of a kneeling zoomorphic figure from Royal Tomb M1001, Xibeigang, Anyang. Liang and Kao 1962: pl. LXXII.

furnishings. Marble, on the other hand, was mainly used for large, three-dimensional, and intricately decorated anthropomorphic and zoomorphic sculptures (Figure 8.11), which appear almost solely in the royal tombs. Marble was also made into small plaques and ornamental pieces to be incorporated into decorative designs on wooden objects and possibly murals inside the royal tomb chambers. The evidence of marble being processed in the Daliankeng area, together with the close proximity to the palace/temple complex, clearly indicates direct control over the workshop by the Shang court.

Bronze Foundry Sites. Bronze ritual vessels were some of the most important prestige items and status markers in Shang society (Chang 1980, 1983; Underhill 2002). In spite of their restricted use, the production of ritual bronzes was not an exclusive craft activity limited only to the royal elite in the core area. Anyang archaeologists have found several locales with foundry remains in the Anyang region (Figure 8.5). Some consist of small deposits in trash pits that indicate nearby casting activities. Some, on the other hand, yielded workshop features and large amounts of foundry debris over extensive areas, marking the actual location of the foundries. These locales also appear to cluster together, but as excavations were often conducted during different seasons with separated excavated areas because of the salvage nature of the operations, the exact relationships between the closely situated locales are not entirely clear. It can nonetheless be argued that these finds may belong to four separate bronze workshops: (1) Xiaomintun, the largest Yinxu bronze foundry known to date, which includes Xiaomintun, Xiaomintun West, and Xiaomintun Southeast (Anyang 2006; IA 1987; Yue and Liu 2006); (2) Xiaotun, which includes Xiaotun Northeast, situated under Structure B5 in the palace/temple complex (Shih 1947, 1955, 1959; see also Li 2003), and Xiaotun Northwest, located outside the palace/temple complex and near the tomb of Fuhao (IA 2004)—there may be an additional bronze workshop at Daliankeng between Architecture Groups A and B at Xiaotun Northeast (Kuo 1933; Li n.d.), which may represent the later component of the Xiaotun Northeast foundry; (3) Miaopu, which includes Miaopu North and Xuejiazhuang (IA 1987); and (4) Dasikongcun South (Kao in press; Li in press).

The scope of the foundry remains discovered at these sites varies considerably. The largest foundries, Miaopu and Xiaomintun, yielded tens of thousands of mold and furnace fragments, whereas at Xiaotun Northwest only 20 mold pieces were found, and at Dasikongcun South only 80 were found. At Dasikongcun South, however, the site's direct association with foundry activities is evidenced by the discovery of mold fragments, furnace pieces, and grinding stones for finishing bronzes (Li in press). The same can be argued for the Xiaotun Northwest site, although it is not clear whether the small numbers of mold fragments represent primary deposits. If sites with foundry remains at Xuejiazhuang and Miaopu North and Xiaomintun West, Southeast, and Xiaomintun were indeed sections of the same foundries, with the bone workshops in the same general vicinities, then this would lead to the conclusion that there were several specialized precincts for craft production at the Shang capital at Anyang, covering areas more extensive than indicated by individual finds (Figure 8.5).

Dates of the Anyang Foundry Sites. Cross-dating pottery from the bronze foundry sites shows that foundry activities began as early as Yinxu Phase I at Miaopu, Xiaotun Northeast, and Dasikongcun South (Figure 8.6).[11] Foundry activities at Xiaotun Northeast, represented by foundry debris under Structure B5, continued to the Late Yinxu Phase II and may have ended sometime during the second half of the Yinxu period. The associated pottery from Xiaotun Northwest is of the Late Yinxu Phase II date, in accordance with the date suggested by the Type I mold found at the site (IA 2004:136–137, fig. 65). Foundry operations at Miaopu North began in Yinxu Phase I, and the scale of bronze production likely reached its peak during Phase II, judging by the number of trash pits containing foundry remains dating to this period; nonetheless, the overall area of the foundry continued to expand during Yinxu Phases III and IV, perhaps because of changes in workshop organization (see below). Foundry operations at Xuejiazhuang may have spanned from Yinxu Phase II to Phase IV (IA 2001:92).

The three locales with foundry remains in and near the modern village of Xiaomintun are not physically connected, and the associated pottery suggests overlapping periods of operation. Bronze production at Xiaomintun West began during Yinxu Phase II and continued to Phase III or IV (IA 1987). Xiaomintun Southeast may have begun operations in Phase II, although most of the foundry activities are dated to the second half of the Yinxu period, and the foundry may have continued to operate to the very end of the Shang occupation at Anyang (Anyang 2006; Li et al. in press). Xiaomintun is reported to be of similar dates (Yue and Liu 2006).

Before more precise dating of the foundry sites becomes available, it can only be said that the foundries at Xiaotun, Dasikongcun South, and Miaopu began producing bronzes in Yinxu Phase I. By Phase II, with the addition of Xiaomintun, all four foundries were in operation. Foundry activities at Dasikongcun South and Xiaotun Northeast at Structure B5 may have stopped sometime during Phase III. Other foundries, however, continued production, with Xiaomintun Southeast and Xiaomintun operating till the end of the Late Shang occupation at Anyang. In other words, at any given time, there were several bronze workshops in production simultaneously.

Production Procedures Represented in the Foundry Remains. In most cases, foundry remains, rather than workshop features, are the only data available for reconstructing foundry activities at Anyang. Thanks to technical studies and replication experiments, the

TABLE 8.2. Foundry Procedures Represented at each Anyang Foundry

FOUNDRY SITE	CLAY PREPARATION	MODEL MAKING	MOLD AND CORE MAKING	METAL MELTING	CASTING AND RETRIEVING CAST BRONZE	RETOUCHING AND FINISHING
Dasikongcun				X	X	
Daliankeng				X	X	
Xiaotun NE		X		X	X	
Xiaotun NW				X	X	
Miaopu N	X?	X	X	X	X	X
Xuejiazhuang				X	X	X
Xiaomintun W				X	X	X
Xiaomintun SE				X	X	X
Xiaomintun	X	X	X	X	X	X

operational sequence of the Shang section-mold casting is fairly well understood, and we are able to examine what production procedures are represented in the foundry remains. Semifinished model fragments clearly indicate model making, and completed model pieces, especially model pieces for vessel appendages, may also represent mold replicating using the model. Furnace and mold fragments were produced in the production steps of metal melting, casting, and bronze retrieving. Grinding stones, on the other hand, represent the polishing and finishing of cast bronzes after they were retrieved from mold assemblies. The production procedures represented at the foundry sites are summarized in Table 8.2. The identification of the foundry procedures observable at Miaopu North, Xuejiazhuang, Xiaotun Northwest, and Xiaomintun West is based on drawings and textual descriptions in related publications (IA 1987). The identification of foundry procedures at Xiaotun Northeast, Daliankeng, Dasikongcun South, and Xiaomintun Southeast is based on my (2003) direct observation of the artifacts. Xiaomintun is not discussed in detail, as the full site report is still forthcoming.

The size of Xiaotun Northeast is relatively small in comparison with Miaopu North or Xiaomintun Southeast (Tables 8.3, 8.4). Nonetheless, the remains from Xiaotun Northeast

TABLE 8.3. Total number of foundry remains from Xiaotun Northeast in the IHP collection

REMAINS	NUMBER
Mold	2739
Model	50
Core	99
Furnace	75
Recycled core	10
Mouth piece	2
Fired clay	2
Clay bars	3
ND	14
Total	2994

exhibit a full range of foundry activities from model making to metal melting, casting, and retrieving cast bronze from the mold. No remains indicating the mold making, retouching, and finishing phases of bronze production were found, although their existence can be inferred from the presence of the other procedures. The foundry remains from Miaopu North reflect a similar range of foundry activities from model making to bronze finishing, with the additional finds of carving tools and unfired mold fragments, which represent the stage of model and mold making. At

TABLE 8.4. Total number of core, mold, and model fragments from Xiaomintun Southeast

REMAINS	2000AG	2001AG	TOTAL	PERCENTAGE
Core	898	212	1,110	3.6
Mold	17,986	11,424	29,410	96.2
Model	25	18	43	0.1
Total	18,909	11,654	30,563	99.9

Note: Number includes remains found in secondary contexts.

Xiaomintun Southeast, however, most of the foundry debris comes from the later stages of bronze production: metal melting, casting, and bronze retrieving and finishing. There is no clear indication of model making, though the presence of a few model pieces for appendages and fired mold pieces not used for casting suggests the presence of mold-making activities. Evidence for clay preparation, model making, and mold drying was also found at Xiaomintun, though the most represented foundry procedures remain casting and bronze retrieving (Yue and Liu 2006). Xiaomintun West and Xuejiazhuang may have been used mainly as bronze-casting and bronze-finishing locales. The foundry remains at Dasikongcun South and Xiaotun Northwest are too few for much to be said about these sites; the remains represent mainly metal melting (furnace pieces) and casting activities (mold fragments). The foundry remains from Daliankeng also indicate similar casting activities.

Although it is possible that at the smaller foundry sites the model-making section of the workshop was not located by the excavators, the absence of any clear indication of model making at Xiaomintun Southeast is significant, considering the size of the uncovered area and the amount of foundry debris found. This forms a strong contrast with Xiaotun Northeast, where foundry remains, though far fewer, represent a considerable range of foundry activities. Miaopu North is the only other foundry where most of the production procedures are represented in the archaeological remains, though little can be further deduced from the published data except for the workshop layout (see below).

Range of Product Types Represented in the Foundry Remains. Comparisons between products manufactured at the different foundry sites can also be made (Table 8.5). At Xiaotun Northeast, the bronzes manufactured range from weapons and horse and chariot fittings (Figure 8.12) to ritual vessels of small, medium, and perhaps large sizes (Figure 8.13a–c), with bronze vessels accounting for most of the model and mold fragments. At Miaopu North, the main category of products seems again to be ritual vessels, though a small number of mold fragments for weapons and chariot fittings was found. The Miaopu foundry was also casting exceptionally large bronze vessels (Figure 8.14; see below). At Xiaomintun West, the identifiable core fragments suggest that the foundry mainly produced weapons and tools. Ritual vessels, probably with fewer varieties, and chariot fittings were also made. Both weapons and ritual vessels were made at Xuejiazhuang. Mold fragments at Daliankeng mainly indicate the production of bronze vessels (Figure 8.15), though a mold piece for arrowheads was also found. Almost all of the identifiable foundry remains at Xiaomintun Southeast are for ritual vessels. The types and sizes of the vessels cast also display wide variation, from small common vessels to large *ding* tripods. Of the identifiable mold fragments, only molds for vessels were found at Dasikongcun South and Xiaotun Northwest. Xiaomintun was again mainly casting bronze vessels, although very few mold fragments for weapons, tools, and chariot fittings were also found (Yue and Liu 2006:363).

Most of the Anyang foundries, located in

TABLE 8.5. Types of Bronzes Manufactured at each Anyang Foundry

FOUNDRY SITE	RITUAL VESSELS	HORSE AND CHARIOT FITTINGS	WEAPONS	TOOLS
Dasikongcun	X			
Daliankeng	**X**		X	
Xiaotun NE	**X**	X	X	X
Xiaotun NW	**X**			
Miaopu N	**X**	X	X	
Xuejiazhuang	X		X	
Xiaomintun W	X	X	**X**	**X**
Xiaomintun SE	**X**			
Xiaomintun	**X**	X	X	X

Note: The most common type of evidence is represented by a bold X.

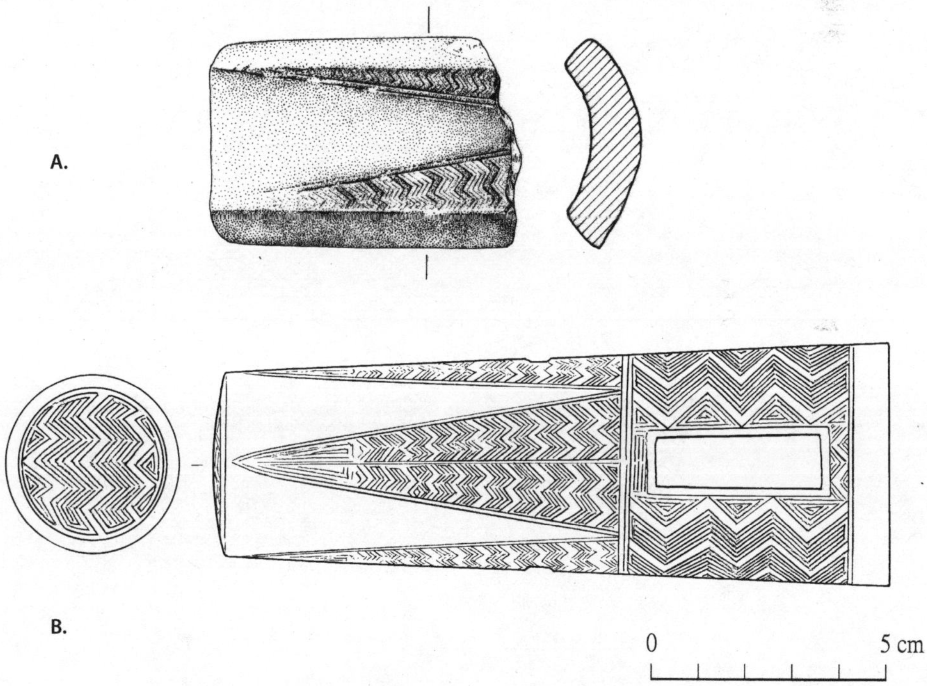

FIGURE 8.12. (a) A mold fragment from Pit YH42, Xiaotun Northeast, for a bronze axle cap. Original drawings by Lai Shu-li; copyright © the Institute of History and Philology, Academia Sinica. (b) An axle cap with identical decoration from Chariot Pit 1136-1137, Xibeigang. Original line drawing by Li Xiating; copyright © the Institute of History and Philology, Academia Sinica.

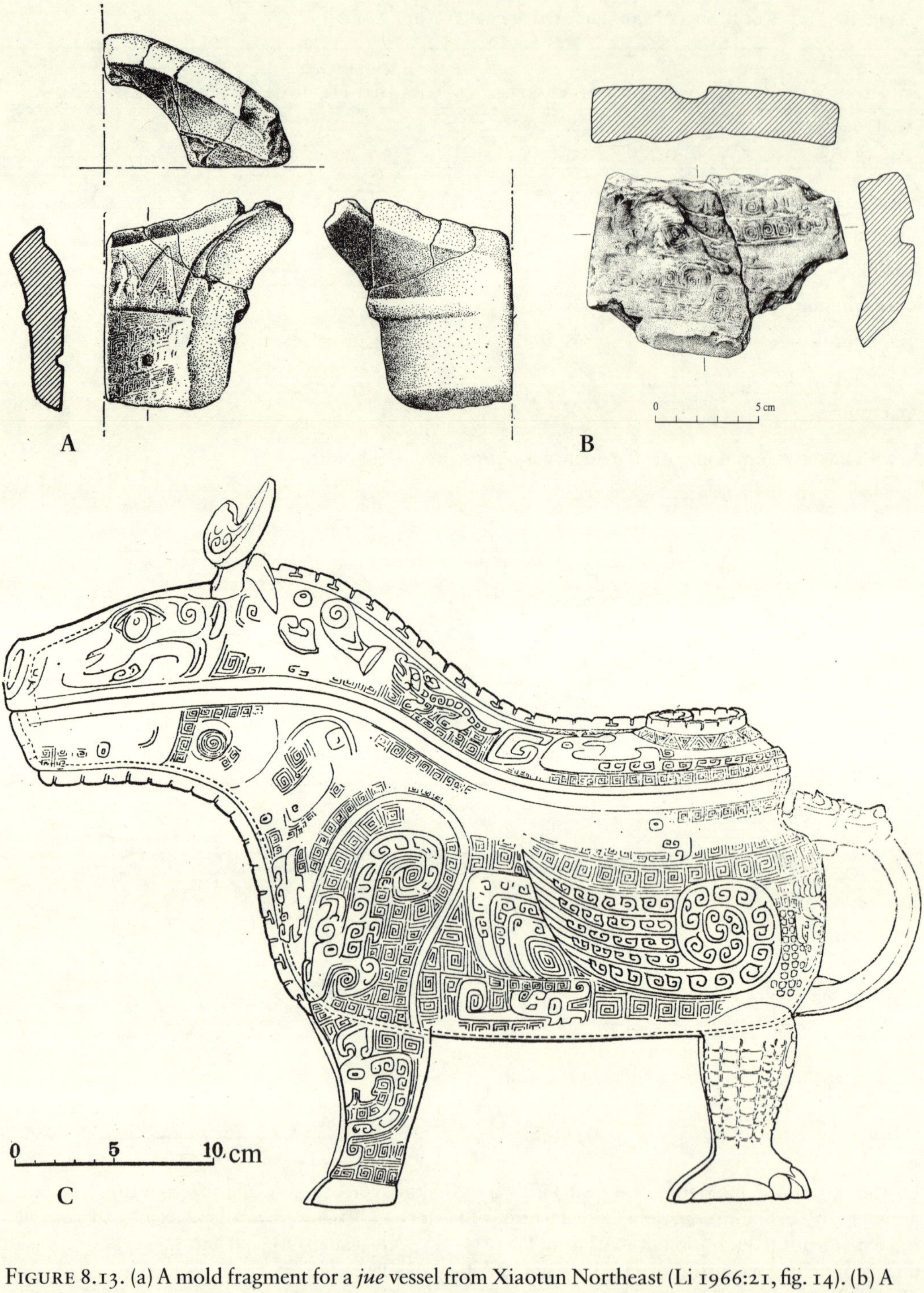

FIGURE 8.13. (a) A mold fragment for a *jue* vessel from Xiaotun Northeast (Li 1966:21, fig. 14). (b) A mold fragment from YH83, Xiaotun Northeast (original drawings by Lai Shu-li; copyright © the Institute of History and Philology, Academia Sinica), that was probably used to cast a vessel similar to (c) the one found in the tomb of Fuhao (Institute of Archaeology, Chinese Academy of Social Sciences 1985: fig. 5).

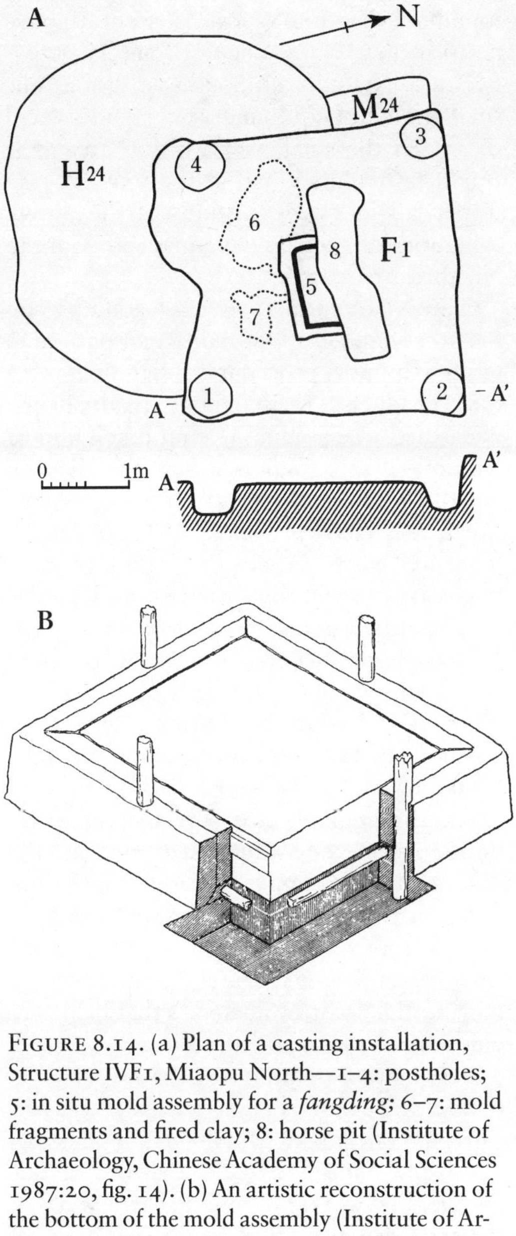

FIGURE 8.14. (a) Plan of a casting installation, Structure IVF1, Miaopu North—1–4: postholes; 5: in situ mold assembly for a *fangding*; 6–7: mold fragments and fired clay; 8: horse pit (Institute of Archaeology, Chinese Academy of Social Sciences 1987:20, fig. 14). (b) An artistic reconstruction of the bottom of the mold assembly (Institute of Archaeology, Chinese Academy of Social Sciences 1987:49, fig. 35).

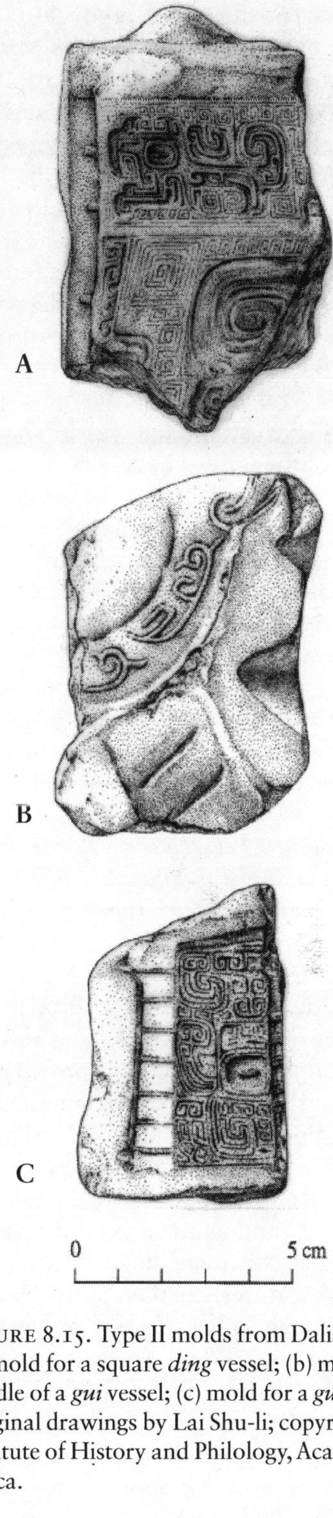

FIGURE 8.15. Type II molds from Daliankeng: (a) mold for a square *ding* vessel; (b) mold for the handle of a *gui* vessel; (c) mold for a *gu* vessel. Original drawings by Lai Shu-li; copyright © the Institute of History and Philology, Academia Sinica.

the core area or not, were producing high-status bronzes. For instance, at the Miaopu foundry, an attempt was made to cast a vessel very close in size to the Si Mu Wu *fangding*, the largest Shang bronze known (Figure 8.14; IA 1987). At Xiaotun Northeast, mold fragments for large ritual vessels were identified, including a vessel that may be similar to the zoomorphic *zun* unearthed from the tomb of Fuhao (Figure 8.13b–c). Horse and chariot fittings were another type of high-status bronze. Although technologically simpler to cast, the use of horse and chariot fittings must have been highly exclusive, as access to chariots with bronze fittings must have been limited to the elite of the highest rank. The foundry at Xiaomintun Southeast cast a number of rarely seen bronzes with a new decorative style (Anyang 2006; Li 2003; Li et al. in press), and the most recent finds at Xiaomintun village include the remains of a casting installation for a large *ding* (Yue and Liu 2006:360, fig. 2). The diameter of the installation suggests that the vessel would have been larger than any known bronzes from the Yinxu period.

From Holistic to Prescriptive? Reconstructing the Production Organization of the Anyang Foundries. Franklin's thesis on holistic and prescriptive technologies has long been the only existing model for craft production and production organization in Bronze Age China. Nonetheless, her model remains descriptive and is not intended to provide further analytical means to study the archaeological remains. The parameters identified in the comparative study of pottery production at Jingdezhen and Yixing, that is, the range of products and production procedures represented at a given location and the distribution of workshop features, can therefore be used to examine the archaeological remains from the Anyang bronze workshops to identify and reconstruct the organization of production (Li 2003, 2007).[12]

Of all the Anyang foundry sites, Xiaotun Northeast and Xiaomintun Southeast form the most interesting contrast in terms of foundry procedures and the range of products observed in the foundry remains. The bronze foundry at Xiaotun Northeast seems to have performed a wider range of foundry activities than Xiaomintun Southeast despite the smaller quantity of foundry debris uncovered there and the smaller excavation area. The products manufactured at the foundry also display a wider variety. Almost all the known categories of bronzes were made at Xiaotun Northeast.

The Xiaomintun Southeast foundry, on the other hand, displays a different pattern in terms of foundry procedures and the manufactured products. The amount of foundry debris unearthed at Xiaomintun Southeast is almost ten times greater than that found at Xiaotun Northeast (Tables 8.3–8.4). Nevertheless, the site seems to have been involved mainly in the latter half of the bronze production process, from casting to finishing the cast bronzes. The manufactured items are overwhelmingly ritual vessels of various kinds, from the more common types of gu, *ding*, *you*, and *gui* to the more elaborated forms of bronzes with flamboyant appendages that may have marked the beginning of a new style (Li et al. in press).

The different degree of specialization manifested in their operational sequence and the different ranges of manufactured items found at Xiaotun Northeast and Xiaomintun Southeast deserve explanation. Returning to the working models of Yixing and Jingdezhen, we may be able to draw analogies that help provide tentative interpretations.

The material patterns in the foundry debris from Xiaotun suggest that workers performed model making, mold making, casting, and bronze finishing in close vicinity to one another and in a more confined area and that the same group of workers was in charge of the entire production process from the stage of making the model to the end of finishing the bronze. In comparison, workers at the particular location at Xiaomintun Southeast were mainly in charge of a relatively limited range of tasks in the workshop, from casting to retrieving and finishing bronzes. The wide diversity represented in the vessel forms and styles in the mold fragments, which must have required different skills and techniques in model and mold making, may be another indication that workers at that particular locale were mainly involved in the casting operation and not the vessel-forming part of the production sequence. The fact that the Xiaomintun West foundry focused mainly on tools and weapons further suggests specialized production among different workshops at Xiaomintun.

The tentative conclusion that can be drawn from the above discussion is that the Xiaotun Northeast foundry exhibits features that may more closely resemble the holistic production model and may have had a form of production organization based on cellular work units, with a group of workers performing a fuller range of the production procedures. At Xiaomintun Southeast, it is likely that workers were divided into specialized units according to the tasks they performed, and different operational sequences were probably performed at different locales within the workshop. In other words, the production organization at Xiaomintun Southeast seems more akin to a form of flow-line production and closer to a prescriptive technology. As Xiaotun Northeast was in operation mainly in the first half of the Yinxu period, and Xiaomintun Southeast, the second half, it is likely that the differences in the range of production procedures and finished products represent a temporal change in the production organization.

Additional evidence for changes in the workshop organization of Shang bronze foundries comes from Miaopu North, the only foundry site with published information on workshop facilities. The Miaopu North foundry was in operation throughout the Yinxu period. Although the range of production procedures and product types suggests a more holistic form of production organization, the paucity of published data about the foundry debris does not allow further analysis. More information, however, can be obtained from the workshop layout. The foundry features at Miaopu include structures with rammed-earth foundations and walls, sheds for casting activities, and working areas with prepared floors, some of which may have been equipped with underground water pipes (IA 1987, 2001; for an English summary, see Li 2003). Although the exact function of many of these features cannot be established because of poor preservation conditions, reconstructed maps of the foundry features indicate that the workshop expanded southward over time, and the distribution of the rammed-earth structures may have changed from a more clustered pattern to a more dispersed one, a shift that also took place between the first half and the second half of the Yinxu period (Figure 8.16). When considered together with the changing patterns in the range of products and production procedures, the change in the spatial layout may also suggest a shift from a more holistic production organization to a more prescriptive one. The presence of not one but two forms of production organization at Anyang suggests that Franklin's theory of Shang bronze production needs further revision. Although Franklin's model is instructive, the archaeological data present a more complex picture.

Revising Franklin's Hypothesis. The more holistic form of production organization may have already existed in the earlier Erligang bronze foundries, Nanguanwai and Zijingshan (for the two foundry sites, see Henan Provincial Institute of Cultural Relics and Archaeology 2001; see also Liu and Chen 2003: 95–96; Thorp 2006:85–99). The excavators argue that there may have been a division of production between the two Erligang bronze workshops. The foundry at Nanguanwai specialized in making tools, and the foundry at Zijingshan produced both weapons and tools. Both workshops, however, were also casting ritual bronzes, even though in very small quantities.[13] The range of products at the two foundries is therefore closer to what we have observed at Xiaotun Northeast at Anyang, with craftspersons manufacturing a full repertoire of bronzes at a single location. The situating of bronze workshops outside of the walled enclosure and away from the main settlement can also be contrasted sharply with the Anyang examples and may indicate a more centralized control over bronze production.

Although Franklin's distinction between holistic and prescriptive production emphasizes the absence or presence of managerial

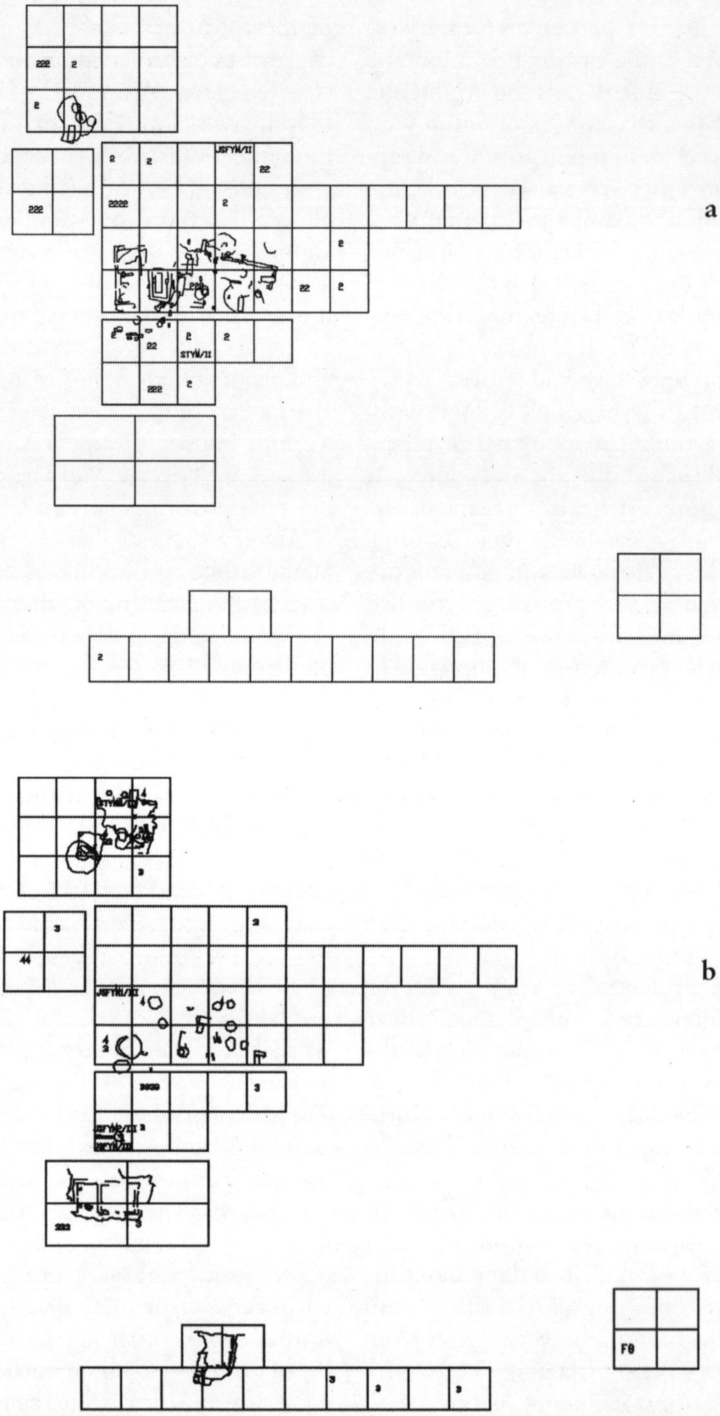

FIGURE 8.16. Distribution of foundry features during (a) Yinxu Phase II and (b) Yinxu Phase III and IV at Miaopu North. Numerals represent trash pits found in the excavation trenches.

control, the changes in production organization observed between Xiaotun Northeast and Xiaomintun Southeast should not be seen as indications for the rise of managerial control during the latter half of the Yinxu period. In the case of bronze production in early China, managerial or elite control was already present before the Yinxu period (Bagley 1999; Liu and Chen 2003). As has been argued by Bagley (1999:137–142), the qualitative change in bronze production in terms of elite control had already occurred in the Erlitou culture, when the small-scale metal production of tools, mirrors, and ornaments of the previous Qijia culture (2000 B.C.) was replaced by the large-scale production of ritual bronzes at Erlitou.

The changes in production organization at Anyang bronze foundries may in fact have been a response to increased demand. Anyang archaeologists have argued that the area of occupation at Anyang doubled during the second half of the Yinxu period and that the distribution of settlements and cemeteries expanded westward (IA 2001:42). The number of burials in the larger cemeteries such as those in the western sectors and at Miaopu North also increased during the latter half of the Yinxu period (Anyang 1979; IA 1987). It is conceivable that with the increase in population, including the size of the elite, and the expansion of the Shang occupation at Anyang, the internal demand for bronzes also increased.

The presence of Anyang-style bronzes outside the Anyang region is usually seen as an indication of the Shang state's political influence. However, when compared with the Erligang period, the political influence Anyang had over its neighboring regions had apparently diminished (Bagley 1977, 1999). The oracle bone inscriptions of the Late Shang period indicate that the Shang court constantly engaged in military actions with or against the neighboring regions (e.g., Keightley 1983, 1999). In its military campaigns, it was often aided by allies, who could also become enemies at another time. Bronzes—as political paraphernalia necessary for the legitimization of power—could also be used as a diplomatic tool through which to establish and maintain political alliances. A decline in Shang power during the Anyang period might have meant that more bronzes were needed to establish outside links and alliances. The Shang court might have needed to produce more bronzes to pacify the political and military foes around its border, such as the increasingly powerful Zhou polity in the west. Within the Shang state, too, the Shang court might also have needed to grant more political gifts to its secondary elite in the core area and to subordinate polities outside Anyang to consolidate existing political ties.

Who Ran the Workshops? Control and Patronage of Craft Production at Anyang

To summarize points made in the previous discussion, craft production at Anyang was characterized by the presence of multiple and contemporaneous workshops for the same crafts, the close proximity of multiple craft-producing activities, and the juxtaposition of craft production in both the core area and the local craft production precincts. The decentralized and dispersed pattern of craft-producing activities at Anyang raises questions about the nature of the control and patronage of craft production. As will be shown below, the Shang court employed different forms of control not only over different workshops but also over different crafts.

Depending on the forms of political and social structure the Shang court maintained at its capital, craft production at the Shang capital could have been organized and conducted in one of the following ways:

1. The Shang court kept highly centralized control over all aspects of the craft production precincts, from procuring raw materials and manufacturing to the distribution of finished products.
2. Craft production at Yinxu was conducted through a loosely organized network of production centers controlled by secondary elites at the nearby settlements. The secondary elites were in charge of procuring raw materials and managing workshop operations. Sumptuous items for royal use were supplied to the court by the individual production centers.

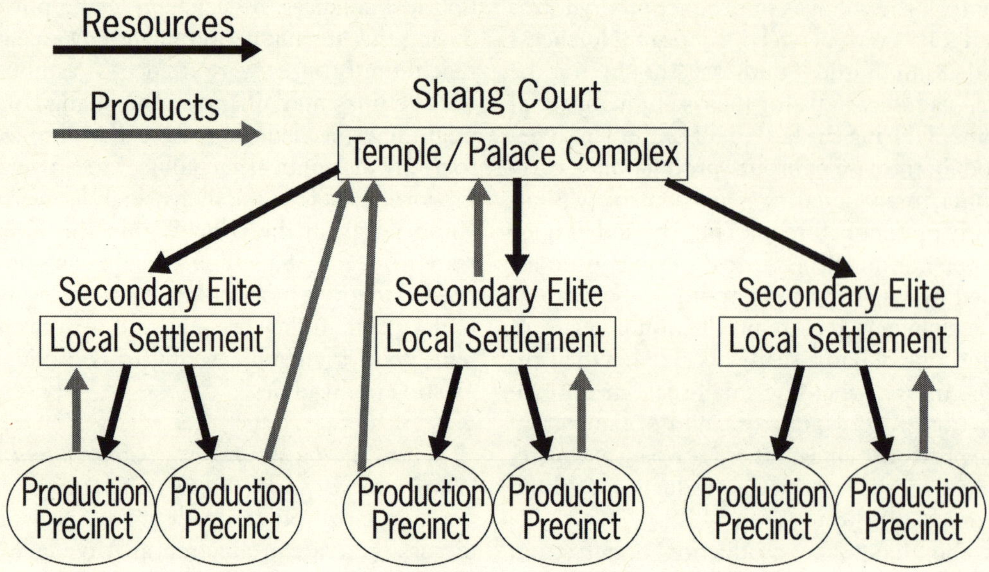

Figure 8.17. A hypothetical model for the organization of craft production at Anyang.

3. The Shang court controlled the procurement and distribution of raw materials for craft production. Each production precinct obtained the necessary resources from the court, with the actual craft production being managed by the secondary elites residing in the nearby settlements. The Shang court also had direct and perhaps total control over the finished products.

Although the current state of research has yet to provide definite answers, I would nonetheless argue that the textual and archaeological data point to the last scenario as most plausible (Figure 8.17).

Oracle bone inscriptions bear surprisingly little information that can be directly linked to craft production. Craft production seems to have existed outside of the main areas of concern for which the Shang king regularly sought divination. Although the outcome of the royal hunt was often more eagerly divined, the number of successful and unsuccessful casting operations in bronze foundries, for instance, does not seem to have been of much interest, and there is only one instance that may represent an inquiry made in relation to bronze casting (Yan 1973). Oracle bone inscriptions also mention the taking of turtle shells, presumably the raw material to be prepared for divination use, and the official title of *gong*, which in the later periods was associated with craft production. Beyond these few and scattered examples, there is little textual evidence about craft production in Anyang.

The Shang king clearly enjoyed absolute power. The large-scale human sacrifices at Xibeigang as well as the ability to mobilize labor for communal work and military campaigns point to political power of a coercive nature that characterizes a powerful kingship (Chang 1980; Keightley 1999). For Keightley, the human sacrifices "indicate that similar demands for service and obedience must surely have been part of the idealized, and in all likelihood actual, political culture of the living" (1999:290). In contrast to the strong kingly power, the bureaucratic system of the Shang state is often described as "incipient," and official titles seem to have been flexible, with their function unspecialized (Keightley 1999:286–288; see also Zhang 1986). The many officials and princes mentioned in oracle bone inscriptions also suggest "a lack of routine administrative delegation" in the Shang court (Keightley 1999:290). This combination of a strong kingship and a loosely organized incipient bureaucratic system observed in oracle bone inscriptions seems to have been paralleled in the realm of craft production.

Of all the workshop sites, only the foundries at the palace/temple complex, the Huayuanzhuang bone refuse pit, and the lithic workshops are situated within the Anyang core area surrounded by the moats. The central location suggests that the operation of these workshops was closely related to the Shang court, a view supported by the production of high-status bronze vessels and horse and chariot fittings at Xiaotun Northeast and by the jade and marble objects produced at Xiaotun North and Daliankeng.

However, foundries outside of the core area, such as Miaopu North, Xiaomintun, and Xiaomintun Southeast, were also making exceptionally large and stylistically striking vessels. These "hypertrophic" vessels (Clark and Parry 1990) could not have been used by members outside of the royal court. The manufacturing of weapons, which is the ultimate means for exercising coercive force, must also have been carefully overseen by the ruling elite, as exemplified by the later Warring States period practice of inscribing the names of craftspersons and supervising officials on bronze weapons (Huang 1974; Li 1985). The fact that the precinct workshops outside the core area were manufacturing such objects shows that the distribution of the finished products must have been controlled by the ruling elite to ensure their exclusive use of prestige items. In fact, the privilege of the lesser elite to operate bronze foundries may well have come with the prescribed obligation to produce prestige goods for royal use.

High-status bronzes, however, formed only a very small fraction of the products produced at the precinct workshops. Mold fragments from the precinct foundry sites indicate that vessels of common types and simple decoration, such as gu, jue, gui, and *you*, were the main products (Li 2003). Bronze production at the precinct workshops was therefore designed mainly to meet the needs of the lesser elite in the nearby settlements.

The dispersed pattern of bronze production may seem perplexing, as the technology and the procurement of raw materials would have stipulated tight administrative supervision and control. However, the extensive use of bronze ritual vessels among the Shang elite in feasting and mortuary practices (e.g., Underhill 2002) may have generated a large enough client base whose demand could have necessitated the more dispersed distribution of bronze production. At a foundry the size of Xiaomintun, the large-scale workshop operation would have created a noxious, hazardous, and unsanitary environment. The massive dump of foundry debris and the constant heat and smoke may have rendered the production of bronzes an undesirable craft activity that needed to be located away from the core area. This may also explain the small scale of foundry operation in the core area. Given the nature of the Shang rulership, bronze production in the precincts could still be guarded and controlled by the court, and one of the means for such control would have been through the administered distribution of the raw materials (Figure 8.17).

Bone production at the precincts, on the other hand, was oriented to meet only local demands. Although yielding large amounts of production debris, the precinct bone workshops consistently produced products of less prestigious and even utilitarian nature. It is conceivable, as in the case of lapidary working, that the skilled artisans for producing highly decorative bone objects were retained in the core area, under the direct supervision of the Shang court.

The lithic workshops present yet another picture. The fact that the only two known lithic workshops were located in the core area suggests that the Shang king was more concerned with control over lapidary production than bronze or bone production. Compared with the case for bronzes, the intended clients and the demand for jades and especially marble were more restricted, and once the raw materials were quarried and transported to the capital, lithic production would in fact require less labor and workshop space than bronze production. The scale and size of lithic production, therefore, would have been more manageable if placed in the core area. The elaborately constructed structure and the central location also indicate that the workers enjoyed higher social status.

The type of control the Shang king exercised over craft production at the capital of Anyang therefore seems to have been determined by the exclusiveness of the prestige goods, the demand, and the logistics of workshop operation. The production of the less prestigious items, such as hairpins, was consigned to the secondary elite and took place outside of the core area. The closely supervised craft-producing activities, that is, jade and marble working, were those that produced objects that in some way conveyed the most culturally and politically charged messages by way of excessive labor investment in the procuring and especially the manufacturing stages (Clark and Parry 1990; DeMarrais et al. 1996; Schortman and Urban 2004). Even though bronzes were an integral part of the Shang elite culture, the production organization of bronzes at Anyang may have been a compromise between exercising control and meeting the logistics of workshop operations.

CONCLUDING REMARKS

The importance of craft production as a research topic has yet to be widely recognized in the field of Anyang, if not Chinese, archaeology. The current archaeological data at Anyang therefore do not provide the same resolution for studying craft production as seen in other ancient civilizations. The notion of "multicraft production" has been explored here through the dual nature of section-mold casting technology, which can better be termed as "co-craft" production, and in the way the Shang court organized different craft production activities in its capital. Emphasis has also been placed on bronze production at Anyang, for despite its technological complexity, bronze production remains the best-understood craft of the Shang civilization.

It is as yet unclear how raw materials for craft production at Anyang were acquired and distributed. The extraordinary amount of bones found in Huayuanzhuang in the core area suggests some form of centralized mechanism for, and control over, the distribution of raw materials for bone working. The procurement of ores and fuels for bronze production must have involved even more coordination and managerial efforts among different sectors of the Shang society. Acquiring raw materials for marble and lapidary work also required administered and coordinated efforts, although the current state of research allows little further analysis. Systematic studies on the sourcing of exotic materials imported to Anyang have just begun, and the results should enhance our understanding of the nature of the Shang court's interaction with its neighboring regions.[14]

Admittedly, the discussion of Anyang craft production in this chapter is still limited to the traditional "top-down" approach that focuses mainly on the elite aspects of production. Many scholars in the field, including authors in the current volume, have stressed the importance of conducting research at the levels of households, individual craft producers, and social agents (e.g., Costin 1998; Dobres and Hoffman 1994; Feinman 1999; Schortman and Urban 2004; Sinopoli 2003; Wright 1998). To respond to such theoretical approaches, those of us who focus on Bronze Age China will need to reorient field strategies and consciously seek information that leads to an understanding of not just the elite, who composed only a small fraction of the Shang society, but also the artisans and the commoners, the populace that have been largely overlooked in the archaeological data.

Notes

1. Dates for the early Chinese Bronze Age cultures in this chapter follow those in Bagley 1999, which tend to be more conservative than dates given in the Chinese archaeological literature.
2. For a more detailed discussion of replication experiments on section-mold casting, see Li 2003.
3. The breakdown of the manufacturing steps in Figure 8.5 is based on Tan's textual descriptions. However, Tan's descriptions of the manufacturing steps for each vessel type are not standardized. In addition, the number of

workdays Tan provides is sometimes a lump sum of the amount of time spent on several steps in the manufacturing sequence. For instance, though model making involves stages such as preparing the clay, shaping the model, carving decorations, and firing the model, only the total time of model making is given, whereas mold making is broken down into finer manufacturing steps for the calculation of invested time.

4. A comprehensive summary of Anyang archaeology can be conveniently found in Institute of Archaeology 2001 (in Chinese). English readers can find summaries of data and in-depth discussions on Anyang archaeology in Chang 1980, 1986; and Bagley 1999. The recent book by Thorp (2006) offers concise English summaries of Anyang archaeological research that also include the more recent finds. For studies of oracle bones in English, see various works by Keightley (1978, 1999, 2000). The now-outdated book by Li Chi (1977) nonetheless provides a firsthand account of the Institute of History and Philology, Academia Sinica, Anyang excavations from the perspective of the pioneers of Anyang archaeology.

5. The recent discovery of a walled settlement, Huanbei, located north of the Huan River, has significantly changed the picture of Anyang archaeology. It is now known that the Huanbei walled settlement predates the Anyang or Yinxu settlement south of the river. See Thorp 2006 and Tang et al. 2000 for English publications on the find.

6. In the Chinese archaeological literature, these walled settlements are often termed *cheng*, and the enclosing rammed-earth walls, *cheng qiang*. The common English translation of these terms as "city" or "urban center" and "city walls" can be misleading, for in many cases the settlement layout and the nature and function of the settlement within the enclosing rammed-earth walls are not clearly understood. For a discussion of walled settlements in ancient China, see Yates 1997.

7. See, for instance, Ding 1959; Lin 1979; and Qiu 1983. A detailed synthesis in Chinese of the available epigraphic, textual, and archaeological data can be found in Zhu 1990. See Chang 1976 and Keightley 1999, 2000, for English discussions on the issue.

8. An earlier version of the following discussion was presented at the "Workshop on Early Chinese Civilization: Social, Cultural, and Historical Dynamics of Early Chinese Civilization in Archaeological, Palaeographical, Philological, and Historical Perspectives," held at the University of British Columbia, British Columbia, March 10–12, 2005.

9. The fact that all craft production sites found at Anyang so far are attached workshops with large outputs may be related to the prevalence of salvage archaeological projects, which focus mainly on burials and sites with rich cultural deposits. Until recently, there have been relatively few excavations focusing on residential areas at Anyang. We therefore know very little about craft production at the household level.

10. For discussions of Type I and Type II molds from Anyang, and their possible dates, see Li 2003.

11. The foundry remains from Xiaotun Northeast and Dasikongcun can only be approximately dated based on the remaining complete pottery vessels and the stylistic attributes of the mold fragments now housed in the Institute of History and Philology (Li 2003, in press).

12. It must be noted, however, that as yet there is no clear picture of what types of installations and facilities existed in all the bronze workshops at Anyang (Li 2003). Miaopu North and Xiaomintun are the only two foundry sites containing evidence of workshop features. The formation process of the workshop refuse deposits at sites such as Dasikongcun South and Xiaotun Northwest has also not been clearly established. The current data consist mainly of foundry debris unearthed at refuse areas or trash pits that may have been near the production area. They are therefore rather limited for the purpose of studying the production organization.

13. The excavators report finding mold fragments for chariot fittings. The line drawings and photos in the site report, however, do not provide conclusive answers.

14. Current projects include sourcing for pottery (Jing Zhichun, personal communication, 2006) and jades (Chen Kwang-tzu, personal communication, 2006).

References Cited

Aitchison, Leslie

1960 *A History of Metals: Volume One*. MacDonald and Evans Ltd., London.

An Chin-huai (An Jinhuai)

1986 The Shang City at Cheng-chou (Zhengzhou) and Related Problems. In *Studies of Shang Archaeology*, edited by K. C.

Chang, pp. 15–48. Yale University Press, New Haven.

Anyang Archaeological Team, Institute of Archaeology, Chinese Academy of Social Sciences

1961 1958–1959 nian Yinxu fajue jianbao (Preliminary Report of the 1958–1959 Excavations). *Kaogu* 2:63–76.

1976 1975 nian Anyang Yinxu de xin faxian (New Discoveries at Yinxu, Anyang in 1975). *Kaogu* 4:263, 264–272.

1979 1969–1977 nian Yinxu xiqu muzang fajue baogao (Excavations of the Cemeteries in Western Yinxu between 1969 and 1977). *Kaogu xuebao* 1:27–146, pls. 1–20.

1992 1986–1987 nian Anyang Huayuanzhuang nandi fajue baogao (Excavation Report of the 1986–1987 Seasons at Huayuanzhuang South). *Kaogu xuebao* 1:97–128, pls. 15–20.

2006 2000–2001 nian Anyang Xiaomintun dongnandi Yin dai zhutong yizhi fajue baogao (Excavation Report of the 2000 and 2001 Seasons at the Shang Bronze Foundry at Xiaomintun Southeast, Anyang). *Kaogu xuebao* 3:351–384, pls. 9–16.

Bagley, Robert W.

1977 P'an-lung-ch'eng: A Shang City in Hupei. *Artibus Asiae* 39(3–4):165–219.

1987 *Shang Ritual Bronzes in the Arthur M. Sackler Collections*. Arthur M. Sackler Foundation, Washington, D.C.

1990 Shang Ritual Bronzes: Casting Techniques and Vessel Design. *Archives of Asian Art* 43:6–20.

1993 Replication Techniques in Eastern Zhou Bronze Casting. In *History from Things: Essays on Material Culture*, edited by S. Lubar and W. D. Kingery, pp. 231–241. Smithsonian Institution Press, Washington, D.C.

1995 What the Bronzes from Hunyuan Tell Us about the Foundry at Houma. *Orientations* 26(1), January: 46–54.

1996 Debris from the Houma Foundry. *Orientations* 27(9), October: 50–58.

1999 Shang Archaeology. In *The Cambridge History of Ancient China: From the Origins of Civilization to 221 B.C.*, edited by M. Loewe and E. L. Shaughnessy, pp. 124–231. Cambridge University Press, Cambridge.

Barnard, Noel

1961 *Bronze Casting and Bronze Alloys in Ancient China*. Australian National University and Monumenta Serica, Canberra.

1986 A New Approach to the Study of Clan-Sign Inscriptions of Shang. In *Studies of Shang Archaeology*, edited by K. C. Chang, pp. 141–206. Yale University Press, New Haven.

Chang, Kwang-chih

1976 *Early Chinese Civilization: Anthropological Perspectives*. Harvard University Press, Cambridge.

1980 *Shang Civilization*. Yale University Press, New Haven.

1983 *Art, Myth, and Ritual: The Path to Political Authority in Ancient China*. Harvard University Press, Cambridge.

Chase, W. Thomas

1983 Bronze Casting in China: A Short Technical History. In *The Great Bronze Age of China: A Symposium*, edited by G. Kuwayama, pp. 100–123. Los Angles County Museum of Art, Los Angles.

1991 *Ancient Chinese Bronze Art: Casting the Precious Sacral Vessel*. China Institute in America, New York.

Chen Guoliang

2006 Erlitou wenhua tongqi zhizuo jishu gaishu (Bronze Technology of the Elitou Culture). *Sandai kaogu* 2:183–220.

Clark, John E., and William J. Parry

1990 Craft Specialization and Cultural Complexity. *Research in Economic Anthropology* 12:289–346.

Costin, Cathy L.

1998 Introduction: Craft and Social Identity. In *Craft and Social Identity*, edited by Cathy L. Costin and Rita P. Wright, pp. 3–18. Archaeological Papers of the American Anthropological Association. Arlington, Virginia.

DeMarrais, Elizabeth, Luis J. Castillo, and Timothy K. Earle

1996 Ideology, Materialization and Power Strategies. *Current Anthropology* 37(1): 15–32.

Ding Shan

1959 *Jiaguwen suo jian shizu ji qi zhidu* (*Clan System Seen in Oracle Bone Inscriptions*). Science Press, Beijing.

Dobres, Marcia-Anne

1999 Technology's Links and *Chaîne*: The Processual Unfolding of Technique and Technician. In *The Social Dynamics of Technology: Practice, Politics, and World Views*, edited by Marcia-Anne

Dobres and Christopher R. Hoffman, pp. 124–146. Smithsonian Institution Press, Washington, D.C.

2000 *Technology and Social Agency: Outlining a Practice Framework for Archaeology*. Blackwell Publishers, Oxford.

Dobres, Marcia-Anne, and Christopher R. Hoffman

1994 Social Agency and the Dynamics of Prehistoric Technology. *Journal of Archaeological Method and Theory* 1(3):211–258.

Fairbank, Wilma

1972 [1965] Piece-Mold Craftsmanship and Shang Bronze Design. In *Adventures in Retrieval: Han Murals and Shang Bronze*, by Wilma Fairbank, pp. 181–201. Harvard University Press, Cambridge. (Original: with Wan Chia-pao, Shang Bronze Chüeh Legs: Carved in the Mold, in *Symposium in Honor of Dr. Li Chi on His Seventieth Birthday*, pt. 1, edited by Li Fang-kuei, pp. 275–280, Academia Sinica, Taipei.)

Feinman, Gary M.

1999 Rethinking Our Assumptions: Economic Specialization at the Household Scale in Ancient Ejutla, Oaxaca, Mexico. In *Pottery and People: A Dynamic Interaction*, edited by James M. Skibo and Gary M. Feinman, pp. 81–98. University of Utah Press, Salt Lake City.

Feng Fugen, Wang Zhenjiang, Bai Rongjin, and Hua Jueming

1980 Shangdai qingtongqi shi zhu jianbao (Preliminary Report of Replication Casting of Shang Bronzes). *Kaogu* 1:91–94.

Feng Fugen, Wang Zhenjiang, Hua Jueming, and Bai Rongjin

1982 Yinxu chutu Shang dai qingtong gu zhuzao gongyi de fuyuan yanjiu (Reconstruction Studies of Shang Bronze *Gu*-Vessels from Yinxu). *Kaogu* 5:527, 532–541.

Fong Wen

1980 *The Great Bronze Age of China: An Exhibition from the People's Republic of China*. Metropolitan Museum of Art, New York.

Franklin, Ursula M.

1983a The Beginnings of Metallurgy in China: A Comparative Approach. In *The Great Bronze Age of China: A Symposium*, edited by G. Kuwayama, pp. 94–99. Los Angeles County Museum of Art, Los Angeles.

1983b On Bronze and Other Metals in Early China. In *The Origins of Chinese Civilization*, edited by D. N. Keightley, pp. 279–296. University of California Press, Berkeley.

1992 *The Real World of Technology*. Anansi, Ontario.

Gettens, R. J.

1967 Joining Methods in the Fabrication of Ancient Chinese Bronze Ceremonial Vessels. In *Application of Science in Examination of Works of Art: Proceedings of the Seminar, September 7–16, 1965. Conducted by the Research Laboratory, Museum of Fine Arts, Boston*, pp. 205–217. Museum of Fine Arts, Boston.

1969 *The Freer Chinese Bronzes, vol. 2: Technical Studies*. Smithsonian Institution, Washington, D.C.

Guo Moruo

1961 Yin yi zhong tuxing wenzi zhi yi jie (One Interpretation for the Pictographs on Shang Bronzes). In *Yinzhou qingtongqi mingwen yanjiu* (*Studies of Yin and Zhou Bronze Inscriptions*), by Guo Moruo, pp. 11–17. Science Press, Beijing.

Henan Provincial Institute of Cultural Relics and Archaeology

2001 *Zhengzhou Shang cheng: 1953–1985 nian kaogu fajue baogao* (*The Shang City at Zhengzhou: A Report on the 1953–1985 Excavations*). 3 vols. Wenwu Press, Beijing.

Hua Jueming

1986 Shila fa zai Zhongguo de qiyuan he fazhan (The Origin and Development of Lost-Wax Casting in China). In *Zhongguo yezhu shi lunji* (*Collected Essays on the History of Metallurgy in China*), edited by Hua Jueming et al., pp. 226–247. Wenwu Press, Beijing.

Hua Jueming, Feng Fugen, Wang Zhenjiang, and Bai Rongjin

1981 Fuhao mu qingtongqi qun zhuzao jishu de yanjiu (Study of Casting Methods of Bronzes from the Tomb of Fuhao). *Kaoguxue jikan* 1:244–272.

Huang Long

1993 Sunan chuantong zhuzao diaocha yu qishi (Investigation of Traditional Casting Techniques in Southern Jiangsu). *Wenwu baohu yu kexue kaogu* 5(1):31–34.

Huang Shengzhang

1974 Shilun San Jin bingqi de guobie he niandai ji qi xiangguan wenti (Dates,

Origins of Manufacture, and Other Related Issues of Weapons from the Three Jin States). *Kaogu xuebao* 1:13–44.

Institute of Archaeology, Chinese Academy of Social Sciences

1980 *Yinxu Fuhao mu* (*The Tomb of Fuhao, Yinxu*). Wenwu Press, Beijing.

1985 *Yinxu qingtongqi* (*Yinxu Bronzes*). Wenwu Press, Beijing.

1987 *Yinxu fajue baogao: 1958–1961* (*Excavation Report of Yinxu, 1958–1961*). Wenwu Press, Beijing.

2001 *Yinxu de faxian yu yanjiu* (*Archaeological Excavations and Researches at Yinxu*). Science Press, Beijing.

2003 *Zhongguo kaoguxue: Xia Shang juan* (*Chinese Archaeology: Xia and Shang*). China Social Sciences Press, Beijing.

2004 *Anyang Xiaotun* (*Report of Excavations Near the Village of Xiaotun, Anyang*). Shijie Tushu Press, Beijing.

Kao Ch'ü-hsün (Gao Quxun)

In press *Dasikong cun di er ci fajue baogao* (*Excavation Report of the Second Excavation Season at Dasikongcun*). Institute of History and Philology, Academia Sinica, Taipei.

Karlbeck, Orvar

1935 Anyang Moulds. *Bulletin of the Museum of Far Eastern Antiquities, Stockholm* 7:39–60.

Keightley, David N.

1978 *Sources of Shang History: The Oracle-Bones Inscriptions of Bronze Age China*. University of California Press, Berkeley.

1983 The Late Shang State: When, Where, and What? In *The Origins of Chinese Civilization*, edited by D. Keightley, pp. 523–564. University of California Press, Berkeley.

1999 The Shang: China's First Historical Dynasty. In *The Cambridge History of Ancient China: From the Origins of Civilization to 221 B.C.*, edited by M. Loewe and E. Shaughnessy, pp. 232–291. Cambridge University Press, Cambridge.

2000 *The Ancestral Landscape: Time, Space, and Community in Late Shang China (ca. 1200–1045 B.C.)*. China Research Monograph 53. Institute of East Asian Studies, University of California, Berkeley.

Kuo Pao-chün (Guo Baojun)

1933 B qu fajue ji zhi yi (Excavation at Area B, pt. 1). *Anyang fajue baogao* 4:579–596.

Lemonnier, Pierre

1992 *Elements for an Anthropology of Technology*. Anthropological Papers, Museum of Anthropology, University of Michigan, No. 88. Ann Arbor.

Li Chi (Li Ji)

1930 Minguo shi ba nian qiuji fajue Yinxu zhi jingguo ji qi zhongyao faxian (The 1929 Fall Excavation at Yinxu and Its Important Finds). *Anyang fajue baogao* 2:219–252.

1977 *Anyang*. University of Washington Press, Seattle.

Li Chi (Li Ji) (editor)

1966 *Gu qiwu yanjiu zhuankan di er ben: Yinxu chutu qingtong jue xing qi zhi yanjiu* (*Archaeologia Sinica, New Series, No. 2: Studies of the Bronze Chüeh-cup*). Institute of History and Philology, Academia Sinica, Taipei.

Li, Xueqin

1985 *Eastern Zhou and Qin Civilizations*. Translated by Kwang-Chih Chang. Yale University Press, New Haven.

Li, Yung-ti

2003 The Anyang Bronze Foundries: Archaeological Remains, Casting Technology, and Production Organization. Unpublished Ph.D. dissertation, Anthropology Department, Harvard University.

2007 Specialization, Context of Production, and Alienation in the Production Process: Comments and Afterthoughts. In *Rethinking Specialized Production in Archaeological Contexts*, edited by Rowan Flad and Zachary X. Hruby, pp. 169–180. Archaeological Papers of the American Anthropological Association, No. 17. University of California Press, Berkeley.

In press Shiyusuo 1936 nian Dasikong cun di er ci fajue de Yin dai yicun ji qi xiangguan wenti (The 1936 Excavation at the Dasikong Village by the Institute of History and Philology, Academia Sinica). *Bulletin of the Institute of History and Philology*, Academia Sinica, Taipei.

N.d. A Shang Royal Workshop: Reinterpreting Finds from the 1930s Excavation at Dialiankeng.

Li Yung-ti, Yue Zhanwei, and Liu Yu

In press Cong Xiaomintun dongnandi chutu taofan tan dui Yinxu qingtongqi de ji dian xin renshi (Bronze Foundry at Xiaomintun Southeast and Its Significance in the Study of Bronzes and Casting Technology of the Shang Period). *Kaogu.*

Liang Ssu-jung (Liang Siyong), and Kao Ch'ü-hsün (Gao Quxun)

1962 *Hou Chia Chuang, Vol. 2: HPKM 1001, Pt. 2: Plates.* Institute of History and Philology, Academia Sinica, Taipei.

Liao Yongmin

1957 Zhengzhou shi faxian de yi chu Shang dai juzhu yu zhuzao tongqi yizhi jianjie (Preliminary Report of a Shang Period Settlement and Foundry Site at Zhengzhou). *Wenwu cankao ziliao* 6:73–74.

Lin Yun

1979 Cong Wuding shidai de ji zhong Zi buci shilun Shang dai de jiazu xingtai (Reconstructing Shang Lineages from *Zi* Oracle Bone Inscriptions). *Gu wenzi yanjiu* 1: 314–336.

Liu, Li, and Xingcan Chen

2003 *State Formation in Early China.* Duckworth, London.

Loewe, Michael, and Edward Shaughnessy (editors)

1999 *The Cambridge History of Ancient China: From the Origins of Civilization to 221 B.C.* Cambridge University Press, Cambridge.

Maryon, Herbert

1963 "Bronze Casting and Bronze Alloys in Ancient China," by Noel Barnard, a Review. *Art Bulletin* 45, December: 394–396.

Maryon, Herbert, and H. J. Plenderleith

1954 Fine Metal-Work. In *A History of Technology, Vol. 1: From Early Times to Fall of Ancient Empires*, edited by C. Singer, E. J. Holmyard, and A. R. Hall, pp. 623–662. Oxford University Press, London.

Qiu Xigui

1983 Guanyu Shang dai de zongzu zuzhi yu guizu he pingmin liang ge jieji de chubu yanjiu (Preliminary Studies on the Lineage System and on the Elite and Commoners of the Shang Dynasty). *Wenshi* 17:1–26.

Rice, Prudence M.

1987 *Pottery Analysis: A Sourcebook.* University of Chicago Press, Chicago.

Roux, Valentine, and Pierre Matarasso

1999 Crafts and the Evolution of Complex Society: New Methodologies for Modeling the Organization of Production, a Harappan Example. In *The Social Dynamics of Technology: Practice, Politics, and World Views*, edited by Marcia-Anne Dobres and Christopher R. Hoffman, pp. 46–70. Smithsonian Institution Press, Washington, D.C.

Ruan Chongwu and Huang Guoji

1959 *Nixing zhuzao (Clay Mold Casting).* Industrial Mechanics Press, Beijing.

Schortman, Edward M., and Patricia A. Urban

2004 Modeling the Roles of Craft Production in Ancient Political Economies. *Journal of Archaeological Research* 12(2):185–226.

Shih Chang-ju (Shi Zhangru)

1947 Yinxu zuijin zhi zhongyao faxian fu lun Xiaotun diceng (Recent Important Discoveries at Yinxu and a Discussion of the Stratigraphy at Xiaotun). *Tianye kaogu baogao* 2:1–82.

1955 Yin dai de zhutong gongyi (Bronze Casting Techniques of the Shang Dynasty). *Bulletin of the Institute of History and Philology* 26:95–130.

1959 *Yinxu jianzhu yicun (Architectural Remains at Yinxu).* Institute of History and Philology, Academia Sinica, Taipei.

Sinopoli, Carla M.

2003 *The Political Economy of Craft Production: Crafting Empire in South India, c. 1350–1650.* Cambridge University Press, Cambridge.

Smith, Cyril

1962–1963 "Bronze Casting and Bronze Alloys in Ancient China," by Noel Barnard, a Review. *Harvard Journal of Asiatic Studies* 24:284–286.

1981 *A Search for Structure: Selected Essays on Science, Art and History.* MIT Press, Cambridge, Massachusetts.

Su Rongyu, Hu Zhisheng, Lu Liancheng, Chen Yuyun, and Chen Yiwei

1988 Yu guo mudi qingtongqi zhuzao kaocha he jinshuqi jiance (Technological and Metallurgical Studies of Bronzes from the Yu-State Cemetery). In *Baoji Yu guo mudi (The Cemetery of the State of Yu at Baoji)*, edited by Lu Liancheng and Hu Zhisheng, pp. 530–638. Wenwu Press, Beijing.

Tan Derui

1999 Zhongguo qingtong shidai taofan zhuzao jishu yanjiu (Study of Bronze Age Clay Mold Casting Techniques in China). *Kaogu xuebao* 2:211–250, pls. 13–24.

Tan Derui and Huang Long
1996 Houma taofan de kexue yanjiu (Scientific Studies of the Houma Molds). In *Fenhewan: Dingcun wenhua yu Jin wenhua kaogu xueshu yantaohui wenji* (*Fenhewan: Collection of Papers Presented at the Conference of the Paleolithic Finds at Dingcun and the Archaeology of the Jin Culture*), edited by the Shanxi Provincial Institute of Archaeology, pp. 218–225. Shanxi University Press, Taiyuan.

Tan Derui, Huang Long, Wang Yongji, Lu Houyuan, and Lu Haibin
1993 Zhiwu guisuanti ji qi zai gudai qingtongqi taofan zhizao zhong de yingyong (Phytolith and Its Application in the Making of Ancient Clay Molds for Bronze Casting). *Kaogu* 5:469–474.

Tang Jigen
1998 Yinxu jiazu mudi chutan (A Preliminary Study of Lineage-Based Cemeteries at Yinxu). In *Zhongguo Shang wenhua guoji xueshu taolunhui lunwenji* (*Papers of the International Symposium on the Shang Culture in Ancient China*), edited by the Institute of Archaeology, Chinese Academy of Social Sciences, pp. 201–207. Encyclopedia of China Publishing House, Beijing.
2004 New Excavations at Anyang and Their Implications for the Study of Shang Civilization. Speech given at the Institute of History and Philology, Academia Sinica, October 20, Taipei.

Tang, Jigen, Zhichun Jing, and G. Rapp
2000 The Largest Walled Shang City Located in Anyang, China. *Antiquity* 74:479–480.

Thorp, Robert L.
2006 *China in the Early Bronze Age: Shang Civilization*. University of Pennsylvania Press, Philadelphia.

Underhill, Anne P.
2002 *Craft Production and Social Change in Northern China*. Kluwer Academic/Plenum Publishers, New York.

Wan Chia-pao (Wan Jiabao)
1972 Shang pian: Anyang qingtong rongqi de zhuzao ji qi jishu de fazhan (Part 1: The Casting and Technical Development of Anyang Bronze Vessels). In *Gu qiwu yanjiu zhuankan di wu ben: Yinxu chutu wu shi san jian qingtong rongqi zhi yanjiu* (*Archaeologia Sinica, New Series, no. 5: Studies of Fifty-Three Ritual Bronzes*), by Li Chi and Wan Chia-pao, pp. 1–54. Institute of History and Philology, Academia Sinica, Taipei.

Wild, Ray
1975 *Work Organization: A Study of Manual Work and Mass Production*. John Wiley and Sons, London.

Wright, Rita P.
1998 Crafting Social Identity in Ur III Southern Mesopotamia. In *Craft and Social Identity*, edited by Cathy L. Costin and Rita P. Wright, pp. 57–70. Archaeological Papers of the American Anthropological Association. Arlington, Virginia.

Xu Jingsong, Dong Yawei, and Li Taoyuan
2001 Panlongcheng chutu da kou tao gang de xingzhi ji yongtu (The Function of Wide-Mouthed Pottery Jars Unearthed at Panlongcheng). In *Panlongcheng*, edited by the Hubei Provincial Institute of Cultural Relics and Archaeology, Vol. 1: 599–607. Wenwu Press, Beijing.

Yan Yun (alias for Lin Yun)
1973 Shang dai buci zhong de yezhu shiliao (Records of Casting in Oracle Bone Inscriptions of the Shang Dynasty). *Kaogu* 5:299.

Yang Hsi-chang (Yang Xizhang)
1986 The Shang Dynasty Cemetery System. In *Studies of Shang Archaeology*, edited by Kwang-Chih Chang, pp. 49–63. Yale University Press, New Haven.

Yang Xizhang (Yang Hsi-chang)
1983 Shang dai de mudi zhidu (The Shang Dynasty Cemetery System). *Kaogu* 10:929–934.

Yates, Robin D. S.
1997 The City-State in Ancient China. In *The Archaeology of City-States: Cross-Cultural Approaches*, edited by D. L. Nichols and T. H. Charlton, pp. 71–90. Smithsonian Institution Press, Washington, D.C.

Yetts, W. Perceval
1929 *The George Eumorfopoulos Collection, Catalogue of the Chinese and Corean Bronzes, Sculpture, Jades, Jewelry and Miscellaneous Objects, vol. 1: Bronzes: Ritual and Other Vessels, Weapons, Etc.* Ernest Benn Ltd., London.

Yue Hongbin, Yue Zhanwei, and He Yuling
2006 Xiaotun gongdian zongmiao qu buju chutan (Preliminary Study of the Layout

of the Palace/Temple Complex at Xiaotun). *Sandai kaogu* 2:328–343.

Yue Zhanwei, and Liu Yu
2006 Yinxu zhutong yizhi zongshu (Overview of Bronze Foundry Sites at Yinxu). *Sandai kaogu* 2:358–374.

Zhang Wanzhong
1962 Houma Dongzhou taofan de zaoxing gongyi (Shape Forming Methods with Clay Molds of the Eastern Zhou Foundry Site at Houma). *Wenwu* 4–5:37–42.

Zhang Yachu
1986 Shang dai zhiguan yanjiu (Study of Shang Official Titles). *Gu wenzi yanjiu* 13:82–115.

Zheng Ruokui
1995 Yinxu "Da Yi Shang" zu yi buju chutan (A Preliminary Analysis of the Lineage-Based Settlement at "Dayishang," Yinxu). *Zhongyuan wenwu* 3:83, 84–93.

Zheng Zhenxiang
1988 Yinxu fajue liu shi nian gaishu (A Summary of the Excavations and Research at Yinxu for the Past 60 Years). *Kaogu* 10: 929–941.

Zhu Fenghan
1990 *Shangzhou jiazu xingtai yanjiu* (*Family and Lineage Structures of the Shang and Zhou Periods*). Tianjin Guji Press, Tianjin.

III

Complementary Approaches to and Perspectives on Multicraft Production

9

Diachronic Change in Crafts and Centers in South-Central Veracruz, Mexico

BARBARA L. STARK

The Mesoamerican record reveals economic changes in crafts linked to the growth of a market system, well documented at the time of the Spanish conquest. This study provides new information about how crafts and polities responded to new political and economic situations, showing that small polity sponsorship or control of crafts (and presumably markets) played a role in the growth of a commercial system after the collapse of large states. The nature of craft organization on a regional scale can be given a first approximation through survey and systematic collections; excavations are difficult to pursue on such a large scale despite their superior potential for disclosing the social contexts and taphonomy related to craft workshops.

In the western Lower Papaloapan Basin (WLPB) of south-central Veracruz, Mexico (Figures 9.1–9.2), multiple crafts show changes toward greater nucleation at major centers and some increases in the intensity of craft work despite continuing dispersed habitation in a tropical lowland environment. I explore the concepts of "multiple crafts" and "multicrafting" with respect to survey data in order to better highlight the strengths and limitations of survey versus excavation for craft studies. I evaluate local competition, increased marketing, economies of scale, economies of place, product transportability, political strategies, and labor costs as causes for change versus persistence in the regional spatial organization of crafts. Different crafts show different trajectories of change or stasis in respect of nucleation and intensification.

One important factor is not discussed in this essay: population growth that may have reduced access to agricultural land and encouraged individuals or families to pursue crafts as an additional or alternative livelihood (Arnold 1985). Comparative population estimates are potentially misleading without an extensive program of testing residential locations in the study region because periods lack comparable arrays of diagnostic ceramics. For example, there may be more diagnostics for one period than another, but there may not have been more people. Perhaps more importantly, settlement data indicate the possibility that seasonal rainfall farming could have drawn on lightly occupied interfluve and upriver areas that acted as a reserve for the populations concentrated along the lower parts of streams and rivers and in low-lying alluvial deltas. If so, any population–resource imbalances cannot be assessed adequately without further survey that concentrates not on the centers and their surroundings but, rather, on environmental zones. Instead, my survey was oriented to clarifying the nature of the settlement hierarchy. My suspicion is that population growth does not readily account for the patterns that will be discussed because the Early and Late Classic periods appear to have the highest occupation evidence,

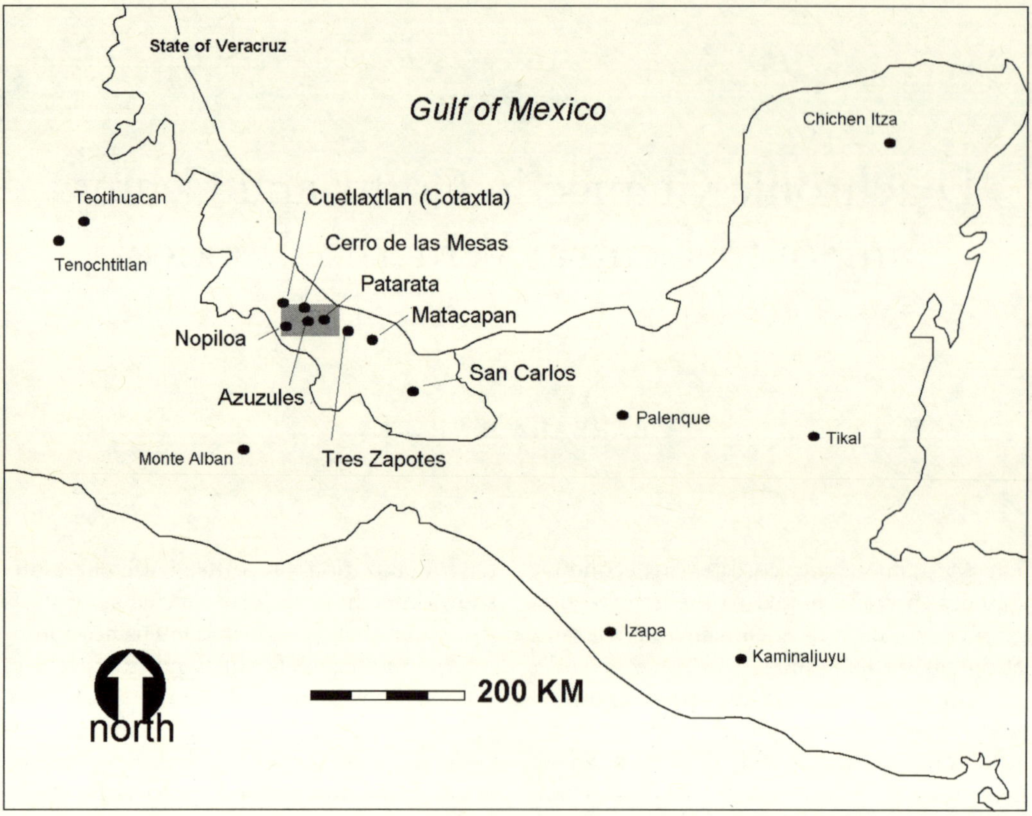

FIGURE 9.1. Map of Mesoamerica with a shaded box indicating the area of the western Lower Papaloapan study region in Veracruz, Mexico.

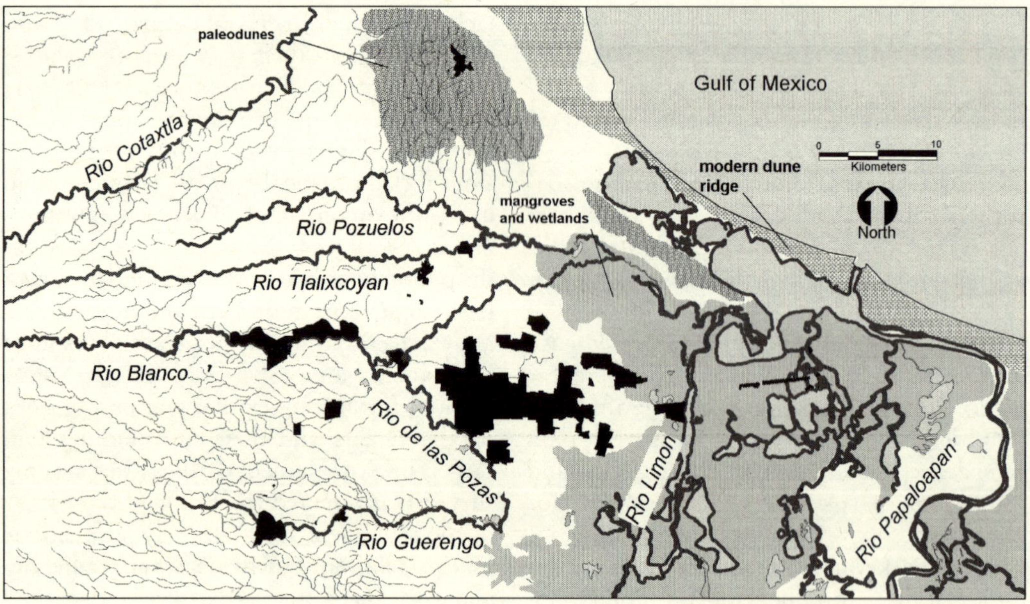

FIGURE 9.2. Map of the study region showing landforms, rivers, and the intensive survey blocks (in black).

but the patterns of craft changes pertain to the Late Classic through Postclassic periods, with population levels likely lower during the Postclassic period.

In the study region, full-coverage survey and systematic collections conducted in two cycles (1984–1988, 1998–2002) provide the primary evidence. A linked survey by Stuart Speaker (2001) provides part of the survey coverage, but some differences in the collection and analysis procedures affect its comparability. Settlement information derives from (1) the delta of the Blanco River, (2) upriver along the Blanco, (3) along the Guerengo River, (4) the mangrove estuarine zone, and (5) the paleodunes. A total of 3,757 features has been recorded, with 2,280 collected. Intensive survey has covered 97 km^2 (plus 43.2 km^2 of reconnaissance survey to locate additional monumental construction). The great majority of features are domestic mounds or nonmound artifact concentrations related to residences. Several complexes of formal, monumental architecture are scattered through the region, some representing the cores of settlements and others, outlying structures that likely represent palace complexes or settlement subdivisions related to the cores.

My analysis will focus on the Classic and Postclassic periods using the following subdivisions: Early Classic, A.D. 300–600; Late Classic, A.D. 600–900; Middle Postclassic, A.D. 1200–1350; and Late Postclassic, A.D. 1350–1521. I have not established a complex pertaining to the Early Postclassic, A.D. 900–1200, although Late Classic materials may continue partly into the Early Postclassic period and materials associated with the Middle Postclassic period may have an earlier inception during the last part of the Early Postclassic period. I do not address the Preclassic period because no evidence of specialized craft production is evident for that period, perhaps in part because most Preclassic settlement locations have subsequent occupation that reduces the chance of detecting Preclassic specialized production.

For the most part, surface evidence is most reliable for the detection of craft activity when production is more intensive and skews surface collections with by-products. Surface evidence is more ambiguous for industries that produce little obvious surface debris (e.g., spinning) or for which production is not sufficiently intensive to skew surface remains. Pottery and blade making are examples of industries in which sufficiently intensive production has a good chance to affect surface collections unequivocally. The use of scrapers, cotton spinning with whorls, and lapidary work are examples of industries in which one or at most a few telltale tools of production or by-products may occur in a given surface collection, but the surface evidence is unlikely to be abundant even if specialists worked at the location. Consequently, we can mainly note if such remains are widely scattered versus concentrated, but we usually cannot distinguish degrees of intensity in the craft.

With a diachronic record, we can examine the changing locations of production in these industries as well as evidence about the intensity of production in some cases. All the evidence suggests production in household contexts, with differences in the intensity of production from a level ancillary to household activities to workshop levels (e.g., Heller 2000).

Five industries provide information about local crafts during the Classic period (A.D. 300–900) and Postclassic period (A.D. 900–1521): (1) prismatic obsidian blade making from imported preformed polyhedral cores; (2) obsidian scraper use, with scrapers mainly from tabular pieces struck transversely from nearly exhausted prismatic cores; (3) pottery making; (4) spinning cotton thread with ceramic spindle whorls; and (5) lapidary activities using imported raw material and tools. At least some figurine production occurred in conjunction with pottery production, so figurines will not be analyzed separately here. An additional possible specialization, making molded ceramic spindle whorls, is signaled by a concentration of 11 whorls and one mold at a mound at Dicha Tuerta (another mold was observed in the possession of a local family there). Because sampling vagaries are a strong possibility with items that occur in extremely low frequencies, such as whorl molds, I am not convinced we have enough information about whorl production to adequately interpret Dicha Tuerta as a location where whorls were made. No other whorl mold was encountered in survey.

In brief, the evidence of more spatially concentrated production increases in the Late Classic period, especially in the vicinity of major centers, and becomes more marked in the Postclassic period. Some of the nucleated crafts represent more intensive production, as well. Concomitantly there are increasing signs of possible multicrafting. The reasons for craft changes and their implications for broader processes in Mesoamerica are my primary concern here. Scenarios regarding the origin and spread of specialization and marketing in Mesoamerica link them to a series of changes related to particularly large states, especially when those states collapsed. The WLPB evidence suggests that a combination of polity sponsorship and marketing that crosscut polities may have stimulated the craft changes. The detection of marketplaces and market institutions is particularly tricky archaeologically, and it is not the focus here (Hirth 1998). Rather, I focus on craft activity and locations, with markets assumed as a likely institution at least by Late Classic times onward.

CONCEPTS OF MULTIPLE CRAFTS AND MULTICRAFTING

I discuss primarily multiple crafts, that is, different crafts that occur in proximity to each other, but the possible relationships to multicrafting warrant consideration, especially with survey evidence. Here I use *craft* to indicate a line of technological production yielding a particular kind of artifact or other object. Households supplying their own artifacts or buildings must necessarily engage in multiple unspecialized crafts. I do not treat household autoproduction as multicrafting even though a single household member may fabricate many different items, as this does not represent craft specialization. Households may have "embedded specialists" as members, however (Ames 1995).

In craft activities, I use *specialization* for production beyond the needs of the domestic group but also in greater quantity than represented by occasional excess production for socially motivated gift giving. *Multicrafting by specialists* refers to two or more craft activities conducted by a craftsperson or social unit. Below I discuss further the issue of what constitutes distinct crafts. Note that the "social unit" can have analytical ambiguity because of different levels of inclusiveness (e.g., from nuclear family, to extended family, to residential district and so forth). We have to clarify particular cases to avoid comparisons at incommensurate scales. In particular, ambiguities in the scale of social units point to potential relationships between multiple crafts and multicrafting, as discussed below. Note that Shimada (this volume) sees some multiple crafts as instances of multicrafting. The links or cooperations that may be entailed among clustered or adjacent yet different crafts are a key factor.

Setting aside the broad range of manufacturing activities in household autoproduction, we can distinguish multicraft situations along a "scale of difference." Many craft specialists narrow their range of products among items that are linked by a related technology. For example, pottery specialists may make a range of forms or sizes or may use varied materials for different kinds of vessels. In one sense, such specialists may be seen as producing multiple products, such as differently sized jars or bowls and jars; the basic technology is the same or quite similar, however. The person(s) who manufactures the objects does not necessarily change for different kinds of vessels. It is not convenient to define this situation as multicrafting because it is, more simply, multiple products within a craft.

The concept of multicrafting may be apropos when quite functionally distinct products are fabricated within the same broad technological procedures; one example is ceramic figurines used in domestic ritual plus pottery vessels used in serving, storing, or cooking foods. Some carryover of skills occurs within a broad technological domain like ceramics, such as experience in the treatment of clays, guiding firing temperatures, and other technical matters. The procurement of raw materials can represent an economy of scale because an infrequently produced item does not require a special trip. For example, figurines in Mesoamerica are ordinarily small and do not consume as much clay as pots. It would be economical for potters also to make figurines because they could fold raw material procurement for the smaller (and often less frequently produced and consumed) figurines into clay procurement for pottery. Nevertheless, products quite different in function and form, such as figurines and pots, *may* involve distinct production skills (e.g., the use of intricate molds for some figurines or the making of intricate molds as well as using them). The person(s) manufacturing the objects might still be the same, however. Cases of quite contrastive products within a technological domain require case-specific consideration to determine if the products demand different skills and procedures.

Multicrafting may involve distinct sets of technological activities, such as lapidary and pottery production. These situations are farther along the scale of difference and match our usual sense of multicrafting. In these cases, the skills, scheduling demands, and work spaces may diverge considerably between the crafts. Such distinct technologies (e.g., pyrotechnic versus abrasive or subtractive) may nevertheless be linked economically through cooperation, perhaps at different scales of a social unit. They might also be linked temporally, with one craft pursued during a lull in the other, perhaps conditioned by weather, demand, or labor scheduling for other reasons. Ceramicists and metalworkers, because they both employ pyrotechnic methods, might share fuel acquisition trips even if they use differing grades or amounts of fuel. Potters and lapidaries might share procurement trips for pigments and abrasives, whether to markets or to raw material sources. Work spaces may partly overlap, for example, parts of an open-air patio or the open space around a domicile.

Although the same person(s) might perform both crafts in the case of distinct technologies, very different technologies and their skills begin to challenge the training and time of individuals and pose a dilemma in personnel allocation. It becomes more likely that different persons will perform the tasks, even if there is a social unit to which all belong.

Whether the different craftspersons are usefully viewed as engaging in examples of multicrafting or of multiple crafts depends on the linkage discerned through the social unit. Otherwise, multiple specialized crafts are a common occurrence within urbanized settlements. Each different craft specialization may be separately organized in its own social units.

This does not imply that critical economic and social co-operations do not link multiple crafts, however, a point directly pertinent to my study. One example is coproduction, such as Bohemian jewelry makers who produced glass ornaments and were economically complementary to metalworkers, who made chains and settings but generally were members of different households and workshops (Jargstorf 1993: esp. 17–20). In Mesoamerica, we have some indications of this kind of coproduction between or among crafts, for example, between Maya masons and wooden lintel carvers at Tzibanche (Harrison 1972, cited in Reents-Budet et al. 1994:219–222). In coproduction, multicrafters engage in distinct technologies that are united in particular objects, such as the "horizontal integration" (an instance of coproduction) described by Topic (1990:164–165) at a workshop at Chan Chan in the Andean area, in which woodworking and metalworking were joined in the production of a carved hand with copper inlays. Shimada (1996) notes multiple combinations of technologies in elite objects from Middle Sicán tombs. Reents-Budet et al. (1994:218–219) argue for multicraft coproduction among potters who made certain Classic Maya vases and the painters who created elaborate scenes on the vessels.

At Otumba, a Late Postclassic city-state in the Basin of Mexico, a different cooperation of "chained" relationships is exemplified by obsidian lapidary work, in which exhausted or nearly exhausted prismatic cores from blade making were used by lapidaries as tools and as raw material for ear spools and other ground and polished products (Otis Charlton 1993, 1994). This functional chaining may have been arranged strictly through market exchanges, but other direct arrangements may have occurred to assure a steady supply. In my study region, chained relationships may have linked obsidian blade makers and obsidian scraper users. Both coproduction and "chained" production are aspects of Shimada's (this volume) concept of multicrafting.

Thus, I apply *multicrafting* to a range of situations involving marked variation in the products or technologies of specialized producers who shared a social unit that had a direct bearing on their craft work. I use *multiple crafts* for the settlement scale when multicrafting is not indicated or is uncertain but there are technologically diverse crafts. Although some of my surface collections possibly show multicrafting, they are more securely regarded as representing multiple crafts.

We should not be too quick to judge that multiple crafts within a settlement lacked the interesting social and economic relationships of multicrafting. Craft activity may be indicated from surface data or testing, but without excavation we cannot be entirely sure how waste disposal locations and original manufacture locations link with a set of nearby residences. The point is that multiple crafts in some reasonable propinquity in a settlement *might* be linked through a social unit, and we should not simply assume that closely positioned crafts were entirely independent of one another socially.

Even if different crafts pertain to separate social units within a settlement, they may enjoy mutual benefits in the procurement of raw materials or access to consumers through multipurpose trips (i.e., trips conducted though pooling efforts by the specialists, or provided by "middlemen," or through multipurpose trips made by consumers to the specialists' settlement). Consequently, one of the broad questions we can explore involves the reasons for multiple craft concentrations, whether at the intermediate levels of small social units (multicrafting) or at a larger scale within settlements (multicrafting and multiple crafts).

One's tactics for analyzing multiple craft concentrations can emphasize "internally" directed studies of the production locations or may include more "outwardly" directed examination of regional patterning in craft production and distribution. Widmer (1991) provides a good example of an internally directed analysis of multicrafting that exposes several potential scales of social units in apartment compounds in the urban center of Teotihuacán in the Basin of Mexico. He suggests three social levels in the Tlajinga 33 compound, the nuclear family, a larger lineage scale, and the still larger scale of the entire apartment compound. At a still larger scale the Tlajinga district has additional compounds besides Tlajinga 33 with evidence of specialist pottery production (Sullivan 2006), so that a "barrio" is indicated within which cooperation or multipurpose trips may have advantaged potters. Tlajinga 33 is interesting in respect to multicrafting because it shows the different scales at which we can consider the implications of multicrafting. Initially, pottery making was evident from surface collections, and multicrafting was not obvious, in part because of the nature of lapidary technology, taphonomic patterns, the curation of tools, and slower consumption and production rates for stone and shell items. Thus, multicrafting may be more common than we imagine from survey data. An appreciation of possible multicrafting is dependent in part on surface conditions and the intensity of surface collections. For example, Nichols et al. (2000) analyzed a maguey fiber processing workshop through intensive surface collections that also revealed the production of whorls, figurines, basalt items, and lapidary products. The authors make a reasonable case on the basis of surface collections for multicrafting in a district of Otumba.

A first step for the analysis of multiple crafts (and multicrafting) from surface evidence is observation of craft production indicators in close spatial proximity. In the WLPB, which has a predominantly dispersed settlement pattern, "close proximity" can mean collections from a scattering of residential mounds in the immediate environs of the formal monumental core of a center or collections from a scattering of dispersed residential mounds in the countryside that are near one another. The nucleation of multiple crafts at a settlement increases the probability of multicrafting relationships but does not guarantee it. Even if multiple crafts are indicated in the same surface collection, we cannot be sure of multicrafting because the craft activities may have shifted over time, leaving a palimpsest rather than a record of contemporaneous crafts. Thus, inferences from systematic surface collections differ from those from excavations of workshops. Survey information excels at establishing a large-scale picture of craft locations and changes for multiple crafts, whereas excavation of workshops (various authors, this volume) can better determine social relationships among producers and the ways that space was allocated or materials were shared, necessarily at a more localized scale.

This essay considers reasons for craft nucleation in the WLPB as well as some craft intensification in a region where dispersed occupation and household-scale specialization did not show craft nucleation initially. Craft nucleation became more characteristic through several periods, but one craft, cotton spinning, seems immune to this process.

THE WESTERN LOWER PAPALOAPAN BASIN SURVEY AND ANALYSIS PROCEDURES

The WLPB has rich aquatic and agricultural resources. Low-lying lands and areas near rivers were favored for occupation throughout the sequence. The settlement record shows some variation over time in the degree of clustering of domestic mounds (and domestic surface artifact concentrations), but settlement is generally dispersed. Even major public construction groups are scattered widely in the region. Survey routinely emphasized domestic features, the most ubiquitous settlement remains, in order to track the chronology of occupation and analyze economic and social differences.

Survey methods used "full coverage," with 20 m spacing of survey personnel when surface artifacts could be observed but up to 50 m if the ground cover prevented observation of artifacts, with the spacing dependent on topographic visibility. Surface collections were systematic over a measured area if ground visibility permitted, but opportunistic collecting was employed when only burrows, looters' holes, or paths exposed artifacts. Even opportunistic collections were made of all of the same designated artifact categories. Thus, artifact analyses can use densities of items from the systematic collections, but other measures, such as proportions or quantities, can be applied to systematic and opportunistic collections alike. Several rare artifacts were collected whenever found, including outside of collection units or features. These "isolated finds" cannot be analyzed on a comparable basis to the surface collections, but they can contribute to spatial analyses. For the crafts to be discussed below, all observed artifacts were collected, except in the case of pottery, where all rim sherds and any decorated or unusual sherds were collected.

To examine spatial patterning in production evidence, I employ thematic maps. To have a satisfactory scale to show distributions, I constructed a compressed format. Survey blocks appear in their actual locations in Figure 9.2 but are positioned in Figure 9.3 close to each other but roughly respecting their relative positions. My analysis concentrates on major centers, not on lesser centers, in order to accommodate the space limitations of the chapter. On the thematic maps, the monumental cores of centers are indicated by a surrounding line, with individual features omitted to improve the clarity of the maps. Pertinent individual collections in the thematic maps are shown with a triangle indicating the center point of the feature.

Because settlement is dispersed, there is no obvious boundary for occupation at the centers. Therefore, as an arbitrary device to provide comparable information, Table 9.1 lists craft indications within a half-kilometer buffer around each major center. The major centers examined are Cerro de las Mesas for the Early Classic period, Azuzules and Nopiloa for the Late Classic period, El Sauce for the Middle Postclassic period, and Callejón del Horno for the Late Postclassic period. Because

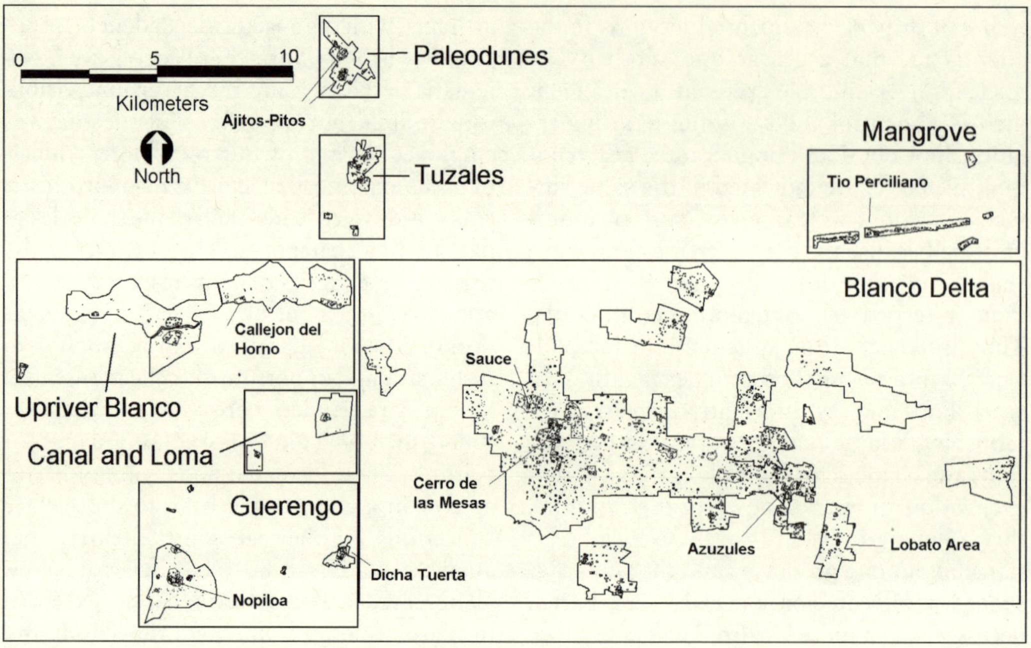

FIGURE 9.3. Compressed base map showing archaeological features. A line surrounds each monumental complex, and centers mentioned in the text are labeled.

of variations in the area of the monumental cores and in the survey blocks, each buffer covers a different amount of area. Some buffers, for example, are truncated by the survey limits. A half-kilometer buffer is adequate in most cases to capture the patterns discussed, but lapidary craft remains at Nopiloa all occur between 0.5 and 1 km from the core. Distribution maps will show craft indications at any distance, so a broader spatial view is provided by the thematic maps than by Table 9.1.

For reasons of space, Table 9.1 omits collections within the buffers that only produced a spindle whorl(s) as evidence of craft activity. On this basis, 36 collections are omitted for Cerro de las Mesas; 18, for Azuzules; 2, for Nopiloa; 13, for El Sauce (plus four in the buffer overlap with Cerro de las Mesas); and none from Callejón del Horno. As discussed below, cotton spinning was widespread and may have been an activity at most households, both for domestic use and possibly for tribute obligations and market opportunities.

Craft indications such as scrapers, lapidary tools or materials, pottery molds, wasters, kiln fragments, and whorls are usually represented in collections by a single artifact. For such relatively rare artifacts, this is not surprising, as the median number of rims per collection is 68, and the collections are typically from small individual domestic mounds. Obsidian and pottery evidence is more abundant, and each requires special steps in the analysis. In particular, the prismatic blade–making evidence ranges from a single artifact to nearly 200 indicators. Because blade-making debitage can be used or modified, such pieces might circulate outside the locale of blade making. Consequently, with blade manufacture, collections with larger quantities of blade-making evidence provide more assurance that the collection is related to blade making in the vicinity. Blade manufacturing evidence raises several interpretive issues requiring discussion.

The recovery of blade production debris on mounds does not pinpoint the exact location of blade making, as debris was likely cleaned up and disposed. I assume that blade-making activities were conducted either at mounds where the debris was encountered or nearby. Aside from the more intensive situations, blade-making by-products occur in small

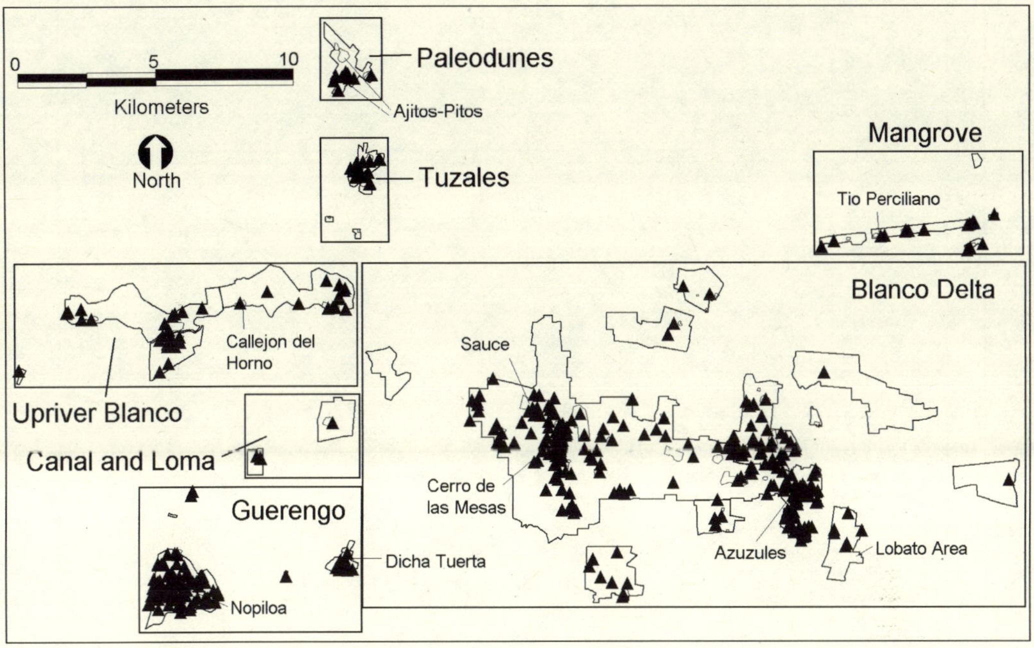

FIGURE 9.4. Compressed distribution map showing any occurrence of a black obsidian blade production indicator.

amounts widely, with a possibility that at least some of the fragments were acquired from a blade-making location and used as tools. Figure 9.4 shows collections with any occurrence of black to dark gray obsidian blade production indicators to illustrate this point.[1]

To isolate more intensive, workshop-related blade-making concentrations, I have examined three different approaches to isolate more intensive activity: (1) the ninetieth percentile of production indicator densities (this measure has the disadvantage of ignoring collections that lack a measured area), (2) the ninetieth percentile of production indicator counts (this index is sensitive to the varying collection sizes), and (3) the ninetieth percentile of the ratio of production indicators to rim sherds (this index can be skewed if pottery varies independently in amount). Each measure has advantages and disadvantages, but consistent spatial patterning near a center for all three indexes is a good guide to more intensive blade production there. However, for reasons of limitations of space, the second index, the ninetieth percentile of counts, is not presented in thematic maps, as it is the most subject to extraneous fluctuations. The densities of blade production indicators per m^2 or their ratios to rim sherds do not identify workshops per se, but they do segregate more intensive blade-making locations. The densities per m^2 of blade production indicators for the ninetieth percentiles range from 0.01 to 0.03, depending on obsidian color. These are comparable to the lower values observed at systematic collection units at the Late Postclassic town of Cotaxtla, Veracruz, by Ohnersorgen (2001:214). For the ratio to rim sherds, WLPB values range from 0.02 to 0.04, depending on obsidian color. Again, these ratios are similar to the lower values in Cotaxtla systematic collections. At Cotaxtla, as at mound 1756 at El Sauce in the WLPB during the Middle Postclassic period (Heller 2000), collections labeled as workshop debris have much higher densities. The intensity of production activity and the amount of by-products are best viewed as gradients, however, not threshold phenomena.

To isolate more intensive pottery production, two measures have been applied concurrently, the ninetieth percentile of density for

Table 9.1. Presence of Obsidian Craft Production Artifacts from Collections within a 0.5 km Radius around Major Centers

Center and Coll. No.	Obsidian Prismatic Blade Production Indicators, 90th Percentiles									Obsidian Scrapers		
	Black Count	Black Density	Black Ratio to Rims	Clear Gray, Count	Clear Gray, Density	Clear Gray, Ratio to Rims	Green, Count	Green, Density	Green Ratio to Rims	Black	Clear Gray	Green
Cerro de las Mesas												
817			x									
63	**x**											
Clear Gray Obsidian, Postclassic; Comal, Postclassic												
71											x	
326				x		x						
369					x							
566					x							
1183					x							
64				**x**								
834											**x**	
Azuzules												
1125										x		
1126		x										
1192										x		
1272		x										
1278		x										
1301										x		
1531										x		
1273		**x**								**x**		

Table 9.1. (continued)

Center and Coll. No.	Obsidian Prismatic Blade Production Indicators, 90th Percentiles									Obsidian Scrapers		
	Black Count	Black Density	Black Ratio to Rims	Clear Gray, Count	Clear Gray, Density	Clear Gray, Ratio to Rims	Green, Count	Green, Density	Green Ratio to Rims	Black	Clear Gray	Green
Nopiloa												
949	x											
953	x		x									
6278	x											
6352	x	x										
6353	x											
6383	x		x									
6386	x		x									
6387	x											
6398										x		
6406										x		
6445			x									
6279	x	x	x									
6303	x		x									
6304	x	x	x							x		
6305		x	x									
6319	x		x							x		
6325	x									x		
6360	x		x							x		
6407	x									x		
6456										x		

Table 9.1. (continued) Presence of Obsidian Craft Production Artifacts from Collections within a 0.5 km Radius around Major Centers

	Obsidian Prismatic Blade Production Indicators, 90th Percentiles									Obsidian Scrapers		
Center and Coll. No.	Black Count	Black Density	Black Ratio to Rims	Clear Gray, Count	Clear Gray, Density	Clear Gray, Ratio to Rims	Green, Count	Green, Density	Green Ratio to Rims	Black	Clear Gray	Green
El Sauce												
116				x		x						
119											x	
127											x	
132				x		x						
842											x	
848											x	
1152					x							
1172						x						
1754				x	x	x						
1817				x	x	x						
124				x								
209										x	x	
214				x							x	
224						x					x	
1176				x		x					x	
1751				x		x						
1756	x			x	x	x					x	
1818					x						x	

TABLE 9.1. (continued)

Center and Coll. No.	Obsidian Prismatic Blade Production Indicators, 90th Percentiles									Obsidian Scrapers		
	Black Count	Black Density	Black Ratio to Rims	Clear Gray, Count	Clear Gray, Density	Clear Gray, Ratio to Rims	Green, Count	Green, Density	Green Ratio to Rims	Black	Clear Gray	Green
Buffer overlap between Cerro de las Mesas and El Sauce												
113				x								
115				x		x					x	
202											x	
353				x		x					x	
354										x		
356						x						
Callejón del Horno												
947											x	x
6002												x
6004												x
6011												x
946				x			x				x	x
6012									x			
6081				x	x	x	x				x	
6082				x			x	x	x	x	x	
6083				x					x			x

Note: Bold entries indicate collections with more than one craft category.

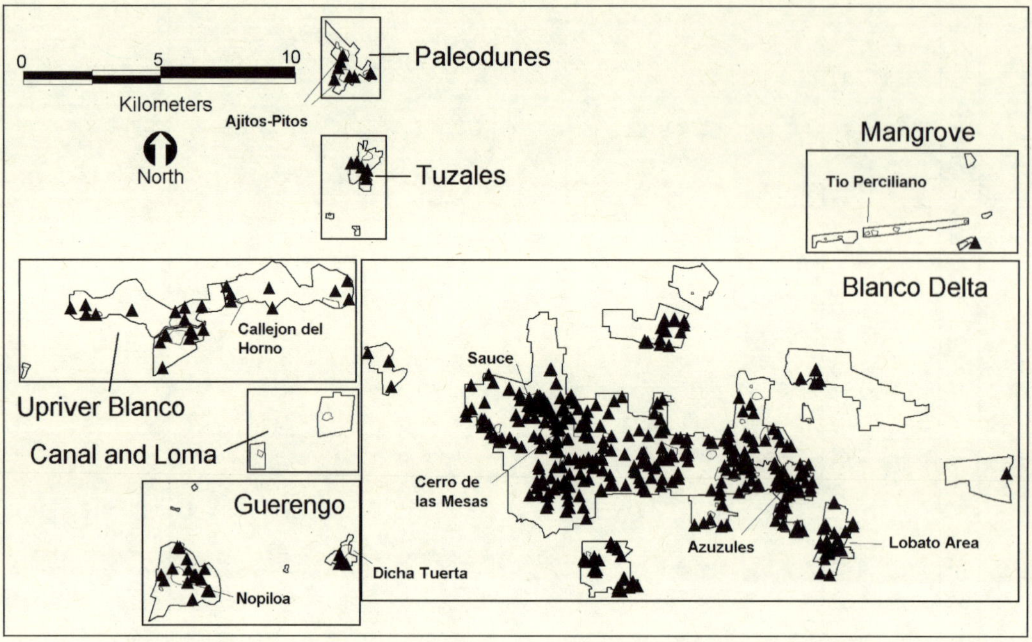

FIGURE 9.5. Compressed distribution map showing collections with one or more spindle whorls (isolated finds excluded).

the type of pottery and the ninetieth percentile of percentages for that type among all the pertinent collections (Stark and Garraty 2004). These measures are designed to detect de facto wasters, that is, sherds at production locations from vessels that broke but do not exhibit obvious misfiring characteristics. Other evidence of pottery production, such as molds, obvious wasters, and kiln fragments, occurs rarely but also offers clues about production. Because of space limitations, pottery production, already analyzed in other publications (Stark 1989, 1992; Stark and Garraty 2004), is not treated in detail here. Likewise, cotton spinning has been treated more thoroughly elsewhere (Stark et al. 1998).

In the subsections that follow, I note briefly the craft profile for the dominant center(s). Table 9.1 summarizes the craft tools and recognizable by-products found in individual collections. As will be demonstrated, most centers show evidence of multiple crafts in their immediate vicinity, and, in a few instances, there also are residential mounds with definite or possible indicators of more than one craft activity. In some of the latter possible multicrafting instances, the scale of some activities remains uncertain between household self-provisioning and specialized production.

EARLY CLASSIC PERIOD (A.D. 300–600)

Cerro de las Mesas, located in the Blanco delta, was the dominant center in the WLPB during the Early Classic period (Figure 9.3; Stark 2003). Imported black to dark gray obsidian (referred to as "black" below), mainly from Zaragoza-Oyameles, Puebla, is the main raw material used for the Classic period prismatic industry (Stark et al. 1992). During the Early Classic period obsidian blade production (Heller and Stark 1998) shows spatial dispersion and a low level of production at any one location. Cerro de las Mesas and its immediate environs have somewhat more evidence of blade production than distant residences, but the difference is not marked (Heller and Stark 1998).

During the Early Classic period as well as later, cotton thread spinning also shows a dispersed pattern without marked concentrations of whorls (Figure 9.5; Stark et al. 1998). Pottery production throughout the Classic period remains dispersed among households for most vessel types (Stark 1992). Lapidary tools are not particularly associated with Early Classic Cerro de las Mesas, unlike later major centers; nor are obsidian scrapers concentrated there, as will be shown below. In sum, there is no indication from surface evidence of a concentration of crafts at or in the immediate vicinity of Cerro de las Mesas, despite the unusually large size of the center and the likelihood that it held sway over much of the WLPB (Stark 2003).

Cerro de las Mesas is close to the Middle Postclassic center of El Sauce, and, as a consequence, Postclassic occupation remains are scattered over the Cerro de las Mesas area. The half-kilometer buffers used for the analyses of craft remains for the two centers overlap. Cerro de las Mesas entries in Table 9.1 that involve Postclassic craft artifacts are grouped under a separate subheading, and buffer overlap collections are grouped under a subheading for El Sauce because they predominantly involve Postclassic production evidence, for example, clear gray obsidian rather than the black obsidian of the Classic period.

Two collections from Cerro de las Mesas show the co-occurrence of a whorl with some other craft activity, black blade making or lapidary work. Whether this is a sign of multicrafting remains uncertain, as a whorl could be related to household provisioning of domestic textiles. Although textiles are a likely element in tribute or marketing, we cannot say if a specific collection represents specialization on the basis on one or two whorls. In aggregate, the abundance of whorls in the region and the indications from ethnohistoric records of cotton cloth and cotton fiber tribute suggest that some spinning was not solely household provisioning.

LATE CLASSIC PERIOD (A.D. 600–900)

In the Late Classic period, Cerro de las Mesas was not abandoned, but it ceased to function as the paramount center. A triad of Late Classic centers divided up the old realm and held smaller territories. At the east end of the Blanco River delta, Azuzules succeeded Cerro de las Mesas as the focus of public construction in the Late Classic period. Comparable large centers include Nopiloa along the Guerengo River to the southwest and Ajitos-Pitos on the paleodunes to the north. With this political change, we see considerable economic change.

A variety of evidence suggests more nucleation of some crafts at the centers. Two major centers provide craft evidence, Nopiloa and Azuzules. The third major center, Ajitos-Pitos, is located in an area today predominantly devoted to pineapple and pastures, making surface collections more difficult and spotty. Perhaps if we had more comparable collections, some craft production would be evident, but at present there is no indication of crafts in the vicinity of Ajitos-Pitos. The major center identified in the mangrove swamp, Tio Perciliano, is smaller than the other three, although it is the largest identified in the mangrove swamp. Tio Perciliano is located approximately in the middle of the linear, lagoon-edge Nacastle-Patarata settlement and constitutes its monumental core. Nacastle-Patarata produced evidence of pottery production (Stark 1989).

Prismatic Blade Production

With all three blade production indexes, the immediate environs of Nopiloa stand out as a concentrated area of black blade production debris, whereas greater Cerro de las Mesas does not stand out by any of the three measures (Figures 9.6 and 9.7 show the density and rim ratio indexes). A secondary center in the Nopiloa settlement system along the Guerengo River, Dicha Tuerta, has a more modest indication of blade production involving fewer collections. Azuzules and the outlying groups at Azuzules South and Azuzules Southeast have a more fitful and scant indication of blade making that varies in location according to the measure applied, suggesting some blade production near public buildings but not at the scale of concentrated, more intensive activity characteristic of Nopiloa.

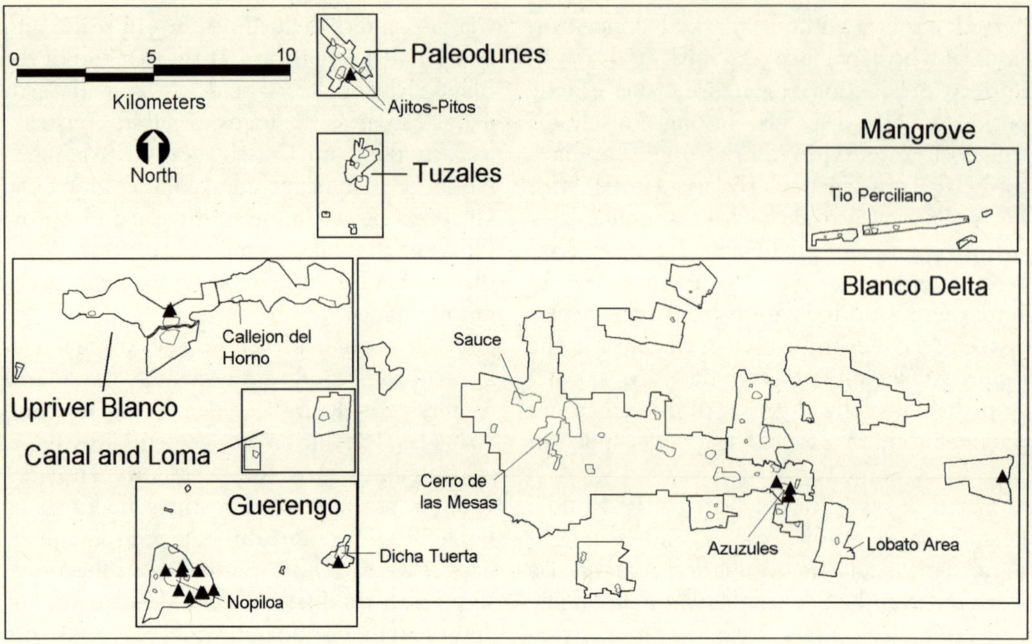

FIGURE 9.6. Compressed distribution map showing collections at or above the ninetieth percentile of densities per m2 of black prismatic blade production indicators (≥0.02667).

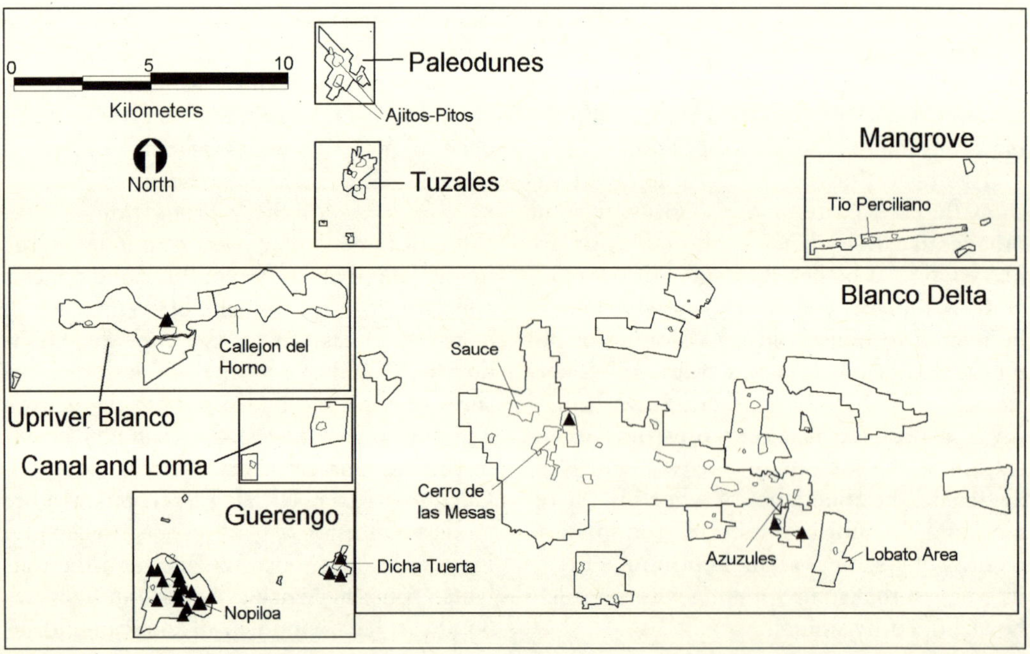

FIGURE 9.7. Compressed distribution map showing collections at or above the ninetieth percentile of the ratio of black prismatic blade production indicators to rim sherds (≥0.04444) for collections with 68 or more rims, the median.

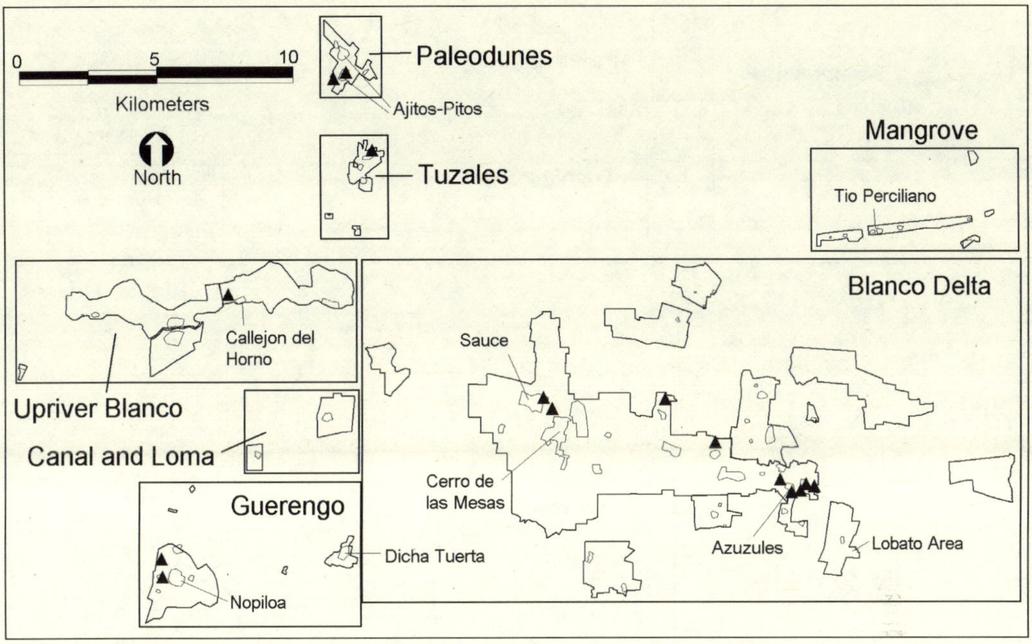

Figure 9.8. Compressed distribution map showing black obsidian scrapers that do not occur in collections that meet any of the three ninetieth percentile blade production indexes.

Obsidian Scrapers

I have included obsidian scrapers as a category for analysis because in the subsequent period, the Middle Postclassic, a mound near the center of El Sauce has an extraordinary concentration of scrapers, suggesting a specialized location employing them in a craft (Lewenstein 2001). Obsidian scrapers are scarce (110 from collections and 69 isolated finds, including all visual color groups and periods).

The scrapers were primarily manufactured on tabular pieces removed from reduced prismatic cores. Most scrapers were produced by detaching a flake that crossed the core transversely, with the tabular flake sometimes subsequently retouched along its margins. We do not know yet how black obsidian scrapers were employed during the Late Classic period, although Middle Postclassic gray scrapers show evidence of woodworking (Lewenstein 2001). They may have originated as a byproduct of prismatic core rejuvenation, but it is more likely that they are a separate product, removed from nearly exhausted prismatic cores specifically to serve as scrapers.

A few black obsidian scrapers do not occur in collections that meet any of the ninetieth percentile blade production indexes; they are concentrated around Azuzules, although only five collections are involved there, each with one scraper (Figure 9.8). Near Azuzules only one scraper occurs in a collection that reaches one of the ninetieth percentile criteria for blade production (Figure 9.9). Apart from the ninetieth percentile collections, most collections with scrapers have one blade production indicator, and the scrapers might have been made at these locations.

At Nopiloa, eight collections in the half-kilometer buffer have black scrapers, and most of these also are among the ninetieth percentile blade-making collections, which raises the possibility that they, like the prismatic blades, were a product of the workshops. If we consider a slightly greater distance, 13 collections around Nopiloa have scrapers.

If utilized, Nopiloa and Azuzules scrapers may indicate multicrafting because most co-occur with blade production evidence. Lewenstein (2001) found wear indicating that clear gray obsidian scrapers were used on wood or woody plants during the Middle Postclassic

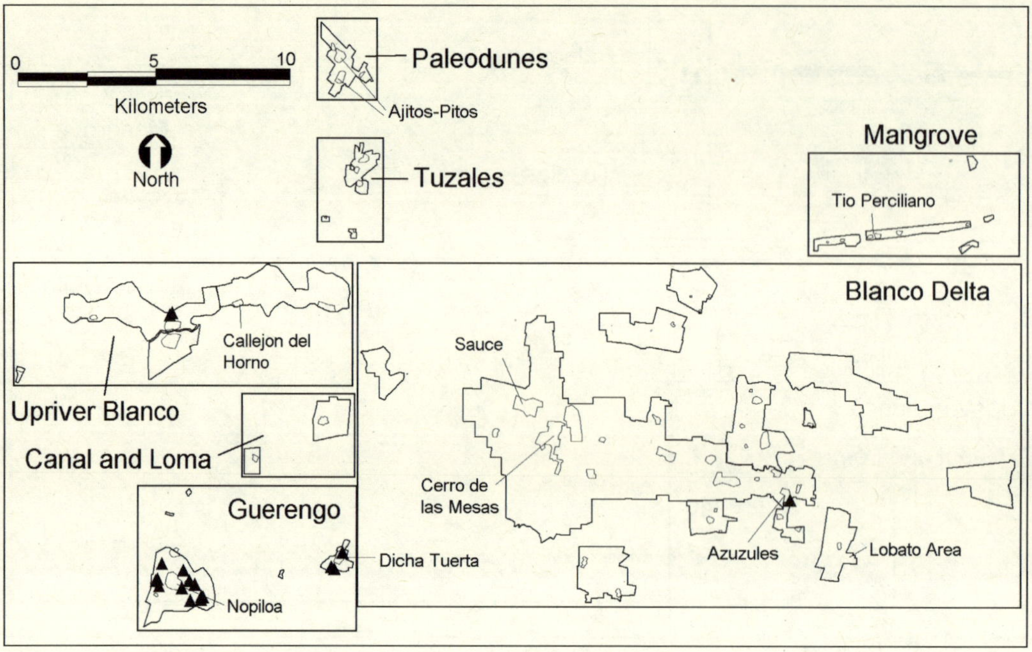

FIGURE 9.9. Compressed distribution map showing black obsidian scrapers that occur in collections that meet any of the three ninetieth percentile blade production indexes.

period, including those from an area related to a blade-making workshop. A use-wear analysis of black scrapers from Azuzules and Nopiloa has not yet been performed, but Amy Jo Vonarx (personal communication, 2004) recently examined a sample from Nopiloa and determined that they do have scraper wear. The WLPB is unlikely environmentally as an area of maguey processing, which is one activity suggested in the Basin of Mexico as a reason for the use of heavy obsidian scrapers (Brumfiel 1980:462). Small tanged obsidian scrapers, perhaps used in *aguamiel* (agave sap) collection in the highlands, are similar in size to those in the WLPB, but possible pre-Hispanic agave fiber–scraping tools illustrated by Parsons and Parsons (1990:291–292, 301–305, 311) are larger. The WLPB lacks a complete tool kit suited to the range of activities in agave processing, however.

Lapidary Items

The lapidary artifacts cannot themselves be separated according to major periods, and the timing of lapidary activities can be judged only by context. The evidence involves rare raw materials (e.g., from which greenstone segments were removed) and small abraders that could have served in fine lapidary work. Small abraders may have had other uses than ornament production, such as polishing celts. Lapidary items are found near Nopiloa to a greater extent than any other Classic center (Figure 9.10), but all five occur between 0.5 and 1 km from the center and are not listed in Table 9.2.

Pottery

There is no surface evidence for a spatial concentration of pottery production during the Early Classic period, and for the most part this continues to be the case during the Late Classic period (Stark 1992). However, the density and frequency of orange bowls in some collections near Azuzules suggest that they were produced there in Late Classic times (Stark and Garraty 2004). Ultrafine orange pottery production is a specialty in the Late Classic Nacastle-Patarata settlement in the mangrove swamp where the center of Tio Perciliano is

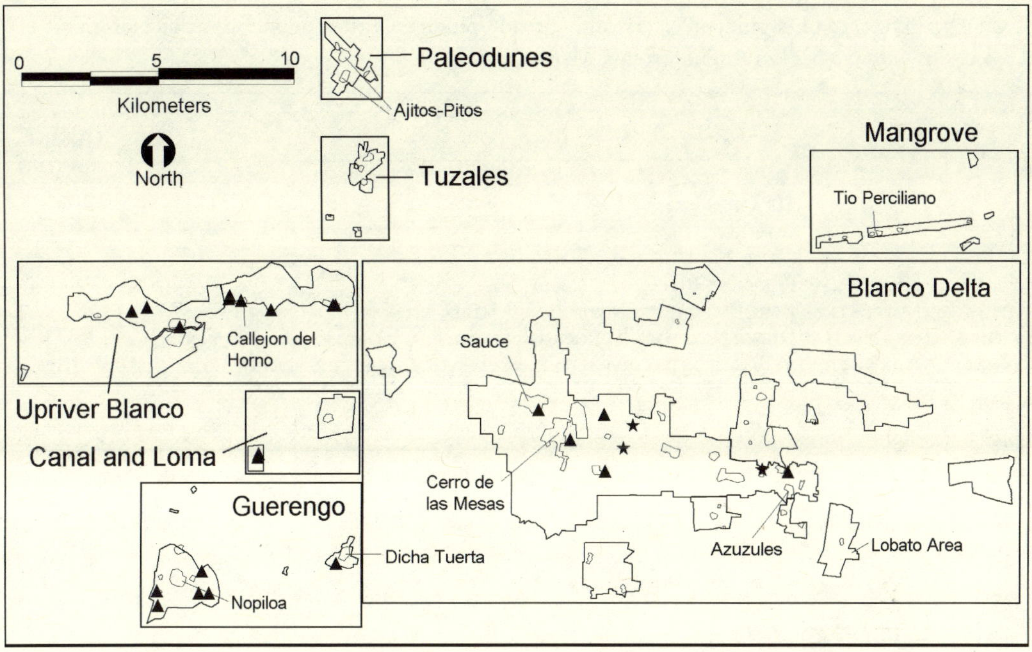

FIGURE 9.10. Compressed distribution map showing lapidary raw materials and abraders. Isolated finds are shown with a star.

located (Stark 1989, unpublished field data). Thus, although most pottery production presumably continued in household workshops without leaving a strong surface signature, some Late Classic pottery production was at a more intensive scale and associated with centers.

Multicrafting

Among collections that have multiple craft indicators, three at Azuzules have a whorl co-occurring with indications of orange bowl production. One more collection has these two characteristics plus a scraper and one of the ninetieth percentile indexes suggesting black prismatic blade production. As discussed with Cerro de las Mesas, a whorl co-occurring with one other category of craft evidence is not particularly convincing as multicrafting, but the last case with the co-occurrence of three distinct craft activities is more promising. Nopiloa has three collections with three activities indicated and four in which only two are suggested, one of them spinning. Although none of the three instances of possible multicrafting provides a strong possibility, we cannot rule it out, and even the four cases that involve a spindle whorl remain possibilities.

MIDDLE POSTCLASSIC PERIOD (CA. A.D. 1200–1350)

During the Middle Postclassic period, only one center is known from the survey, El Sauce, just northwest of the core of Cerro de las Mesas. Ceramic cross-dating places it in this period, and the strikingly different material culture likely indicates the intrusion of people with a highland cultural affinity (Curet et al. 1994; Garraty and Stark 2002; Stark 1995, in press). We do not know if earlier centers had already been abandoned or if newcomers hastened their decline. Although El Sauce itself has more concentrated residential remains, domestic occupation continued to be scattered out around the center and into the countryside. Among the economic changes is a marked decline in the use of black obsidian, with clear gray obsidian, apparently mainly from Pico de Orizaba, entirely or largely replacing it (Stark et al. 1992).

Table 9.2. Presence of Lapidary, Pottery, and Thread-Spinning Craft Production Artifacts from Collections within a 0.5 km Radius around Major Centers

		Pottery						Thread-making
Center and Coll. No.	Lapidary Material or Abraders	Orange Bowls, code 17a, Percent 90th Percentile Density and by Visibility	Mold, Texcoco Molded	Mold, Fondo Sellado	Comal Waster	Kiln Frag.	Waster, Poss. Fine Gray	Whorl
Cerro de las Mesas								
348							x	
356						x		
1433		x						
63								x
375	x							x
Clear Gray Obsidian, Postclassic; Comal, Postclassic								
554					x[a]			
605					x			
64								x
834								x
Azuzules								
1225		x						
1230		x						
1272		x						
1501	x							
1226		x						x
1273		x						x
1302		x						x
1304		x						x
Nopiloa								
6279								x
6303								x
6305								x
6325								x
6360								x
6407								x
6456								x
El Sauce								
116								x
119								x

Table 9.2. (continued) Presence of Lapidary, Pottery, and Thread-Spinning Craft Production Artifacts from Collections within a 0.5 km Radius around Major Centers

		Pottery						Thread-making
Center and Coll. No.	Lapidary Material or Abraders	Orange Bowls, code 17a, Percent 90th Percentile Density and by Visibility	Mold, Texcoco Molded	Mold, Fondo Sellado	Comal Waster	Kiln Frag.	Waster, Poss. Fine Gray	Whorl
124	x							x
209								x
214								x
224								x
1176								x
1751								x
1756								x
1818								
Buffer overlap between Cerro de las Mesas and El Sauce								
113								x
202								x
354								x
356								x
Callejón del Horno								
6014			x					
946	x		x	x[b]				
6012								x
6081	x							
6082	x		x					x

Note: Bold entries indicate collections with more than one craft category.
[a] possible comal waster
[b] possible Fondo Sellado mold

Prismatic Blade Production

In the Middle Postclassic period, blade workshop debris for clear gray obsidian at mound 1756 in El Sauce is accompanied by additional scattered residential features with evidence of production (Heller 2000). Figures 9.11 and 9.12 show the collections that meet or exceed the ninetieth percentile for two indexes of production indicators (density and the ratio to rims).

Obsidian Scrapers

Clear gray obsidian scrapers are concentrated in the immediate vicinity of El Sauce (Figures 9.13–9.14; Lewenstein 2001). Scrapers are particularly abundant at mound 353, where 52 were recovered; mound 353 has evidence of blade production (three pieces) and reaches the ninetieth percentile of clear gray blade production indicators in regard to the ratio to rims and for blade count. Four scrap-

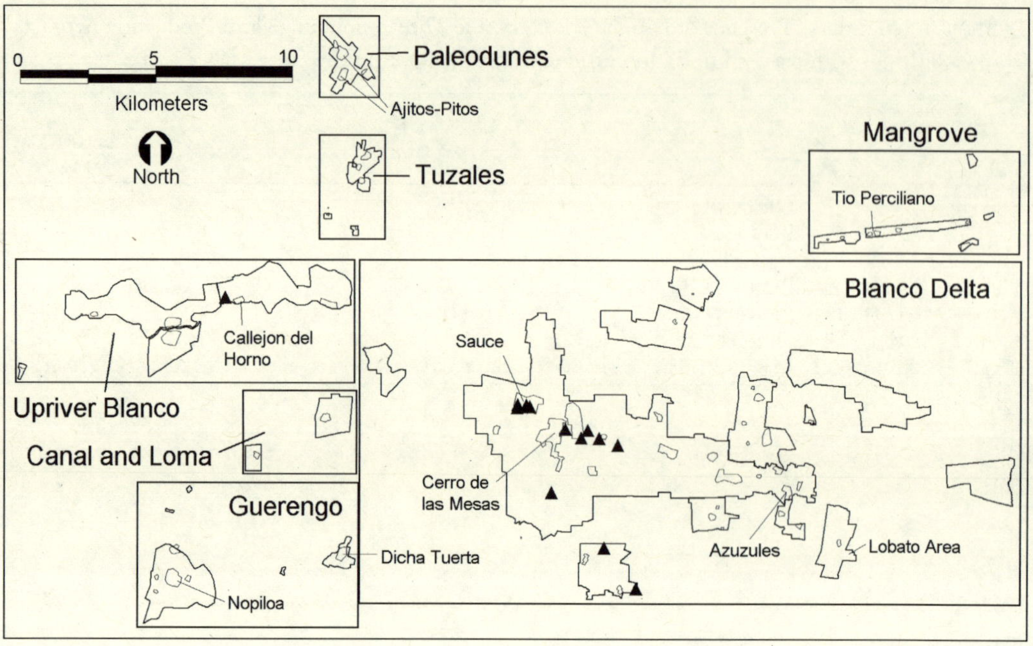

FIGURE 9.11. Compressed distribution map showing collections at or above the ninetieth percentile of densities per m² of clear gray obsidian blade production indicators (≥0.014085).

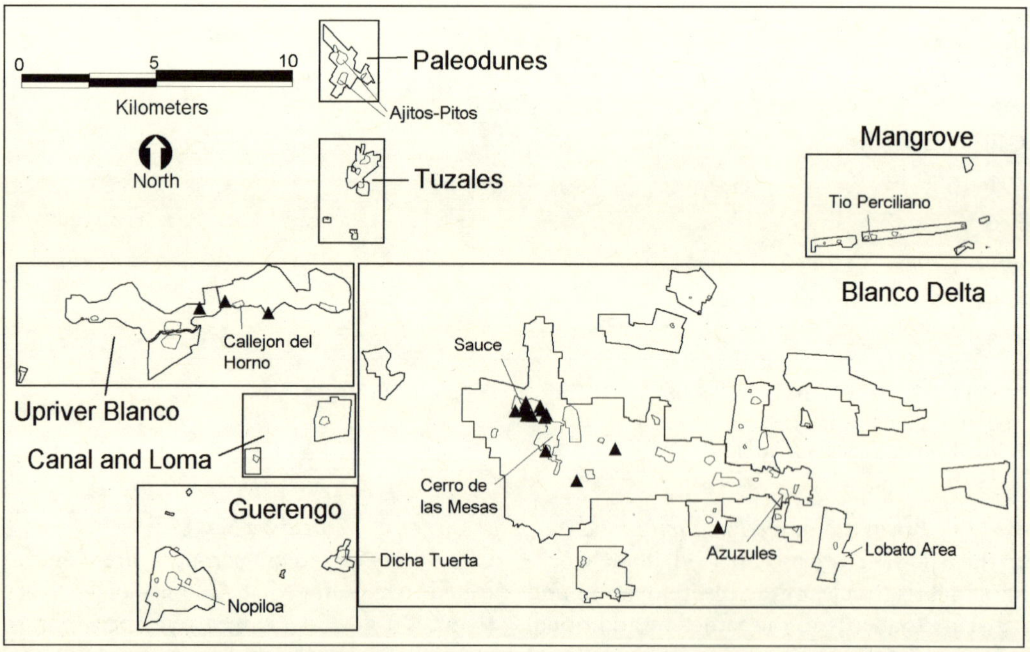

FIGURE 9.12. Compressed distribution map showing collections at or above the ninetieth percentile of the ratio of clear gray obsidian blade production indicators to rim sherds (≥0.022727) for collections with 68 or more rims, the median.

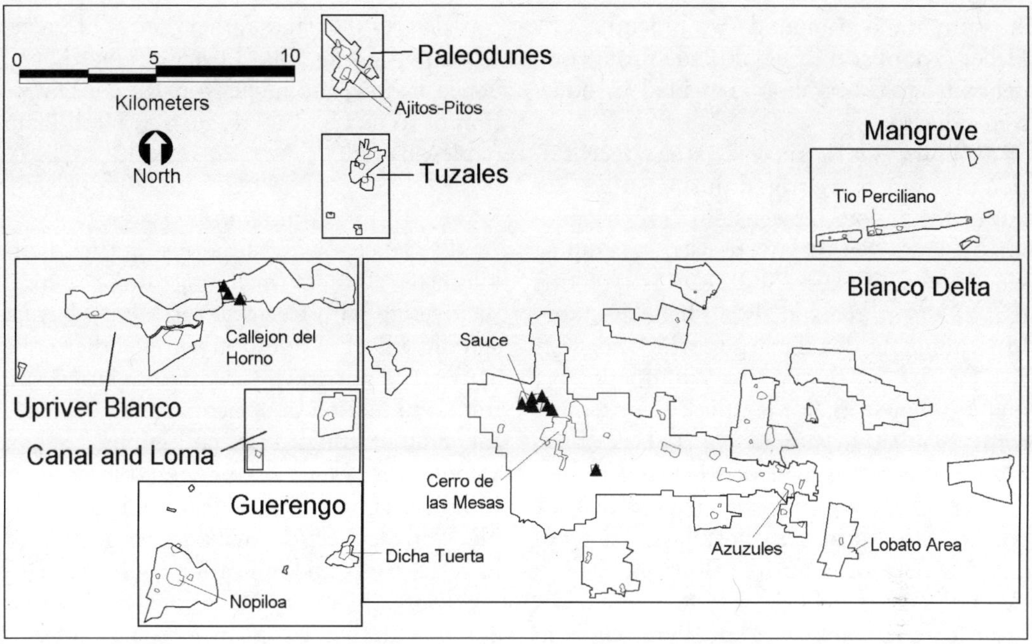

FIGURE 9.13. Compressed distribution map showing clear gray obsidian scrapers that occur in collections that meet any of the three ninetieth percentile blade production indexes.

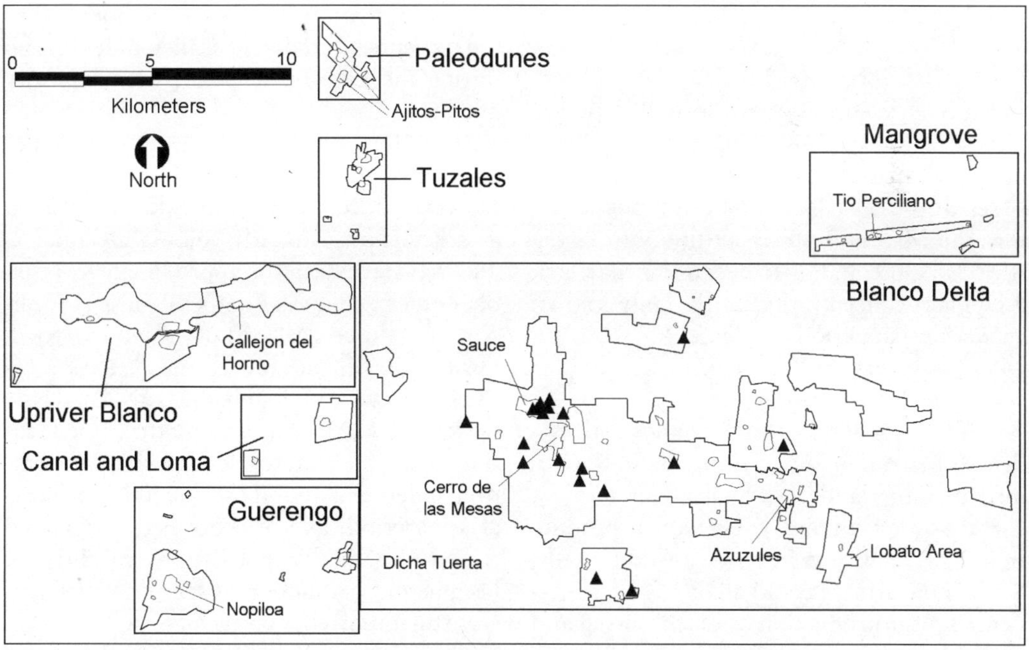

FIGURE 9.14. Compressed distribution map showing clear gray obsidian scrapers that do not occur in collections that meet any of the three 90th percentile blade production indexes.

ers were found at mound 1756, identified by Heller (2000) as a locus of blade production debris from a workshop (195 blade production indicators).[2]

Mound 353 is the most striking concentration of scrapers recovered during survey for any of the periods. Lewenstein's (2001) analysis suggests that they were used in scraping wood or woody plants. She (2001:174) reports that all the scrapers analyzed have a scraper locus along the distal margin, which suggests that multicraft work occurred at mounds 353 and 1756 or in their vicinities. One scraper from the adjacent mound 354 is black obsidian and may be a Classic period artifact.

Clear gray scrapers occur in a less marked concentration at the Late Postclassic center of Callejón del Horno upriver along the Blanco, as will be discussed below. However, Middle Postclassic remains are almost entirely confined to the Lower Blanco delta, which suggests that clear gray scrapers in the delta are a palimpsest of Middle Postclassic and Late Postclassic artifacts, with more from the former than the latter period.

Lapidary Items

El Sauce lacks any concentration of lapidary pieces (Figure 9.10). Although lapidary items are too scarce overall to make much of this observation, it at least raises the possibility that the Middle Postclassic differs from the periods before and after because it may lack a lapidary concentration at the only known center for that period.

Pottery

A small cluster of residential mounds with evidence of *comal* manufacturing lies in the immediate hinterland of El Sauce (Curet 1993), slightly over 2 km away. The comal production was relatively intensive. Additionally, Stark and Garraty (2004) analyze surface evidence for the production of black-on-red and local black-on-orange bowls at a more distant location, the Lobato area 12 km from El Sauce. The Lobato area was recorded by Speaker (2001). It is apparently a small settlement with dispersed residences. These examples suggest that some crafts were concentrated at rural villages rather than at El Sauce. Some pottery may have been produced at El Sauce, however, to judge from the concentration of some types there, such as Hard Plain (code 58).

Multicrafting

At El Sauce, six collections in the half-kilometer buffer (including those within the overlap with Cerro de las Mesas) have a spindle whorl in addition to one other craft activity. As remarked for Cerro de las Mesas, this combination is not convincing evidence for multicrafting. Five collections have a spindle whorl and two other craft indicators, usually a scraper and evidence of blade making. Three collections have no whorl but do have a scraper and indications of blade making. Especially the combinations of two other crafts with whorls provide somewhat improved chances that multicrafting occurred.

Late Postclassic Period (A.D. 1350–1521)

El Sauce appears to have been abandoned, with a new center founded at Callejón del Horno during the Late Postclassic period (Garraty and Stark 2002). Callejón del Horno, located upriver along the Blanco, was the only center discovered for this period during the survey. However, Tlalixcoyan to the north is named as a head town in early colonial documents and may have been the approximate location of another center. Alanna Ossa's initial examination of the central area of the modern town during 2002 did not reveal Late Postclassic ceramics, however. Possibly the Late Postclassic settlement was not at exactly the same location as the modern town. Obsidian importation underwent another shift in the Late Postclassic period, as green obsidian, presumed to represent the Pachuca source in Hidalgo, became more abundant. Clear gray obsidian was still imported, however.

Prismatic Blade Production

Green blade production (and some clear gray blade production) was concentrated near Callejón del Horno. Figures 9.15 and 9.16 show the collections meeting the ninetieth

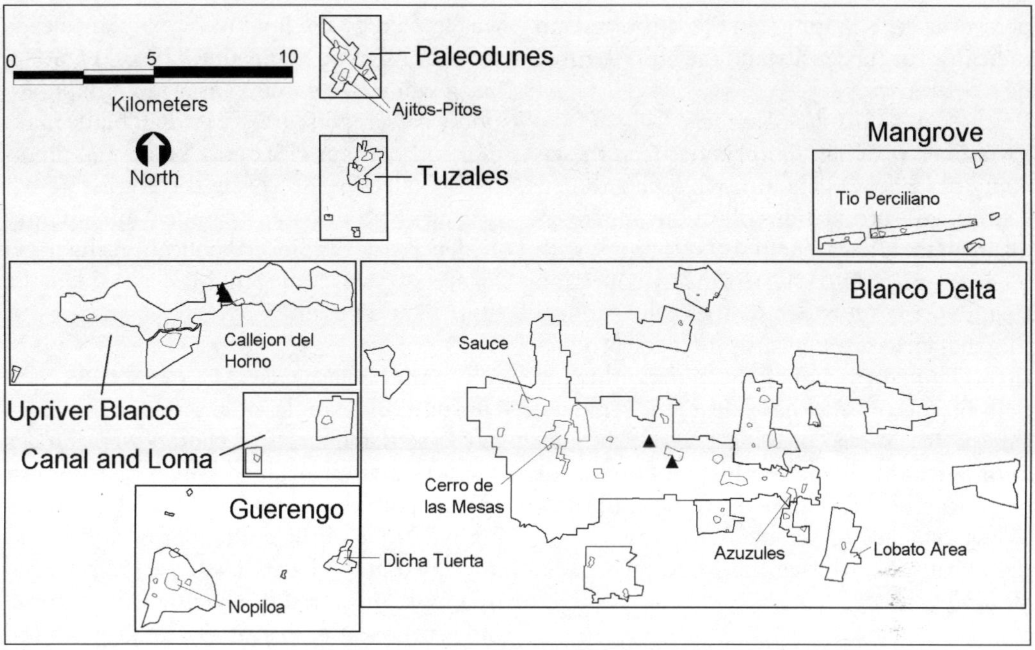

FIGURE 9.15. Compressed distribution map showing collections at or above the ninetieth percentile of densities per m² of green obsidian blade production indicators (≥0.0125).

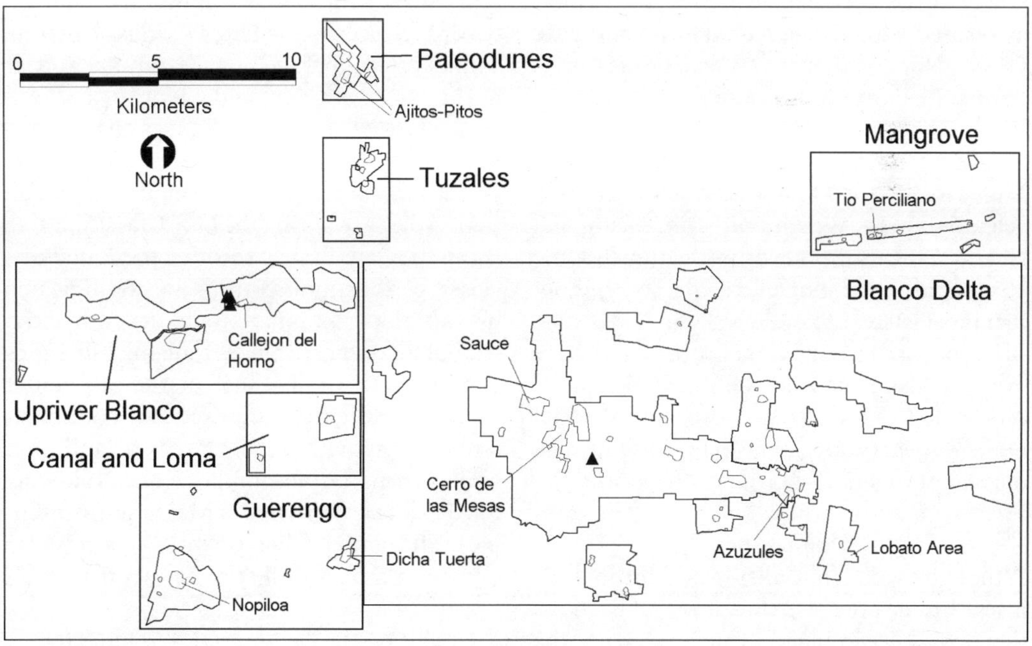

FIGURE 9.16. Compressed distribution map showing collections at or above the ninetieth percentile of the ratio of green obsidian blade production indicators to rim sherds (≥0.025862) for collections with 68 or more rims, the median.

percentile criteria for green blade production indicators using density and the ratio to rims.

Obsidian Scrapers

Two green obsidian scrapers occur in the vicinity of Callejón del Horno in ninetieth percentile blade production collections, and overall green scrapers occur in six collections, with a maximum of two scrapers in any collection (Figure 9.17). Only one scraper is in a collection that lacks any indicators of blade production (Figure 9.18). Four collections within the half-kilometer buffer have clear gray scrapers, three of them also meeting one of the ninetieth percentile blade production criteria for clear gray blades (Figure 9.13). Overall there is less evidence of scrapers around Callejón del Horno than during the Middle Postclassic at El Sauce.

Lapidary Items

Three collections have lapidary pieces near or in the core of Callejón del Horno (Figure 9.10). Although in general lapidary evidence is not entirely clustered near centers, it is more associated with Callejón del Horno and Late Classic Nopiloa than with Cerro de las Mesas during the Early Classic period.

Pottery

Molds for Texcoco Molded censers and Fondo Sellado bowls were found at Callejón del Horno, indicating pottery production in some residential areas. Skoglund et al. (2006) show that most pastes of Fondo Sellado, Aztec-style Black-on-orange, and Texcoco Molded were different from those at the Aztec provincial center of Cotaxtla investigated by Michael Ohnersorgen (2001), in keeping with the idea of local production at Callejón del Horno.

Multicrafting

Although overall for Callejón del Horno there is less buffer area and thus fewer collections that can be examined compared with the other centers, there is some evidence of multicrafting. Two collections have four craft activities indicated, but one of these (946) is an unusually large, emergency supplemental collection made after most of Callejón del Horno was leveled by the landowner to plant beans the year after we mapped it. The other, 6082, has a whorl, a ceramic censer mold, lapidary evidence, scrapers, green obsidian blade making, and perhaps clear gray blade making as well. Collection 6081 has three craft activities, blade production, a scraper, and lapidary evidence. Two additional collections have evidence of two craft activities, provided the scrapers were used.

SUMMARY OF CRAFT CHANGES

During both Postclassic periods for which I have a settlement record, centers were a focus of craft activity. Harbingers of Postclassic craft spatial patterns appear in the Late Classic period. None of the crafts that became more nucleated in the Late Classic period reversed to a more dispersed or less intensive pattern. In fact, nucleation became more pronounced for pottery making during the Postclassic period, although there are important variations. In the Middle Postclassic period, some pottery-making concentrations appear in outlying villages, probably in addition to the center of El Sauce. By the Late Postclassic period, some pottery-making concentrations appear at the center of Callejón del Horno, and outlying craft villages are no longer characteristic. Postclassic times show evidence of more vessel categories with nucleated production than is the case during the Late Classic period. From the Late Classic through the Late Postclassic periods, multicrafting may be indicated in collections with more than one craft activity, but all of the instances remain ambiguous pending evaluation of such matters as scraper use and excavation contexts that could better reveal activities, their intensity, and their scale. In general, most of the centers following Early Classic Cerro de las Mesas show multiple crafts, with a few residential mounds also admitting the possibility of multicrafting.

EVALUATION OF FACTORS AFFECTING THE NUCLEATION AND INTENSIFICATION OF CRAFT PRODUCTION

Several possible reasons can be advanced to account for growing craft specialization starting in Late Classic times and culminating in

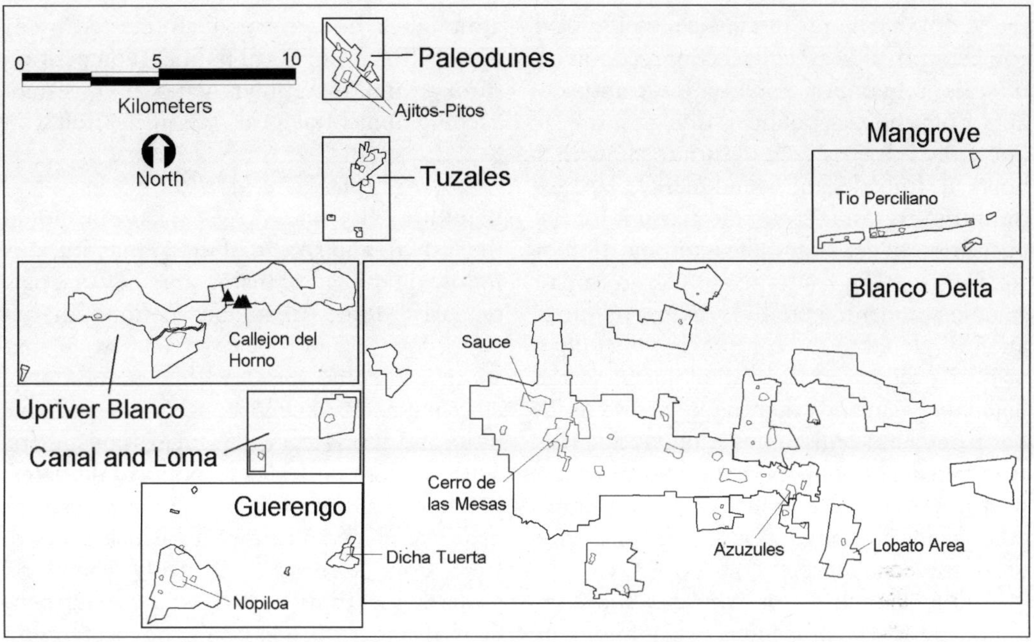

FIGURE 9.17. Compressed distribution map showing green obsidian scrapers that occur in collections that meet any of the three ninetieth percentile blade production indexes.

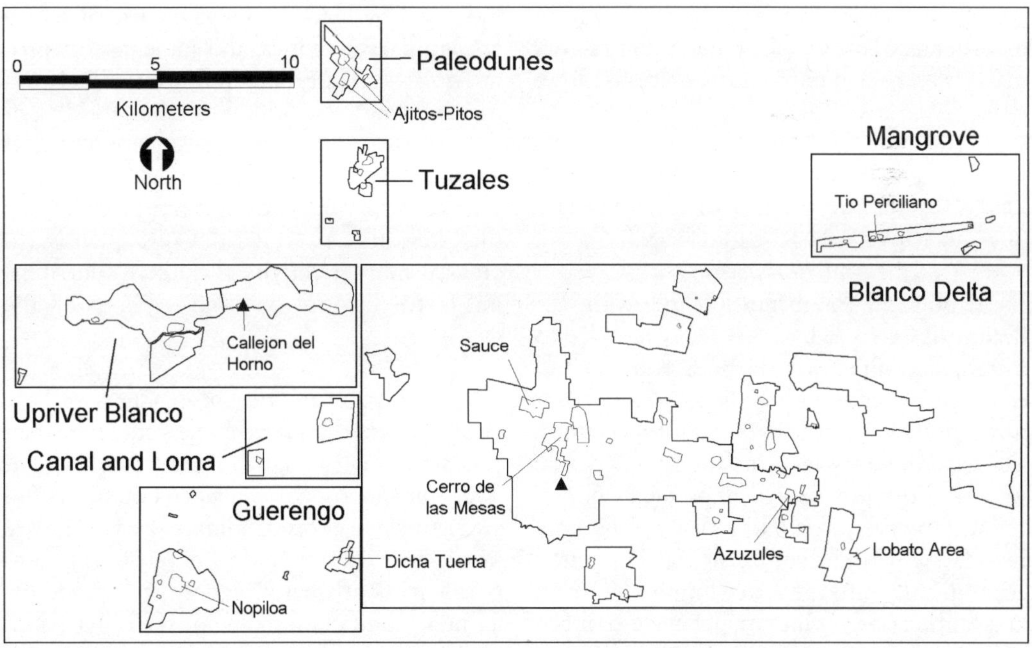

FIGURE 9.18. Compressed distribution map showing green obsidian scrapers that do not occur in collections that meet any of the three ninetieth percentile blade production indexes.

the Middle and Late Postclassic periods: local competition, increased marketing, economies of scale, economies of place, the transportability of products, political strategies, and labor costs. I discuss each in turn to show that political strategies and economies of place are the most promising factors to account for the nucleation of crafts and the intensification of some crafts. The resistance of cotton thread spinning (and inferentially weaving) to nucleation is related to the continuing direct links of fiber production to landholdings and the high labor demands that made the use of labor from rural households economical. This structure does not imply that cotton products were not part of a centralized economy, however, as they may have been demanded as tribute or traded in markets.

Although each factor warrants consideration, the most compelling explanations of continuing processes will be those that are consistent across all three periods. Craft nucleation started during the Late Classic period and persisted through two successive periods, each with substantial reorganization in settlement patterns. Less persistent factors may play a role in particular situations, however.

FACTORS THAT DO NOT PROVIDE A CONSISTENT EXPLANATION OF CRAFT NUCLEATION

Local Competition

Local political fragmentation following the decline of Cerro de las Mesas may have led to increased competition on both political and economic fronts, with polities seeking economic advantage by fostering craft specialties. The triad of Late Classic major centers can be viewed as offering new local political competition. However, it is likely that by the Late Postclassic period, the region was integrated economically into the Aztec Empire (Garraty and Stark 2002), which might have reduced political competition. The collapse and disappearance of the Classic period centers and cultural traditions led to only one Middle Postclassic center (El Sauce), not a situation with local competition. Likewise, in the Late Postclassic period only one local center is known, although a head town likely existed at or near Tlalixcoyan. Local political competition through multiple centers is not a consistent accompaniment of local craft nucleation.

Increased Marketing

Blanton et al. (1993:212–214) have proposed the growth and spread of market institutions following the fall of major state powers, particularly Monte Albán and Teotihuacán, between A.D. 600 and circa 750. They argue that specialists could thrive within major states, enjoying pacified conditions among communities and access to a wide range of consumers. With the loss of extensive political domains, specialists and consumers alike have reason to promote market institutions, which also are freed from strong state controls. Therefore, as market institutions became more independent of state control, we might expect to find as much or more evidence of craft specialization.

The escalation of intensified craft production in the WLPB matches this expectation, as indicated by blade making, the use of scrapers, lapidary activities, and some pottery production—but not for spinning (and likely not for weaving). Nevertheless, the WLPB evidence shows not just an intensification of craft activities but also increased localization at or near centers. This pattern is not predicted by the Blanton et al. scenario. Consequently, increased market activity is a likely contributing factor but not one that explains all the evidence.

Transportability of Products

If we ask whether more nucleated, intensified production is related to products with relatively higher costs in transport, pottery would seem to be a good candidate, as vessels are breakable, weighty, and relatively bulky compared with prismatic blades, greenstone ornaments, and cloth. Some of the other products are breakable, and greenstone is heavy, but pottery combines weight, breakability, and moderate to great bulk, depending on the vessel form. The spatial concentration of pottery specialists at a center means that a range of vessels will be accessible for consumers or "middlemen," and from the perspective of the producers, transport costs are shifted to others.

At least during the Middle Postclassic period, however, some specialist potters were concentrated in villages outside of centers. Because both pottery and obsidian blade production show some intensification and nucleation, product transportability alone does not account for the differential patterns of change. Obsidian blades are small and lightweight, which reduces transport costs, yet production was nucleated. Cotton spinning shows no signs of nucleation although cloth and thread are lightweight. The transportability of products does not provide a consistent explanation of craft concentrations.

Economies of Scale

Economies of scale refers to the increased size of a production unit or rate of production (e.g., through investment in a bigger labor pool or tools) that yields increased returns (utility or profit) and lower costs of products (Schneider 1974:234). Economies of scale are a possible reason for intensification and nucleation. With regard to pottery, economies of scale might relate to the use of kilns for improved control of firing for sizable loads. However, because kilns could be built near centers or far from them, the use of kilns cannot account for the greater nucleation of production at centers.

Because obsidian was imported as preformed blade cores during the Classic and Postclassic periods, there was no scale benefit to bringing the cores to a center versus an outlying location for processing, except for ready access to consumers, which leads to the related topic, economies of place, discussed below. In sum, economies of scale do not appear to account for the nucleation of production at centers.

FACTORS THAT PROVIDE A CONSISTENT EXPLANATION OF CRAFT NUCLEATION

Economies of Place

I use *economies of place* to refer to a range of economic advantages that accrue to the selection of one place versus another for (in this case) craft production. Economies of place include such traditional considerations as (1) multipurpose trips (which make retail marketing or exchange advantageous in higher-order service centers where more consumers are attracted because they can reduce overall transport costs), (2) location near the resources needed for craft activities to reduce trips to obtain materials, (3) administrative protection at centers so that products and facilities enjoy greater security, and (4) easy consultation with patrons at centers to appropriately design commissioned products. Many of the locational issues regarding transport costs have been examined extensively in economic geography (e.g., Bradford and Kent 1977; Haggett 1966:142–152). Although the third point regarding security may be important in some situations, unless the population in general is nucleated, any safer position of craftspeople does not represent protection of other crucial elements in the economic system.

Potters might reduce transport costs with locations close to clay or timber sources, but suitable clay deposits and stands of timber were probably not highly restricted in the region. Specialist pottery production, unlike the other crafts examined, did not always nucleate at centers, and perhaps Middle Postclassic rural specialists were separated from centers because of localized resource advantages, among other factors. The mangrove swamp production at the Nacastle-Patarata community would certainly be in an excellent location for clay and timber, but the same is true of other mangrove locations. Even though none of the centers appears to represent an advantage in terms of natural resources, some potters and other specialists became more concentrated at centers.

More compelling than economies of place related to natural resources are economies linked to transport and access to consumers. Readier access to consumers would occur at centers during market days or festival assemblies. In addition to such regular congregations, other more sporadic trips to centers (e.g., connected to governance) would provide improved craft access if producers (or vendors) concentrated at centers. Only readier access to consumers and markets would account for craft locations near centers, unless we assume that centers were placed near clay sources and wood fuel for pottery making, for which there is as yet no supporting evidence.

A third aspect of economies of place might apply to luxury crafts commissioned by elites (e.g., lapidary work) who were concentrated in and around centers. This production can more readily be supervised and tailored to the commissioners' tastes if the craft work occurs near elites at centers. If support of a craftsperson involved subsistence, a location at or adjacent to elite residences would require less movement than residence at a distance.

Political Strategies

Craft proximity to centers might improve the chances for regulation and taxation, and consequently, insistence on or facilitation of proximity might be an overt political strategy. During the Late Classic period, the triad of smaller polities lacked the extent of territory and tributary clients enjoyed before by Cerro de las Mesas. Fostering crafts for taxation may have been attractive (e.g., via regulation of a market). Eisenstadt (1963:24–28) stresses the importance of "free-floating" resources for central rule, not embedded in the rights and practices of traditional social groups, such as landowning elites. Additionally, craft and possible market services at centers may have helped cement links to the client population and their loyalty. These new centers did not enjoy the historical pedigree of Cerro de las Mesas, which seems to have had a head start during Preclassic times and thus a longer role as a governmental and sacred place.

Obsidian, as an imported material possibly subject to interpolity negotiations, could be more readily centrally controlled compared with pottery or cotton, which draw on widely available local resources. Lapidary materials and tools also require importation and thus might have been easier to control by personnel at centers than pottery or cotton. However, the nucleation of crafts extends beyond those dependent on imported materials, which suggests nucleation of crafts as a broad political strategy.

Labor Costs

If access to consumers (related to transport) and political factors are important, then how do we explain the continued dispersion of whorls, which indicate a low scale of production at many domestic locations, scattered throughout the area? Elsewhere Stark et al. (1998) have argued that pre-Hispanic cotton fiber processing and weaving were not subject to noteworthy economies of scale and were extremely labor intensive. If much of the farming population continued to be dispersed in the countryside, then both cotton growing and a labor pool for thread and cloth production would remain dispersed, particularly if spinning and weaving were integrated into the timetable of domestic activities conducted by women. Cloth production activities can start and stop, taking advantage of various intervals of time, as George Cowgill (personal communication, 2003) has pointed out. Most pottery production and blade production, in contrast, do not depend on an extremely labor-intensive process. Each of these two industries can be centralized more readily. If textiles were a common tribute demand, then this obligation could have mobilized households widely in spinning and weaving.

Spinning and textile weaving *can* be localized as specialties. Nichols et al. (2000) suggest a maguey textile workshop area at Otumba on the basis of a concentration of whorls, but they also note that whorls continue to be widely distributed in households. Sinopoli (1988) argues for textile specialists at the Indian capital of Vijayanagara from textual evidence. Thus, the "inertia" of textile production in respect to centralization and greater intensity can be overcome under particular historical conditions; one possibility suggested by George Cowgill (personal communication, 2003) is that some luxury textiles might have been more spatially concentrated because they required special knowledge or skills linked to elites at centers. In general, however, the spatial concentration of textile production appears to have been uncommon prior to the mechanization of lint processing and weaving.

One other markedly labor-intensive industry is lapidary work. Some lapidary evidence is dispersed, but overall the evidence was recovered at or near monumental structures (primary centers, lesser centers, and outlying groups) by a 3:1 margin. Raw materials, as in the case of obsidian, would have been imported and therefore perhaps subject to central control. Unlike obsidian, lapidary raw materials are mainly fine stone, especially greenstone, that would have been exceptionally valued and likely destined for prestige goods. Except for the labor demand in production, reasons of political or social control would suggest that lapidary production would have been relatively close to centers, especially as some lapidary work may have been performed by attached specialists. Perhaps these two considerations, labor needs versus elite control, produced the somewhat mixed spatial pattern observed. Some abraders, however, may have been employed in polishing celts used in quotidian activities, contributing to a mixed distribution.

DISCUSSION OF THE NUCLEATION OF CRAFTS

The reasoning outlined above accounts for a process of nucleation and intensification in some crafts under conditions of more vigorous marketing and greater reliance by political powers on craft taxation and controls. The indications of multiple crafts near centers may be only part of the process of change, as multicrafting may have applied in some instances, such as woodworking and knapping. Feinman and Nicholas (2000) suggest that multicrafting is one response to the Mesoamerican limitations of transport that constrained income from a particular craft. Without a program of craft-focused excavations, we cannot yet gauge multicrafting, but Ohnersorgen (2001) notes that some collection locales at the Late Postclassic Aztec provincial center of Cotaxtla, Veracruz, that have evidence for obsidian blade production also include indications of pottery and figurine production. Thus, multicrafting may have been part of a wider escalation of craft specialization and one form of craft intensification.

Access to consumers, population growth (not addressed here), and transport technology are also profoundly important in craft patterns. Many craft specialists can be viewed as individuals with poor agricultural opportunities who turn to crafts to supplement or create income (e.g., Arnold 1985; Smith and Heath-Smith 1994:368–369). Among other factors, growing population may produce poor access to land for some individuals. In Mesoamerica striking transport limitations restrict cost-effective access to consumers and may dampen household investments in craft production in favor of a mix of farming and crafts or a mix of crafts (Nichols 1994:185). Although concentrations of demand accompanying urbanization increased craft opportunities, craft nucleation and intensification in the WLPB are a feature of small polities and centers with only modest nucleation of population. Additionally, the Middle Postclassic period appears to show a population decline in view of the more localized remains confined to the Blanco delta, yet craft nucleation continued unabated. The particulars of the WLPB sequence call attention to the likely importance of periodic markets at central places and governmental strategies. Economies of place made craft nucleation at centers efficient in terms of transport and access to consumers, and labor demands help make sense of the lack of nucleation of some crafts. Further, the nucleation of crafts may have been encouraged by ruling authorities to enhance revenues. Other factors do not fully account for the patterns of change when several industries are examined or several periods are considered.

The nucleation of crafts is not necessary for vigorous marketing, even if it offers some transport advantages. Periodic markets offer a mechanism to concentrate demand and vendors. Periodic markets allow dispersed consumers and producers at different villages or residences to make weekly or less frequent trips to markets without living immediately in the market center. For this reason, increased marketing alone seems unlikely to account for the persistent trend toward more craft specialization and more intensive production in close proximity to centers. For example, specialists during Early Classic Cerro de las Mesas times could have brought products into the center, and consumers could have arrived on market days. Thus, marketing cannot be ruled out in the Early Classic period. Certainly with dispersed production, travel is increased in a nonnucleated scenario, but its effects seem modest in view of the possibilities of multipurpose trips that include social and ritual activities as well. The distances from outlying residences to the main centers are not more than a day's walk, and some can be shortened by canoe travel.

With the breakup of the Cerro de las Mesas realm around the close of the Early Classic period, the smaller successor polities had a significantly decreased pool of labor and land on which to draw. Enhancement of exchange at a center would attract consumers and allow more reliable taxation of commercial transactions. In keeping with the interpretation advanced by Blanton et al. (1993:212–214), I conclude that marketing likely became more prevalent and institutionalized in Mesoamerica starting in the Late Classic period.

Contrary to their scenario, however, marketing was not largely liberated from state control in all regions. Rather, in the WLPB crafts at centers seem to have thrived as a part of central policy. Statecraft for smaller polities would likely employ a range of techniques to enhance revenue and power. The encouragement of markets and more intensive specialization that could attract consumers was a reasonable response by more fragmented polities in the Late Classic period. This strategy became more prominent with succeeding small Postclassic polities that did not have close competing neighbors (Middle and Late Postclassic periods) but may have been more closely tied with more distant markets. This is especially true for the Late Postclassic period when the Aztec Empire subdued tributary head towns near the WLPB (Garraty and Stark 2002). Therefore, the process of change was complex, and political controls or sponsorship appear to have been a catalyst for the combination of greater nucleation and intensification in crafts, even if marketing institutions became more widespread and vigorous in Mesoamerica in a transregional process.

The key to the analyses of processes of change that I have conducted using survey data is the systematic collections from individual features. With such data, spatial patterns can be examined for both industries with strong surface signals for craft production (e.g., blade manufacture) and those that are sampled more fitfully (e.g., lapidary work). By focusing on the major centers through time, I have shown a shift to multiple crafts at centers and more intensive craft activity. Several potential instances of multicrafting were identified. The excavation of selected craft locations against this backdrop of regional change can provide a next step to understanding the activity profiles at residential mounds.

The survey design was intensive and specifically geared to enable social and economic interpretations, such as those presented here. Some of the chapters in this volume provide examples of the contextual detail obtained through excavations. In craft analyses, regionally oriented survey and focused excavations strike a balance, each supplying something the other cannot—a broader view of social and economic change versus more precise contexts for craft actions and their social implications. This complementarity is noteworthy and should dispel the lingering objections that excavation-oriented archaeologists sometimes voice about survey.

Acknowledgments

I appreciate the invitation from Izumi Shimada to present a paper at the symposium "Rethinking Craft Production: The Nature of Producers and Multi-Craft Organization," which he organized at the 68th Annual Meeting of the Society for American Archaeology, Milwaukee, 2003. The Proyecto Arqueológico La Mixtequilla survey was made possible by grants from the National Science Foundation (BNS 85-19167, BNS 87-41867, and SBR-9804738) and from the National Geographic Society. Support was also provided by Arizona State University. The Instituto Nacional de Antropología e Historia (INAH) granted permission for the investigation. The Centro INAH Veracruz, its director, Ing. Daniel Goeritz, and the staff members provided timely assistance at numerous junctures. Any success in the project is entirely dependent on the excellent fieldwork of a number of students and archaeologists over the several seasons of survey and excavation. I thank the local communities for their kind welcome and cooperation. George Cowgill, Christopher Garraty, Ben Nelson, Michael Ohnersorgen, Kristin Sullivan, and Paula Turkon provided valuable suggestions to improve the manuscript but are not responsible for its deficiencies.

Notes

1. Categories included as prismatic blade production indicators include the following (database code number in parentheses): (8) prismatic blades with platform reversal scars; (89) hinge recovery blades; (71) small, delicate, ribbon blades; (19) proximal macropercussion blades (>2.5 cm wide); (21) medial macropercussion blades; (23) distal macropercussion blades; (25) whole macropercussion blades; (27) initial-series blades; (26) second-series blades; (29) ridged blades; (30) secondary ridge blades; (33) plunging blades; (35) whole prismatic cores; (37) proximal prismatic cores; (39) medial prismatic cores; (41) distal prismatic cores; (42) prismatic core shatters; (47) longitudinal blade core fragments; (43) prismatic core rims, transversely struck; (45) prismatic core face flakes, transversely struck; (46) prismatic core face flakes, longitudinally struck; (93) proximal macroflakes (for flakes >2.5 cm wide); (94) macroflake fragments; (95) whole macroflakes; (54) flake fragments, probable platform trimming; (55) platform trimming or faceting flakes; (56) probable platform trimming or faceting flakes; (90) transverse core flake fragments; (91) transverse core flakes with faceted dorsal side; (92) transverse core flakes with nonfaceted dorsal side; (97) distal rejuvenation flakes; and (99) lateral rejuvenation blades.
2. Five scrapers were reported from mound 1756 by Lewenstein (2001), but reanalysis of the 1756 collection by Lynette Heller reassigned one piece, giving a new total of four.

References Cited

Ames, Kenneth M.

1995 Chiefly Power and Household Production on the Northwest Coast. In *Foundations of Social Inequality*, edited by T. Douglas Price and Gary M. Feinman, pp. 155–187. Plenum Press, New York.

Arnold, Dean E.

1985 *Ceramic Theory and Cultural Processes*. Cambridge University Press, New York.

Blanton, Richard E., Stephen A. Kowalewski, Gary M. Feinman, and Laura Finsten

1993 *Ancient Mesoamerica: A Comparison of Change in Three Regions*. 2nd ed. Cambridge University Press, Cambridge.

Bradford, M. G., and William A. Kent

1977 Central Place Theory: Christaller's Model. In *Human Geography: Theories and Their Application*, by M. G. Bradford and William A. Kent, pp. 6–27. Oxford University Press, Oxford.

Brumfiel, Elizabeth M.

1980 Specialization, Market Exchange, and the Aztec State: A View from Huexotla. *Current Anthropology* 21:459–478.

Curet, L. Antonio

1993 Regional Studies and Ceramic Production Areas: An Example from La Mixtequilla, Veracruz, Mexico. *Journal of Field Archaeology* 20:427–440.

Curet, L. Antonio, Barbara L. Stark, and Sergio Vásquez Z.

1994 Postclassic Change in South-Central Veracruz, Mexiço. *Ancient Mesoamerica* 5: 13–32.

Eisenstadt, Shmuel N.

1963 *The Political Systems of Empires*. Free Press of Glencoe, Collier-Macmillan Ltd., London.

Feinman, Gary M., and Linda M. Nicholas

2000 High-Intensity Household-Scale Production in Ancient Mesoamerica: A Perspective from Ejutla, Oaxaca. In *Cultural Evolution: Contemporary Viewpoints*, edited by Gary M. Feinman and Linda M. Nicholas, pp. 119–142. Kluwer Academic/Plenum Publishers, New York.

Garraty, Christopher P., and Barbara L. Stark

2002 Imperial and Social Relations in Postclassic South-Central Veracruz, Mexico. *Latin American Antiquity* 13:3–33.

Haggett, Peter

1966 *Locational Analysis in Human Geography*. St. Martin's Press, New York.

Harrison, Peter

1972 *The Lintels of Tzibanche*. Atti del 40th Congresso Internazionale degli Americanist 1. Roma, Genoa.

Heller, Lynette

2000 Postclassic Obsidian Workshop Debris from El Sauce, Veracruz, Mexico. *Mexicon* 22:139–147.

Heller, Lynette, and Barbara L. Stark
1998 Classic and Postclassic Obsidian Tool Production and Consumption: A Regional Perspective from the Mixtequilla, Veracruz. *Mexicon* 20(6):119–128. (Figures reprinted in *Mexicon* 21[1 (1999)]: unnumbered page following page 12.)

Hirth, Kenneth G.
1998 The Distributional Approach: A New Way to Identify Marketplace Exchange in the Archaeological Record. *Current Anthropology* 39:451–476.

Jargstorf, Sibylle
1993 *Baubles, Buttons and Beads: The Heritage of Bohemia*. Schiffer Publishing Ltd., Atglen, Pennsylvania.

Lewenstein, Suzanne
2001 Use of Obsidian Blades, Flakes, and Scrapers in the Mixtequilla. In *Classic Period Mixtequilla, Veracruz, Mexico: Diachronic Inferences from Residential Investigations*, edited by Barbara L. Stark, pp. 170–175. Institute for Mesoamerican Studies, Monograph 12. University at Albany, New York.

Nichols, Deborah L.
1994 The Organization of Provincial Craft Production and the Aztec City State of Otumba. In *Economics and Politics in the Aztec Realm*, edited by Mary G. Hodge and Michael E. Smith, pp. 175–194. Institute of Mesoamerican Studies, State University of New York, Albany; and University of Texas Press, Austin.

Nichols, Deborah L., Mary Jane McLaughlin, and Maura Benton
2000 Production, Intensification, and Regional Specialization: Maguey Fibers and Textiles in the Aztec City-State of Otumba. *Ancient Mesoamerica* 11:267–291.

Ohnersorgen, Michael Anthony
2001 Social and Economic Organization of Cotaxtla in the Postclassic Gulf Lowlands. Unpublished Ph.D. dissertation, Department of Anthropology, Arizona State University, Tempe.

Otis Charlton, Cynthia
1993 Obsidian as Jewelry: Lapidary Production in Aztec Otumba, Mexico. *Ancient Mesoamerica* 4:231–243.
1994 Plebians and Patricians: Contrasting Patterns of Production and Distribution in the Aztec Figurine and Lapidary Industries. In *Economies and Polities in the Aztec Realm*, edited by Mary G. Hodge and Michael E. Smith, pp. 195–219. Studies in Culture and Society Vol. 6. Institute for Mesoamerican Studies, University at Albany, State University of New York, Albany; and University of Texas Press, Austin.

Parsons, Jeffrey R., and Mary H. Parsons
1990 *Maguey Utilization in Highland Central Mexico: An Archaeological Ethnography*. Museum of Anthropology, University of Michigan, Anthropological Papers, No. 82. Ann Arbor.

Reents-Budet, Dorie, Ronald L. Bishop, and Barbara MacLeod
1994 Painting Styles, Workshop Locations, and Pottery Production. In *Painting the Maya Universe: Royal Ceramics of the Classic Period*, by Dorie Reents-Budet, pp. 164–233. Duke University Press, Durham.

Schneider, Harold K.
1974 *Economic Man: The Anthropology of Economics*. Free Press, New York.

Shimada, Izumi
1996 Sicán Metallurgy and Its Cross-Craft Relationships. *Boletín Museo del Oro* 41:26–61.

Sinopoli, Carla M.
1988 The Organization of Craft Production at Vijayanagara, South India. *American Anthropologist* 90:580–597.

Skoglund, Thanet, Barbara L. Stark, Hector Neff, and Michael D. Glascock
2006 Compositional and Stylistic Analysis of Aztec-Era Ceramics: Provincial Strategies at the Edge of Empire, South-Central Veracruz, Mexico. *Latin American Antiquity* 14:541–559.

Smith, Michael E., and Cynthia Heath-Smith
1994 Rural Economy in Late Postclassic Morelos: An Archaeological Study. In *Economies and Polities in the Aztec Realm*, edited by Mary G. Hodge and Michael E. Smith, pp. 349–376. Studies on Culture and Society, Vol. 6. Institute for Mesoamerican Studies, University at Albany, State University of New York, Albany; and University of Texas Press, Austin.

Speaker, John Stuart
2001 Settlement and Agricultural Land Use in Ancient Mixtequilla, Veracruz, Mexico. Unpublished Ph.D. dissertation, Department of Anthropology, Tulane University.

Stark, Barbara L.
1989 *Patarata Pottery: Classic Period Ceram-*

ics of the South-Central Gulf Coast, Veracruz, Mexico. Anthropological Papers 51. University of Arizona Press, Tucson.

1992 Ceramic Production in La Mixtequilla, Veracruz, Mexico. In *Ceramic Production and Distribution: An Integrated Approach*, edited by George J. Bey III and Christopher Pool, pp. 175–204. Westview Press, Boulder.

1995 Introducción a la Alfarería del Postclásico en la Mixtequilla, Sur-central de Veracruz. *Arqueología* 13–14:17–36.

2003 Cerro de las Mesas: Social and Economic Perspectives on a Gulf Center. In *El Urbanismo en Mesoamérica: Urbanism in Mesoamerica*, edited by Guadalupe Mastache and William Sanders, Vol. 1:391–422. Instituto Nacional de Antropología e Historia, Mexico City; and Pennsylvania State University, University Park.

In press Archaeology and Ethnicity in Mesoamerica. In *Indigenous Mesoamerica: The View from Archaeology, Art History, Ethnohistory, and Contemporary Ethnography*, by Frances F. Berdan, John K. Chance, Alan Sandstrom, Barbara L. Stark, James Taggart, and Emily Umberger. MS submitted to the University of Utah Press, Salt Lake City.

Stark, Barbara L., and Christopher P. Garraty

2004 Evaluation of Systematic Surface Evidence for Pottery Production in Veracruz, Mexico. *Latin American Antiquity* 15:123–143.

Stark, Barbara L., Lynette Heller, Michael D. Glascock, J. Michael Elam, and Hector Neff

1992 Obsidian Artifact Source Analysis for the Mixtequilla Region, South-Central Veracruz, Mexico. *Latin American Antiquity* 3:221–239.

Stark, Barbara L., Lynette Heller, and Michael A. Ohnersorgen

1998 People with Cloth: Mesoamerican Economic Change from the Perspective of Cotton in South-Central Veracruz. *Latin American Antiquity* 9:1–30.

Sullivan, Kristin S.

2006 Specialized Production of San Martín Orange Ware at Teotihuacan, Mexico. *Latin American Antiquity* 17:23–53.

Topic, John R.

1990 Craft Production in the Kingdom of Chimor. In *The Northern Dynasties: Kingship and Statecraft in Chimor*, edited by Michael E. Moseley and Alana Cordy-Collins, pp. 145–176. Dumbarton Oaks Research Library and Collection, Washington, D.C.

Widmer, Randolph J.

1991 Lapidary Craft Specialization at Teotihuacan: Implications for Community Structure at 33:S3W1 and Economic Organization in the City. *Ancient Mesoamerica* 2:131–147.

10

It's All in a Day's Work: Occupational Specialization on the Peruvian North Coast, Revisited

SUSAN E. RAMÍREZ

ARTISANS AND OTHER SPECIALISTS

References to specialists and, in particular, skilled artisans are common in sixteenth-century manuscript records for coastal Andean populations. Over 20 different types of specialists are identified for the north coast alone in early colonial times. But many questions remain about the production and identity of these specialists and to what extent they existed in the late pre-Hispanic period. I shall revisit the evidence for specialization, highlighting, in passing, the problems of our sources, to finally suggest that relatively small numbers of highly skilled artisans commanded or benefited from the relatively unspecialized activities of large numbers of temporary, rotating workers. I propose a division of labor where commoners (*parques* and *mitayos*) doing their tribute service to the state acquired and transported the raw materials necessary for the highly prized craft production of sumptuary goods necessary to identify authorities, propagate their legitimizing myths, and reward commoners and nobles alike for extraordinary services. This reevaluation of the evidence changes considerably the picture of regional specialization and has larger implications for our understanding of late pre-Hispanic and early colonial native economic and social relations both on the coast and in the highlands.

REFERENCES TO REGIONAL SPECIALISTS

Evidence of coastal specialization can be traced back to the myth of Naimlap as recorded by the Spanish priest Miguel Cabello Valboa. Naimlap, a valiant, captain, caudillo, and "father," arrived by sea from the "supreme part" (south) of the Andes with a principal wife, many concubines, and a multitude of followers. Among these were 40 very valued or esteemed tradespeople (*oficiales*), such as

- Piti Zofi, a trumpeter, who announced the lord's approach;
- Niñacola, a person who cared for his litter and chair;
- Ninagintue, a brewmaster;
- Fongasigde, an aide who scattered powered seashells wherever his lord might pass;
- Occhocalo, a chef;
- Xammuchec, a cosmetician in charge of the lord's face paint;
- Ollopcopoc, an intimate, charged with the lord's hygiene; and
- Llapchiluli, a specialist who produced rich shirts and feathered cloth (Cabello Valboa 1951 [1586]:327).

Such were precisely the specialists that any important ethnic lord needed to prepare, establish, maintain, and propagate his authority in late pre-Hispanic and early colonial times.

Andean chronicles and myths relate the importance of such specialists in supplying the insignia of office. Lords traveled in procession, preceded by a trumpeter who often blew a conch shell to announce his approach and a person who either scattered powdered marine shells or swept the ground and laid down mantles before the retinue. The lord's elaborate dress, the litter on which he rode, and the decorated chair or stool on which he sat, above all others, declared the presence of an authority whose power could be traced back to a divine ancestor. His secondary wives and concubines followed the procession carrying roasted corn and maize beer to distribute to onlookers when he stopped and rested. Personal attendants, like his cook, his cosmetician, and his dresser or valet, were omnipresent (AGI/Patronato, *legajo* [l.] 189, *ramo* [r.] 11, 1566; AGI/Justicia 457: 1066v; AGI/Justicia 461: 854; ART/CoO, l. 154, *expediente* [exp.] 208, 11-VIII-1582, 13v; Hart 1983:256, 262; Martínez Cereceda 1995; Rostworowski de Diez Canseco 1961:15–16, 77).

Such specialization is confirmed in colonial documentation, summarized in Table 10.1. Two dozen types of specialists are listed by pre-Hispanic *curacazgo* (chieftainship) and year of first mention in the primary sources. Some, like the merchants (*mercaderes*) and the adobe maker (*albañil*, discussed below), appear as lone practitioners of their trade or occupation. Others like the litter bearers (*hamaqueros*), potters (*olleros*), fisherfolk (*pescadores*), and metalsmiths (*plateros*) constituted part of a kinship group (*ayllu* or *parcialidad*) specialized in the one activity. The latter passed their knowledge and skills from generation to generation, without the freedom to change their specialization for another, according to Rostworowski de Diez Canseco (1975:317, 1977b:324; see AGI/Justicia 461: 1567; Betanzos 1987 [1551–1557]:115; Espinoza Soriano 1969–1970:9, 12; Morúa 1946 [1590]:329).[1]

Linguistic evidence suggests that some specialists had their own lords. Some lords used a title, often as a surname, that ended in the suffix *siec* (or *ceque*), meaning "lord," after the Spanish arrival, as shown in Table 10.2. Translations of the prefixes suggest their rank or specialty. *Cono*, for example, means "thousand," so *conoseque* translates as "lord of a thousand households." *Calanseque*, which appears most frequently for the communities of Illimo (Túcume; this and subsequent parenthetical names in this paragraph and below refer to the pre-Hispanic polity to which this colonial *encomienda* [the grant of a people and their labor or tribute given to a Spaniard who had served the king] once belonged), Lambayeque (Túcume), Túcume, Jayanca, Chuspo (Collique), and Ferreñafe (Túcume), may refer to the lord of a group of jesters or clowns, because *kall* means "to laugh," as shown in Table 10.3. Pedro Cieza de León (1864 [1553]:219), an early and highly regarded Spanish traveler and chronicler, describes the buffoons and dancers who entertained while others played music and sang. This translation may be too narrow, because, in 1572, a calanseque had charge of a group of litter bearers, suggesting a wider interpretation of the term, perhaps as a parade master or master of ceremonies, a director in charge of the lord's travel in procession or his presence in ritual. Likewise, another title, *purriseque*, may refer to the lords of salt makers, an unlikely status, or, more probably, feather workers who made or procured the much esteemed feathered shirts and sun shades common to local lord and emperor alike. Other titles might refer to confectioners (*senseque*) who worked with supplies of honey. *Tuniseque* may identify religious specialists who had esoteric knowledge of time and the calendar, the universe, and the sky —or astronomy. Or if we accept *tuni* as equivalent to *tumi*, a Quechua term, and translate it after Cabello Valboa (1951 [1586]:269, 506), *tumiseque* could have been the keeper or maker of knives or arms, that is, a metalsmith. These possible translations suggest a tendency seen at the state level for leaders to be valiant in battle precisely for the favor and power bestowed on them by the gods (of the sky, e.g., thunder/lightning, sun) and to demonstrate and confirm this status by the insignia that they kept (AGI/Justicia 418, 1573: 201v; AGI/Justicia 457: 695v; Hart 1983:251; Netherly 1977a:157; Ramírez 1990:510–511, 516,

Table 10.1. Specialists by *Curacazgo* and Date

Specialty	Collique	Jayanca	Pacasmayo	Saña	Sinto	Túcume	Unknown
Albañiles	1566	n.d.					
Cabestreros		1566					
Carpinteros	1566	1566				1573	
Cumbicos		1566					
Chicheros	1566	1540		1566		1560	
Esteros Petateros	1611	1739		1549	1566		1566
Hacheros/ Lenadores		1540					
Hamaqueros	1530s	1566		1566	<1566		1566
Huseros		1566				1566	
Mercaderes	1566	1563			1566	1560	
Olleros	1566	1540	1566		1566	1566	
Ovejeros/ Pastores		1540	1566		1566		
Pescadores	1565	1540	1566	1549	1566	1541	
Pintores (de mantas)	1558	1566			1566	1558	
Plateros	1566	1566			1566	n.d.	
Roperos		1566			1566		
Salineros		n.d.		1549	n.d.	1566	
Sapateros/ Alpargateros		1566				1566	
Sastres		1566					
Silleros						1566	
Tejedores	1582	1540			<1532	1576	
Tintoreros		1566				1566	
Venaderos						1566	

2005; Rostworowski de Diez Canseco 1975: 317, 1977b:221).[2]

Although the data are far from complete and sometimes contradictory, only two of these specialists appear to have used their skills full-time: the merchants and the metalsmiths. Cieza de León writes that in Cajamarca certain metalsmiths were full-time makers of elaborate pieces of gold or large silver vessels ("no entendían en todo el tiempo en más que en labrar ricas piezas de oro o grandes vasijas de plata" [1967 (1550s):65, all translations mine]) in their foundry (*casa de fundición*). To better serve their lord, metalsmiths stayed near his court to engage in the smithing and elaboration of ritual paraphernalia. No platero ever mentions lands (the proxy variable for deciding if a specialist is full- or part-time) or indicates engaging in any other economic activity. Information about the fishermen is also contradictory. Although most do not refer to land, a few report access to some. Fisherfolk did have time for cultivating, in theory, because ocean fishing was a seasonal activ-

Table 10.2. Lords (*Ceques*) on the North Coast, Sixteenth Century

Name	Curacazgo	Date	Source
Pedro Amas xeque	Sinto [Chuspo]	1585	ASFL/Reg. 9, Ms. 7, 1585
Don Diego Calance q[ue]	Chuspo	1563	ART/Lopez de Cordova 29-III-1563
Martin Calanceque	Ferreñafe	1549	AGI/Escribanía 502A, 43v-44
Don Felipe Calanceq[ue]	Ferreñafe	1560	AGI/Audiencíi de Lima 133, 1560
Don Diego Senseq[ue]	Ferreñafe	1560	AGI/Audiencia de Lima 133, 1560
Don Francisco Palarref Conconciqu[e]	Ferreñafe	1560	AGI/Audiencia de Lima 133, 1560
Tuni Ceq[ue]	Illimo	1566	AGI/Justicia 461, 1572
Don Pedro Calanceq[ue]	Illimo	1572	AGI/Justicia 460, 333v
Calanceq[ue], Calanceque	Jayanca	1540	Gama 1974 [1540]: 217, 225
Conoseq[ue]	Jayanca	c. 1540	AGI/Justicia 418, 1573, 256v, 270v
Calanse[que]	Jayanca	c. 1540	AGI/Justicia 418, 1573, 271
Don Joan Ponon cec	Jayanca	1566	AGI/Justicia 461, 1432
Don Joan Pononce	Jayanca	1566	AGI/Justicia 462, 1863
Don Juan Principal de Conosaque	Jayanca	1557	AGI/Justicia 418, 1573, 258
Don Juan Cononceque	Jayanca	1560	ART/CoAsuntos de Gobierno, 13-II-1565
Don Luis Calanceq[ue]	Lambayeque	1566	AGI/Justicia 462, 1867
Don Luis Calanceque	Lambayeque	1572	AGI/Justicia 460, 332v
Don Luis Calanceque	Lambayeque	1572	AGI/Justicia 457, 695v
Don Xpoval Callocic[que]	Lambayeque	1572	AGI/Justicia 457, 714v
Don Gaspar Purriseche	Túcume	1566	AGI/Justicia 462, 1863
Don Andres Conanseque	Túcume	1566	AGI/Justicia 462, 1739, 1863
Don Andres Con nonseque	Túcume	1573	ART/CoO, 3-XII-1575
Conocique	Túcume	1536	AGI/Escribanía 502A, 7
Calanseque	Túcume	1541	AGI/Escribanía 502A, 55
Conoseque	Túcume	1541	ART/CoO, 13-VII-1570, 100v
Locnoserque (el conoserque)	Túcume	1543	ART/Alvarez 28-IV-1543, 33
Don Andres Connonse[que]	Túcume	1569	ART/CoO, 13-VII-1570, 130v
Don Lorenco Puri ce[que]	Túcume	1570	ART/CoO, 13-VII-1570, 166v
Don Andres Cononseq[ue]	Túcume	1571	ART/CoO, 13-VII-1570, 190

ity. Sixteenth-century sources identify at least one potter who had land, but this may have been an isolated case or the scribe may have equated access to clay deposits with possession of land. Salt making, deer hunting, cooking, and pottery and mat making were carried out on a part-time basis (Lothrop 1938:24; Netherly 1977b:15).[3]

The skills of certain of these coastal specialists were so highly prized that, after the Incas conquered the north, some of them were dispersed throughout the highlands. Cieza de León (1967 [1550s]:195) recounts how Tupac Inca Yupanqui took coastal metalsmiths to Cuzco and settled them among other ethnicities. A native chronicler, Santa Cruz Pachacuti

TABLE 10.3. Yunga Prefixes and Their Meaning

PREFIX	POSSIBLE TRANSLATIONS	SOURCE(S)
amas, omas	[?]	
calan, calo, callo	kall = reir [laugh]	Orrego 1958: 91
purri, puri	purri = pluma [feather]	Zevallos Quiñones 1947: 185
	purren = plumas [feathers]	
	purr = salado [salty]	Zevallos Quiñones 1947: 186
	pum, punic = sombre [shade]	Zevallos Quiñones 1947: 187
sen	sem = dulce, suave [sweet, soft]	Zevallos Quiñones 1947: 179
tuni	tumi = lobo marino or mucho [seal, a lot]	Orrego 1958: 85; Zevallos Quiñones 1947: 181, 183, 187
	töni, tüne, tune = huaca [temple, shrine]	Vargas Ugarte 1942 [1939]: 476; Trimborn 1979: 69
	tuni = tiempo, universo [time, universe]	Orrego 1958: 94; Zevallos Quiñones 1947: 187

(1927 [1613]:190), also attributes to the same emperor the distribution of metalsmiths and clothing everywhere that he passed. This may account for about a hundred households or 600 people, specialists in making pottery and formerly subjects of the coastal lord of Collique, who were removed to Cajamarca, where they were rewarded with clothes and given land to grow reeds (*totora*) and straw by the state (Espinoza Soriano 1969–1970:12). Likewise, potters from Jayanca lived and worked as state colonists (*mitimaes*) in Guambos, although they continued to provide tribute labor to their own coastal lords (Gama 1975 [1540]). Hart (1983:271, citing AGI/Justicia 458: 2135) also reports that coastal potters served as mitimaes in the highlands of Huamachuco, making pots for dying yarns or cloth (AGI/Justicia 461: 1452, 1527v; AGI/Justicia 462: 1875; Rostworowski de Diez Canseco 1977a:126).

THE SOURCES

These data are gleaned from such evidence as encomienda grants, individual and group petitions to a Spanish judge, ordinances and royal decrees, wills, court cases, notarial ledgers, and other administrative records, most from the second half of the sixteenth century. The single most important source of data on specialists is the *residencia* (judicial review of a term of office) of a Royal Audiencia (supreme court) judge, Dr. Gregorio Gonzalez de Cuenca. This peninsular Spaniard traveled through the northern valleys and into their highland hinterlands in 1566–1567. Dr. Cuenca came to census and study the northern populations after the threat of rebellion of the Taki Onqoy (1565) movement. He brought royal orders that no native be allowed to ride a horse with saddle and bridle, that lords could not serve their followers maize beer (*chicha*), and that no one be allowed to leave their communities without permission, among others. He also settled disputes, reassessed tribute, resettled scattered populations into easier-to-control nucleated villages (*reducciones*), and wrote ordinances for each community. There were so many complaints about irregularities that his actions were revisited by another Spanish official in 1572–1573. In the thousands of pages of this record are about 200 petitions from natives asking to be exempted from the royal decrees that Dr. Cuenca was sent, in part, to enforce. These, then, identify individuals and whole groups who say that they are engaged in certain types of activities, including trades.

The data these old papers provide are deceptively intoxicating. Scholars, including myself, have been too eager to accept information about local native society without sufficient question and due examination of its biases and flaws. Over the years I have become increasingly critical in my reading of the primary sources and more convinced that scrutiny is necessary before the reliability of the information can be determined and the value of the contents can be known.

I came to this conclusion for several reasons. First, these primary sources were usually written by Spaniards or creoles and, more rarely, by bilingual mestizos or natives. The latter, in acquiring Spanish language skills, accepted, at least in part, European cultural paradigms. They then had the enormous responsibility of translating oral testimony and tradition both literally and conceptually. Complicating their task was the fact that some statements were translated from Mochick to Spanish or from Mochick to Quechua to Spanish. This increased the possibility of error and mistranslation.

Such cultural filters led to the misrepresentation of some institutions. Rostworowski de Diez Canseco (1977a), for example, found, transcribed, and published a copy of what she believes is an early manuscript, which she renamed the "Aviso," about "merchants" in Chincha (on the coast, south of Lima). This now-famous document that Rostworowski de Diez Canseco read in the library of the Royal Palace of Madrid is undated and has no author or provenience. The paper on which it is written, judging from the watermark, was manufactured in the eighteenth century. The handwriting and abbreviations are also of the same era. On the basis of her analysis of this document, Rostworowski de Diez Canseco concludes that there were individual merchants who traveled the coast and into the highlands "selling" or bartering their wares for their own account in the sixteenth century. Yet, another document specifically about Chincha, written in 1558 by the *corregidor* and the local priest (Castro and Ortega Morejón 1968 [1558]), does not mention merchants but, instead, includes information about reciprocity and redistribution. Further suggesting the misconstrued designation of the Chincha "merchants" is a petition from a self-identified fisherman who asks Dr. Cuenca for permission to trade his fish with people of neighboring homesteads. Dr. Cuenca granted his petition but in responding addresses the fisherman as a merchant. This is a case where a high Spanish official reclassified a person, labeling him to fit his own cultural conceptualization of the world. How many other times did this happen (Ramírez 1986, 1997)?[4]

Besides cultural filters, yet another reason for my distrust of a literal interpretation of the documents on hand are the late dates of some of these sources and the ongoing transformation of the native socioeconomic, politico-religious, and cultural institutions and structures. Even the date of the residencia (1572–1573) is already 40 years after initial European contact. Until recently, most students of early colonial history have underestimated the change that occurred. Even by the end of the first three years of contact, the Spaniards had fundamentally reconceptualized the empire and its leadership (Ramírez 1996, 2005). Other changes include the impact of decimating diseases on a native population with no innate immunity to them. Furthermore, starting in 1534–1535, Francisco Pizarro divided up the people of the various curacazgos into as many as four encomiendas (see Table 10.4) and allowed once subordinate lords to assume the airs and acquire the insignia of their former paramount leaders, causing widespread disruption and uncertainty among commoners in terms of service obligations and the like. This division undermined the authority of a native lord whose status was a function of the number of his followers (ART/CoJuez de Comisión 29-III-1558).

Simultaneously, as the native population continued to fall because of disease, overwork, and flight, small groups of surviving relatives moved in with one another. Such informal reconstitution of kin groups in the 1540s continued on a grander scale in the 1560s and 1570s by government fiat that mandated the resettlement of the scattered population in nucleated villages or reducciones. These were usually deliberately located far from ancestral shrines and traditionally worked lands.

TABLE 10.4. Divisions of Pre-Hispanic *Curacazgos*

PRE-HISPANIC CURACAZGO	SPANISH ENCOMIENDAS
Collique	Callanca
	Collique
	Reque
Jayanca	Jayanca
	Pacora
Pacasmayo	Pacasmayo
	Chepen
	Chérrepe
	Moro
Saña	Saña
Sinto	Sinto
	Lambayeque
Túcume	Túcume
	Ferreñafe
	Illimo

Source: Ramírez 1981:286–88

Native residents were compensated for these moves and forfeitures with formal grants of communal lands located nearer their new homes. Meanwhile, the state redefined tribute from labor to goods and currency. As Spaniards moved into the area to populate the new *villa* (Spanish town) of Saña after 1563, they began appropriating vacated lands and other resources. In such a scenario, it should not be surprising that some native specialists adapted their skills to Spanish demand and other natives learned a trade.

A final problem is the nature of the documentation. The major source for coastal data on specialists, as noted above, are the petitions and ordinances written by Dr. Cuenca and his scribes in 1566. It should be noted that all but one of the native petitioners and all the scribes were male, leaving most females unrepresented and all but a single lady's voice silenced. In addition, the circumstances under which Dr. Cuenca granted native requests make the information and particularly the specialty designations therein suspect. I now believe that scribes used some of the artisanal designations in order to get peasants and other commoners exemptions from some of the new royal decrees that Dr. Cuenca came to introduce and enforce. Thus, a peasant, representing himself, his wife, or another family member, who could produce a woven textile or a pot of some sort, might declare himself to be a weaver or a potter at the urging of the scribe to secure permission to leave the homestead or village to which he otherwise might be confined. The production of such petitions was a win-win situation. The scribes earned a fee. Dr. Cuenca's exemptions constituted a license for which he charged and later boasted that he thus generated royal revenues. The natives paid but received the exemption they sought (AGI/Justicia 458: 1319v–1320).

SPECIALISTS AND COMMONERS

Such problems with the evidence and their methodological consequences invite a rethinking of the identities and roles of the specialists to which we have references. Below, I summarize the information on hand for some of the specialists listed in Table 10.1:

- Only one individual adobe maker (albañil), Francisco Quine, appears in the record. Two petitions generated during the Cuenca inspection tour identify him as an employee of the Franciscan fathers at the monastery of Chiclayo (AGI/Justicia 457: 832; AGI/Justicia 461: 1455v–1456).
- *Chicheros* or beer brewers in the 1560s were, as an occupational specialty, in a state of transition. Traditionally, brewing beer was a generalized skill common to women, practiced at both the household and state levels. Unmarried women and widows still brewed the liquor that was so necessary to foster and reward communal labor for their lords in Jayanca in 1566. Dr. Cuenca prohibited this, declaring that each settlement could have only one "tavern." He only relented after receiving numerous petitions explaining that the commoners would not work on communal projects without a liberal distribution of beer, granting exemptions to all who so declared, for a price. A petition of Lambayeque was writ-

ten in the name of a lord, asking for the right to give chicha to his followers; there is no direct indication of who actually did the work. However, Dr. Cuenca's ordinances mention that women still did the brewing for the lords. Another petition from don Pedro Payanpoyfyl, an overseer of the chicheros, specifically states that they made chicha to trade for corn, wool, beads, and other things in the market. They had no lands and firmly reiterate that they are neither farmers nor merchants. These suggest that traditional practices continued in the countryside, whereas others, four decades after contact, had taken up the specialty, catering to urban demand (AGI/Patronato, l. 189, r. 11, 1566; AGI/Audiencia de Lima 28, 1574: 13–15; AGI/Justicia 458: 1779v–1780, 2090v; AGI/Justicia 461: 1567).

- *Cocineros* or cooks are rarely mentioned. In Jayanca they are described as requisitioning animals, like ducks, from the followers of the *curaca* (chief) for his table. In Túcume, a petition indicates a preoccupation with tilling land, but there is no mention of the kitchen (AGI/Justicia 461: 1521; Hart 1983:262–263).
- *Estereros* and *petateros*, or mat makers, by 1582 (50 years after contact) were paid (ANP/Residencia, l. 3, *cuaderno* [c.] 7, 1582, 154).
- *Hamaqueros*, who once transported on their shoulders such personages as the curaca of the Jayanca in the 1530s and 1540s and don Antonio Chayhuaca, curaca of the Chimu, as late as 1559, soon became obsolete. Dr. Cuenca prohibited such carriage because too many natives were so occupied. *Caciques* protested, stating that they could not travel on foot because it would be "damaging" (to their prestige and status?) and expose them to illnesses. They asked to be allowed to ride horses. As a consequence, in the 1560s, hamaqueros reverted to carrying messages for their lords and otherwise acted as their agents. One manuscript describes how they were entrusted with rounding up single and widowed women to serve their lord (AGI/Justicia 404: 370, 453; AGI/Justicia 457: 697v, 1057v; AGI/Patronato, l. 189, r. 11, 1566; ART/Mata: 25-IV-1565; Cieza de León 1984 [1553]:205; Hart 1983:257–258, citing AGI/Justicia 398).
- *Huseros* made a variety of spindle whorls (some for tribute cloth [*hilados de tributo*] and others to spin thin thread) in the 1560s and peddled them among people of various surrounding, but unnamed, encomiendas. Given that the original curacazgos had been divided into various encomiendas, the spindle whorl makers may have been circulating among peoples of their own ethnicity (AGI/Justicia 458: 1934v; AGI/Justicia 461: 1467v; ANCR/1766 for 1677; Netherly 1977a:247).
- *Olleros* or potters of Xequetepeque specialized in making large and medium *tinajones* (large, wide-mouthed jars) to exchange with their neighbors. If they resembled the people of Chincha further south and the ceramicists of Conchopata (described by Isbell [this volume]), however, they only worked part of the year. Other olleros were making adobe bricks for the Convent of San Francisco in Chiclayo in 1581. The *ollero* designation exemplifies how natives occupied in the same task, recall the albañil named above, and working in the same place for the same employer could be variously described in the sixteenth-century documents. This underscores the imprecision of the nomenclature found in the early documents (AGI/Justicia 458: 1933v, 2053v; ASFL/Registro 9, No. 2, Manuscrito 5, 1581; Rostworowski de Diez Canseco 1977a:135).[5]
- *Ovejeros* were shepherds. One man identified as an ovejero, subject to the governor of Moro (Pacasmayo), no longer pastured camelids, if he ever did, but instead cared for hogs as well as imported sheep for a wage in 1582. These shepherds moved animals seasonally, taking them east to higher ground when fodder became scarce on the coast (ART/CoO, l. 154, exp. 208, 11-VIII-1582, 2, 13–14; Spalding 1975:117).
- *Pintores* of Xequetepeque, Collique, Sinto, and Túcume were itinerant. They moved from homestead to homestead in return for

subsistence. Individual men painted, dyed, or otherwise decorated cloth, including ordinary mattress covers of homespun fabric (*colchones* of *ropa de la tierra*), cloaks (*capuces*), carrying cloths (*mantas*), and other items. In the 1560s, don Diego Mochumi declared that his overseers (*mandones*) had 630 skeins (*madejas*) of wool that needed to be dyed, indicating that yarns, in some cases, were colored before weaving. Twenty years later, pintores were being paid to process cloaks (AGI/Justicia 457: 828; AGI/Justicia 458: 1935; AGI/Justicia 461: 1453v–1455, 1457, 1468v; ART/Co-Residencia: 30-VI-1576; ART/CoO, l. 154, esp. 208, 11-VIII-1582, 1, 14v).

- *Plateros* or metalsmiths worked alone or in groups making pins, bracelets, beakers, and other items of both gold and silver (AGI/Justicia 458: 1800v; Calancha 1974 [1638], I: ch. 16, 232 [on Tumbez circa 1638]; Rostworowski de Diez Canseco 1977a:125, 136).
- *Zapateros* or *alpargateros*, or sandal makers, so vital to the redistributive economy of the native state, by the 1560s were equated with shoemakers. By 1560, they had begun to produce for the mass market. In that year, a merchant of Trujillo sold 600 pairs to another Spaniard for 142 *pesos oro* at 450 *maravedís* each (234 pesos 7.5 reals at 8 reals each). By the 1566 visit of Dr. Cuenca, the sandal makers were identified as cobblers who produced, besides shoes, reins, halters (*xaquimas*), trunks (*petacas*), shawls or carrying cloths (mantas), and bags (*bolsas*) of leather on commission and to sell in the market. To complicate matters, one *cumbico* (worker of the finest wool [*cumbi*]) also declared that he made sandals, presumably of the best wool. That these were still being used is evidenced by the will of the governor of Moro, who left three pairs of sandals in his estate: two of red (*colorado*) cumbi and another of cotton (AGI/Escribanía 501a, 102v, cited in Netherly 1977a:245; AGI/Justicia 458: 1935v, 1760; AGI/Justicia 461: 1217, 1223, 1234, 1468v; AGI/Patronato, l. 189, r. 11, 1566; ART/López Cordova, l. 3, 25-X-1560; ART/CoO, l. 154, exp. 208, 11-VIII-1582).[6]

No other detail is given for other specialists, whose mere existence is mentioned in passing in the sources, except for the fisherfolk, treated below.

This information indicates a society and economy in transition. Some specialties were by midcentury obsolete, as was the case for the hamaqueros, who, deprived by Spanish colonial law of their ancestral duty, had become personal attendants and agents of their lord. The cumbicos quickly faded from the scene, too, as fine wool became more difficult or impossible to obtain and shoes replaced sandals as the foot covering of choice for the native elite (see Figures 10.1–10.8).[7] Other specialists, like the alpargateros, adopted their skills to the market to become generalized leatherworkers. Plateros, too, quickly became beholden to the Spanish for their living. They, like some other artisan groups, became so important that Dr. Cuenca exempted them from *mita* (forced, rotating service) labor so that they could work for the Spanish settlers (AGI/Justicia 461: 1514v–1515; AGI/Audiencia de Lima 28, 1574: 6).

New native specialists also appeared in response to colonial demand. Some "learned [their trade] from Spaniards" [an aprendido (su oficio) de españoles] (AGI/Justicia 455: 1684). The ovejeros seem to be one such group. The few camelids, no more than nine, that belonged to a single lord at any given time and that dignitaries kept for ritual purposes, transporting goods, and wool in pre-Hispanic times would not have required more than one or, at most, two shepherds for their care. But the Spanish started breeding and herding European animals on the north coast as early as 1539. This quickly burgeoning livestock industry represented an effective and growing demand for shepherds, who were, in theory, paid for their work both as personal retainers or employees and as mitayo laborers (temporary and rotating forced laborers who worked as part of their obligation to the Inca, curaca, and colonial state). *Cabestreros* (halter mak-

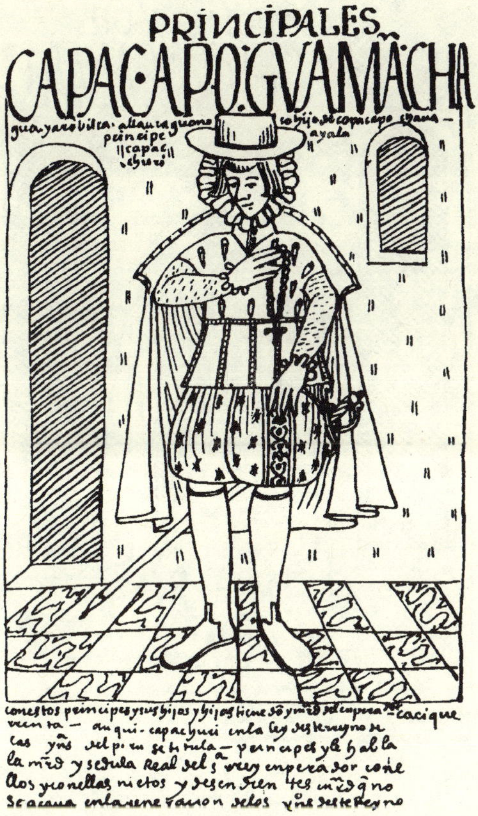

Figure 10.1. Guaman Poma drawing, folio 741. Principales [-] Capac Apo Guama[n] Chagua Yaro vilca.

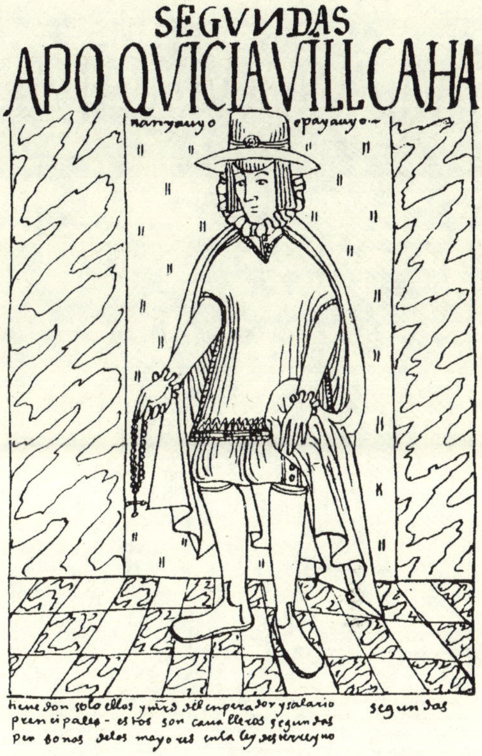

Figure 10.2. Guaman Poma drawing, folio 743. Segundas [-] Apo Qviciavillca.

ers), only mentioned once in the ordinances of Jayanca, were ordered by Dr. Cuenca to sell in the recently established market. Because camelids did not require halters, this specialty indicates a colonial origin and acquired skills. The making of spindle whorls may also have been a growth sector given the Spanish institutionalization of cloth as a common tribute item. Likewise, it is unlikely that mats were ever mass produced until the arrival of the Spanish, given that weaving was a ubiquitous skill and materials were readily available. The chicheros, as noted above, may have been another group that developed as the Spanish outlawed polygamy, turning secondary wives and concubines into "widows," and as a taste for chicha developed in the urban centers.

Merchants, too, may be in this category. I believe that some of the so-called merchants were both retainers and pages of lords who were sent to requisition and barter for goods needed for ceremonial and politico-religious ends. Supporting this view is Cock's (1987) conclusion that sumptuary ware was fabricated for noble use and awarded as gifts. Indeed, one credible source designates such gobetweens as *chasquis* (runners and carriers), as in *mollo chasqui camayoc* (carriers of *Spondylus* shell), transporters of goods from one individual or group to another. The progressive breakdown of the ancient system of reciprocity and redistribution after over 30 years of colonial contact hastened the participation of lords and commoners in the market. This process was beginning to affect the lives of the nobles at midcentury, the era when most of the artisans and specialists first appear in the documentation. Native peddlers became important as personal kinship networks proved insufficient to supply needed items. Trade,

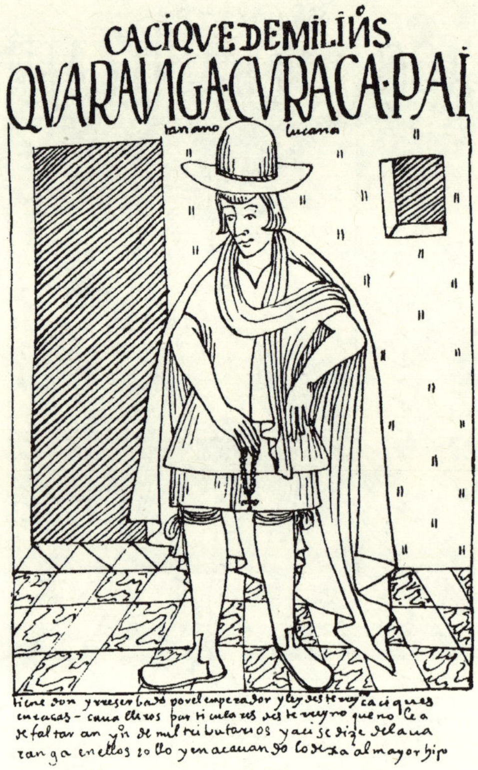

FIGURE 10.3. Guaman Poma drawing, folio 745. Cacique de mil in[di]os Quaranga Curaca Paitan ano lucana.

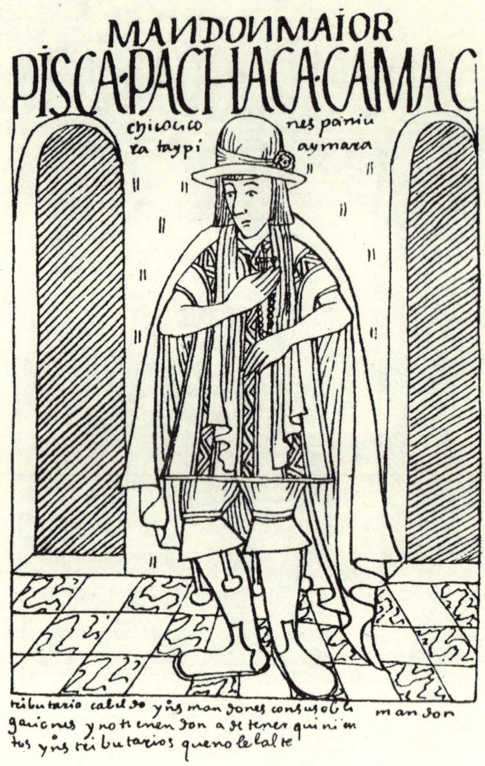

FIGURE 10.4. Guaman Poma drawing, folio 747. Mandon Maior Pisca-Pachaca Camacchicoc.

once established, also called into existence native tailors, shoemakers, carpenters, and the like (AGI/Patronato, l. 189, 4. 11, 1566; ART/Salinas 18-VIII-1539; Falcón 1918, cited in Rostworowski de Diez Canseco 1978:213; Ramírez 1982, 1998).

Cabesteros, *carpinteros*, *hacheros*, *salineros*, *sastres*, and *silleros* are only mentioned in passing in the manuscripts of the era. The absence of additional references to these producers on the north coast suggests that these were not pre-Hispanic specialists but, rather, individuals and groups who developed in response to Spanish demands and tastes or persons who highlighted their skills for the purpose of skirting or avoiding colonial restrictions. Indeed, another source limits specialists to three:

> [Among the natives] there are only three types of specialized tradespersons: potters who make the vessels to brew and store chicha; carpenters who make wooden beakers for drinking...and metalsmiths who in the time of the Inca made vessels and other things of gold and silver. [Solo tres oficios usan entre ellos (los indios): olleros que hazen basijas para hazer y tener chicha; carpinteros que hazen basos de palo para en que beber...y plateros que en tiempo de Ynga hacian de oro y plata las basijas y otras obras.] [British Museum/Add. 13992: 415]

These olleros, carpinteros, and plateros survived 1532, but they "no longer live among the natives because they cannot make enough to eat except in the cities where they earn a living from the Spanish" [ya no viben entre los indios porque no hallan a que ganar de comer sino en las ciudades donde lo ganan entre españoles] (British Museum/Add. 13992: 415).

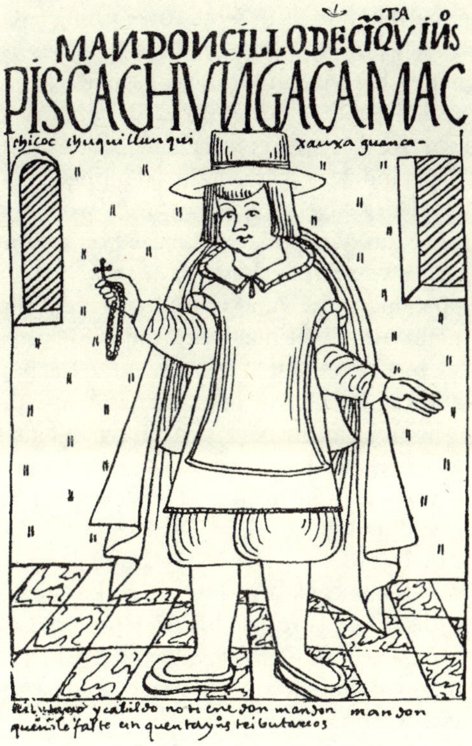

FIGURE 10.5. Guaman Poma drawing, folio 749. Mandoncillo de Ci[n]qu[en]ta In[di]os Piscach-vnga camacchicoc.

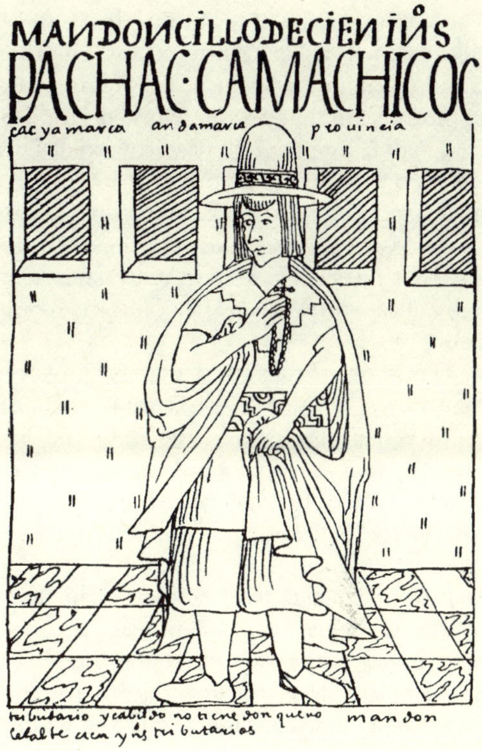

FIGURE 10.6. Guaman Poma drawing, folio 751. Mandoncillo de cien in[di]os Pachac camachicoc.

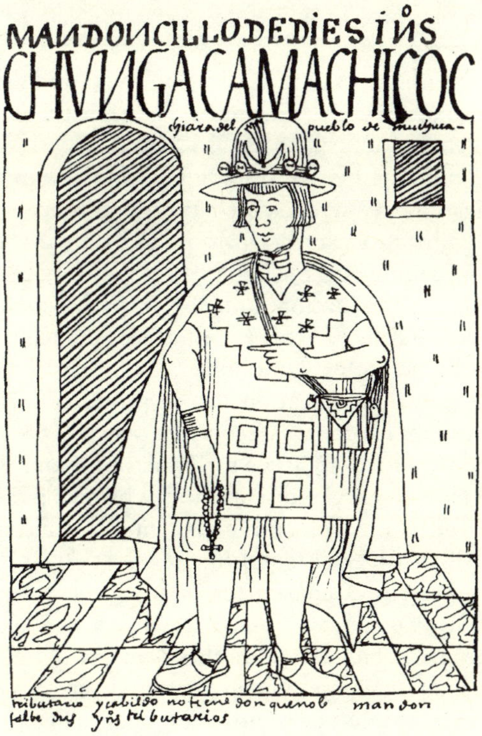

FIGURE 10.7. Guaman Poma drawing, folio 753. Mandoncillo de Dies In[di]os Chvnga camachicoc.

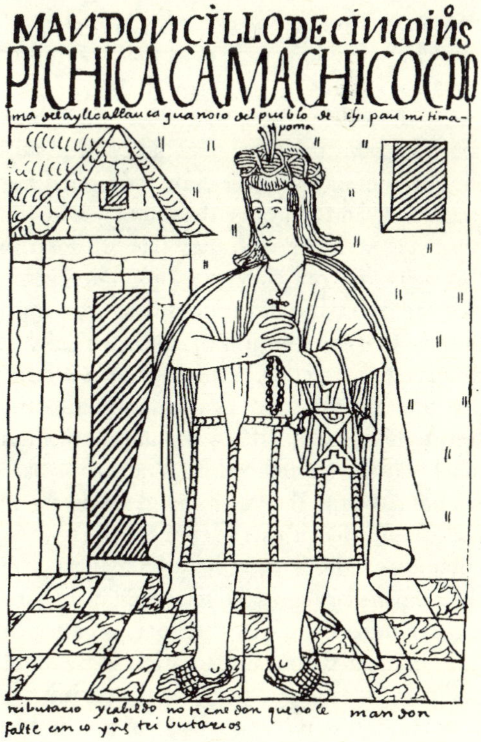

FIGURE 10.8. Guaman Poma drawing, folio 755. Mandoncillo de Cinco In[di]os Pichica Camachicoc.

Those who remained behind in the countryside reverted to being farmers and guarding cattle, often for the Spanish. These statements do not negate that no one worked wood or made salt before 1532. Rostworowski de Diez Canseco found that carpinteros in Chincha made lances and had lands. Sometimes ornate beakers were produced also for the use in ritual (AGI/Justicia 455: 1684; AGI/Justicia 457: 848v; Rostworowski de Diez Canseco 1977a: 126, 135).

The above discussion suggests, however, that only a very few of these mid-sixteenth-century specialists served the state and had pre-Hispanic antecedents. Among these are the chicheros or (female) *chicheras*, hamaqueros, olleros, plateros, and cumbicos/*tejedores* (weavers)/*tintoreros* (dyers)/pintores (*de mantas*). But, as suggested above, the chicheros may have industrialized a process that had been practiced by women at the household and ethnicity level for generations (or longer if we assume continuity with the women who served at Conchopata, studied by Isbell [this volume]). And the hamaqueros had lost their monopoly on transporting their lord to horses and mules (AGI/Justicia 458: 1935, 2090v; AGI/Justicia 461: 1468v, 1567; Diez de San Miguel 1964 [1567]:14, 27, 98; Hamilton 1978:27; Hyslop 1976:174–175).[8]

Of the other three, the sources are most specific about the specialists associated with the garment industry, the cumbicos, tejedores, tintoreros, and pintores (de mantas). The first two were weavers who may have differed in the type of wool they worked. The second two are dyers and decorators. But we cannot be sure that the division of labor was always observed in practice. Agustinos (1964 [1865 (1560?)]:32) recorded the intriguing information that in the northern highlands of Huamachuco the cumbicos had a cult to Quispe guanajai, whom they worshiped under the direction of three priests so that the dyes would come out right. This indirectly suggests that the weavers could have dyed too. The *roperos* are excluded from this category, because the only one for whom information is available had no part in the actual manufacturing process. He states that his duty was to account for or guard hundreds of pieces of clothing (AGI/Justicia 458: 2248v). He thus qualifies for what Father Martín Morúa calls in his chronicle "indios para guardar trajes" [clothes keeping natives] (1946 [1590]:334).[9]

This industry was organized along gender lines. The weavers (tejedores) were women as far back as Mochica times, based on a painting on a vessel that shows women weaving at looms under the supervision of a man. This organization continued into the colonial period. Inspector Gama, who censused the population of Jayanca in 1540, names Ayapoque, a fisherman, as the person in charge of female weavers. Twenty-six years later, the maidens and widows of Jayanca were still cleaning cotton and making garments and sacks (*fardos*) for their lord. Don Diego Mochumi, a principal of Túcume, used widows to make "67 leg-lengths [a measure] of cloaks" [67 piernas de capuzes] (ART/CoResidencia: 30-VI-1576) right up to the end of his life in the mid-1570s. Thus, it appears that women prepared the raw material and wove and made garments, whereas men may also have colored and decorated the cloth (AGI/Patronato, l. 189, r. 11, 1566; AGI/Audiencia de Lima 92; Gama 1975 [1540]:217; Gayton 1978:278; Netherly 1977a:233, 249).

With the exception of the weavers and dyers/painters, the documents are consistently silent about subspecialists. In this case, a nonfinding may be a finding. Other research has already underscored the importance of numbers to a lord's position and status. The more followers, the higher a lord's divine and worldly reputation, and the more he could accomplish. For this reason, lords competed with each other for followers and their labor. A state's organization of the production of this labor force was key to its material prosperity and its ability to satisfy expectations of reciprocity and redistribution, which won the loyalty of its sometimes-recalcitrant subjects (ART/CoJuez de Comisión 29-III-1558).

Most of a coastal lord's labor force was either farmers or fisherfolk. At first contact, peasants and their families lived in homesteads sometimes widely scattered across the landscape. In 1540, natives of Jayanca lived in approximately 250 small hamlets and residential complexes dispersed within a two-league radius of their curaca's abode. But hundreds also lived several days' walk away in the highlands on land first cultivated by the Túcumes, among others, and worked by the Guambos people (Gama 1975 [1540]; Hart 1983:282).

Such relatively isolated agricultural peoples were largely self-sufficient, exhibiting multiple subsistence-guaranteeing skills. One source states that "all generally know how to make the clothing and shoes they need and to cultivate their gardens and build their houses" [todos en general saben hacer lo que an menester para vestir y calçar y cultivar sus labranças y haçer sus casas] (British Museum/Add. 13992: 415). They raised their own foodstuffs and animals, such as the guinea pig. They fished occasionally in the rivers and lakes for additional protein. They constructed their homes. The women spun and wove, cooked, and prepared beer. Children helped care for animals and other younger siblings, chased birds from the crops, and collected firewood.[10]

Fishermen, the second largest group mentioned in the sources, were also capable of multiple tasks. They fished and hunted sea lions and birds on beaches that they had used for generations. They made their own nets and their reed boats and log rafts. Some burned seashells for lime, gathered seaweed as a dietary supplement, and made salt. They and their families dried and preserved their catch. Some learned the currents and navigated to the guano islands and possibly to the *Spondylus* beds nearer the equator. But fishing was a seasonal activity, limited to three months per year. The fishermen entered the sea by "established turns" (*por sus turnos o mitas establecidas*). Therefore, lords used them to carry messages, like chasquis. A very few fishermen from Eten and Túcume also reported having access to cultivatable lands. All, like the farmers, were subject to tribute labor service probably in the off-season. They also were subject to other colonial impositions, like, for example, Cuenca's order for fishermen to supply the towns of Trujillo and Saña with fish and other marine goods (AGI/Justicia 458: 1850, 1934v, 2025v [the last two folios cited in Netherly 1977b:9]; AGI/Justicia 404: 580; AGI/Justicia 457: 1240; AGI/Justicia 461: 1009, 1262v; ANCR/1739; ANP/DI, l. 485, c. 5, 1784; ANP/DI, l. 39, c. 820, 1737; Flores Galindo 1981:162; Guaman Poma de Ayala 1980 [1613]:1131 [1141]; Herrera Calderón 1986:14; Netherly 1977a:243; Pizarro 1978 [1571]:239, 247; Ramírez 1995; Rostworowski de Diez Canseco 1975:312–322, 1977a:137, 1977b:214–215, 1978:212, 1981:15–17).

In late pre-Hispanic times, this combined population of commoners, on request, also put their skills to use when periodically serving the state as mitayos. On a rotating basis, groups of commoners tilled ground to produce cotton, corn, coca, and other goods for their lords (and their ancestral gods). Once these were harvested and processed, workers carried them to state warehouses. They commuted to a ceremonial center to work making adobes for construction. Some grazed camelids for ritual purposes or bred and cared for exotic birds to supply feathers for headdresses and textiles (on llamas, AGI/Patronato, l. 188, r. 22, 1561, 4v, 22–22v; AGI/Justicia 461: 856v; ART/CoO, l. 154, exp. 208, 11-VIII-1582, 12; ART/CoO, l. 157, c. 301, 14-XII-1595, 17; ART/CoResidencia: 30-VI-1576; Cabello Valboa 1951 [1586]:403; Cieza de León 1984 [1553]:21). Calculations based on chroniclers' reports show that about 1 percent of a subject population was periodically recruited to mine ores needed for pottery glazes, paints, and metalsmithing (AGI/Patronato, l. 189, r. 11, 1566; AGI/Audiencia de Lima 28, 1574: 3v, 31; AGI/Justicia 458: 1872; ART/CoAsuntos de Gobierno: 11-X-1585; ART/CoPedimento, l. 280, exp. 3612, 30-X-1583: 1; Espinoza Soriano 1969–1970:39; Falcón 1918 [1567]:149–150; Hart 1983:284; Moseley 1975:196; Netherly 1977a:212; Polo de Ondegardo 1917 [1571]:71; Ramírez 1994:98–100, in press; Rowe 1984).

Mitayo laborers were also used to requisition or otherwise acquire needed supplies, like exotic foodstuffs, shells, precious stones, coca, wood, and ore, to supply the specialists. Sometimes, they traveled great distances for this purpose. Juan de Betanzos writes of personal retainers who were sent on distant missions to procure rare coca leaves for state affairs. They could be sent into the Andean foothills to gather cacti, like San Pedro, for divination and other ritual purposes and dye-stuffs for textile manufacture. They cut trees to make poles to use as shoulder braces for transporting the lord's litter and other objects from stands of wild trees that were open to all. They also collected wood to fire the clay furnaces of the metalworkers and heat the vats for dying yarn and cloth. Thus, it appears that the raw materials and other inputs needed for artisanal production were produced and procured by unspecialized labor doing routine tribute service for their lords. The commoners' occasional need to change location temporarily to comply with this duty is one explanation for enclaves or colonies of one ethnic group living among another and the multiethnicity of some resource sites. It may also explain the kitchen facilities near sacred structures (AGI/Audiencia de Lima 28, 1574: 31; AGI/Justicia 461: 1027v; Betanzos 1987 [1551–1557]; British Museum/Add. 13992: 412; Falcón 1918 [1567]:149; Hyslop 1976:178; Morúa 1946 [1590]:227; Pease 1991:23–24).[11]

AN ARTISAN ELITE

This reconsideration of the data now on hand leads me to the following conclusions. There were far fewer dedicated craft specialists on the north coast in late pre-Hispanic times than the list in Table 10.1 might suggest. The craft producers who did exist constituted a tiny cadre of specialized persons attached to the service of a lord. They included skilled metalsmiths and potters. Women under male supervision made sumptuary wear and brewed chicha to fulfill expectations of state hospitality, to reward subordinates, to sacrifice to their ancestral gods, and to gift the state. The artisans received raw materials and necessary inputs from the government through the labor of the commoners, a finding that seems coincident with Goldstein and Shimada's (this volume) work at the workshop at Huaca Sialupe that was run by groups of artisans working in close proximity but largely independent of each other. The existence of elite craftspeople does not preclude the similar but cruder production of like objects at the household level. The ability to administer the timely production of luxury goods was essential to maintain the prosperity of the ethnicity and the belief in the divine status of the rulers and to foster loyalty among the ruler's followers. To the extent that subordinates drank the lord's beer, ate sacrificial animals, or received a tunic as a reward for extraordinary service, suppliers and producers were also consumers.

On a higher level of abstraction, the finding that there were fewer pre-Hispanic specialists on the coast than previously thought has implications at the imperial level. Scholars have written about the greater variety of specialists on the coast than in the mountain hinterland. One can explain this if one assumes that religio-political leaders representing the older, more evolved states had more followers and proportionately more specialists (e.g., the Chimu [more developed] versus the Inca [the upstart, expanding highland polity]). On this basis and considering the presence of specialists, whom she identifies as merchants, Rostworowski de Diez Canseco postulates the existence of two different economic systems at the time of first European contact. Certain coastal peoples, she states, engaged in trade for personal gain, whereas those in the highlands organized themselves vertically for basic self-sufficiency within their ethnic group. This constituted a less specialized population. But if most of the two dozen specialists developed in the course of the first 30 to 40 years after contact with the Spanish and if those merchants were retainers, then the basis of this argument disappears (Hamilton 1978:21; Murra 1972; Rostworowski de Diez Canseco 1970, 1978:212, 1979).

This leaves us with a greater degree of coincidence between coast and highlands than previously thought. On the coast and in the highlands alike, common laborers put their range of everyday subsistence skills to work procuring or producing the inputs and enabling production for a small elite group of attached craftspersons who manufactured or supervised luxury manufacture for religio-political elite sponsors. The sponsors who directed this effort returned this production at ritual gatherings in the form of beer or distinguishing tunics, cloaks, or bracelets to well-placed and helpful subordinates. All the members of society consumed and benefited to the extent that they were integrated into the political and religious group. Workers benefited directly if rewarded with food or a precious gift symbolizing the leadership's favor. They benefited indirectly when a craft item was presented to the Inca or burned as a sacrifice to the ancestral gods to ensure prosperity and fertility. The distinction, then, between producer and consumer becomes one of quantity and quality, despite the fact that the real measure of a person's status remained the number of persons who would come to his or her aid when asked, the emic definition of "wealth" (which Mills [this volume] also finds in the North American Southwest). It also cautions us that our categories may not coincide with native classifications and that craft distinctions may appear less definitive than we would like.

Notes

1. See also the list of specialists in Falcón 1918 [1567]:149–150 and Guaman Poma de Ayala 1956 [1613]:135.
2. See also Guaman Poma de Ayala (1980 [1613]: [340]) on *farsantes* (tricksters, impostors, perhaps as in or for the purpose of entertainment).
3. Hart mentions cloth painters, chicha makers, and salt makers as without land and, therefore, presumably full-time, as opposed to cooks, mat makers, and deer hunters, who had lands and another basis of subsistence. But she concedes that some nonagricultural activities were performed seasonally (AGI/Justicia 458: 2013; AGI/Justicia 461: 1521; Hart 1983:268–269). Ravines (1978:93, citing Cobo de Ondegardo [lib. 16]) states that metalsmiths and cloth painters also had land. Rostworowski de Diez Canseco (1977a:135) found that Chincha potters worked part of the year.
4. See also Curatola's (1997) analysis of the "Aviso" and the inconsistencies between its information and the historical and archaeological record, which suggests that guano (bird manure) was the basis of prosperity for the people of Chincha.
5. See also Cock 1987.
6. See also Rostworowski de Diez Canseco 1977a:126.
7. Only commoners and *mandoncillos* (native overseers) in charge of five families used sandals by the early seventeenth century. Compare the drawings of natives by Guaman Poma de Ayala 1936 [1613]: folios 741–755.
8. These findings are consistent with the conclusions of Watanabe (2002) and Cornejo (2002).
9. See also Falcón (1918 [1567]:149), who lists *tanti camayoc* and *indios que hacian colores de yerbas*, and Vargas Ugarte (1940:561–562), who suggests that some specialists developed as a result of Spanish demand.
10. Barbara Mills (this volume) writes of household craft producers who may be an analogy to this situation.
11. For another example of multiethnic concentrations of mitimaes, see Murra 1978.

Archives

AGI—Archivo General de las Indias, Seville
Audiencia de Lima
Escribanía
Justicia
Patronato
ANCR—Archivo Notarial de Carlos Rivadeneira, Lambayeque, Peru
ANP–Archivo Nacional del Perú, Lima
Derecho Indígena (DI)
Residencia
ART—Archivo Regional de Trujillo
Corregimiento, Asuntos de Gobierno (CoAsuntos de Gobierno)
Corregimiento, Juez de Comisión (CoJuez de Comisión)
Corregimiento, Ordinario (CoO)
Corregimiento, Pedimento (CoPedimento)
Corregimiento, Residencia (CoResidencia)
López Cordova
Mata
Salinas
ASFL—Archivo de San Francisco, Lima
British Museum, London

References Cited

Agustinos
1964 [1865 (1560?)] Relación de la religión y ritos del Perú. In *Colección de documentos inéditos, relativos al descubrimiento, conquista y organización de América y Oceanía*, edited

by Luis Torres de Mendoza, Vol. 3:3–58. Kraus Reprint Ltd., Vaduz, Germany. (Original: Imprenta de Manuel Quiros, Madrid.)

Betanzos, Juan Diez de
1987 [1551–1557] *Suma y narración de los incas*. Atlas, Madrid.

Cabello Valboa, Miguel de
1951 [1586] *Miscelánea antártica*. Instituto de Etnología, Universidad Nacional Mayor de San Marcos, Lima.

Calancha, Antonio de la
1974 [1638] *Crónica moralizada*. 6 vols. Universidad Nacional Mayor de San Marcos, Lima.

Castro, Fray Cristóbal de, and Diego de Ortega Morejón
1968 [1558] Relación de Chincha. In *Biblioteca Peruana: El Perú a través de los siglos*, Vol. 3: 465–489. Primera serie, 3 vols. Editores Técnicos Asociados, Lima.

Cieza de León, Pedro de
1864 [1553] *The Travels of Pedro de Cieza de León*. Hakluyt Society, London.
1967 [1550s] *El señorío de los incas, Segunda parte de La crónica del Perú*. Instituto de Estudios Peruanos, Lima.
1984 [1553] *Crónica del Perú. Primera Parte*. Pontificia Universidad Católica del Perú, Lima.

Cock, Guillermo A.
1987 The Lord's Control of Artisans, Craftsmen and Labor Specialization in the Jequetepeque Valley, North Coast of Peru. Paper presented at the Mid-West Conference of Latin American Studies, November 6, Chicago.

Cornejo, Miguel
2002 Sacerdotes y tejedores en la provincia Inka de Pachacomac. *Boletín de arqueología (Lima)* 6:171–204.

Curatola, Marco
1997 Guano: Una hipótesis sobre el orígen de la riqueza del señorío de Chincha. In *Arqueología, antropología e historia en los Andes: Homenaje a María Rostworowski*, edited by Rafael Varon Gabai and Javier Flores Espinoza, pp. 223–239. Instituto de Estudios Peruanos, Lima.

Diez de San Miguel, Garci
1964 [1567] *Visita hecha a la Provincia de Chucuito*. Casa de la Cultura del Perú, Lima.

Espinoza Soriano, Waldemar
1969–1970 Los mitmas yungas de Collique en Cajamarca[,] siglos XV, XVI y XVII. *Revista del Museo Nacional [del Perú]* 36:9–57.

Falcón, Lic. Francisco
1918 [1567] Representación hecha por el licenciado Falcón al Concilio Provincial sobre los daños y molestias que se hacen a los indios. In *Informaciones acerca de la religión y gobierno de los incas. Colección de Libros y documentos referentes a la historia del Perú*, edited by Horacio H. Urteaga, Vol. 2:133–176. Sanmartí, Lima.

Flores Galindo, Alberto
1981 La pesca y los pescadores en la costa central (siglo XVIII). *Histórica* 2:159–165.

Gama, Sebastián de la
1975 [1540] Visita hecha en el Valle de Jayanca [Trujillo]. *Historia y cultura* 8:215–228.

Gayton, Ann
1978 Significado cultural de los textiles peruanos: Producción, función y estetica. In *Tecnología andina*, edited by Rogger Ravines, pp. 269–297. Instituto de Estudios Peruanos, Lima.

Guaman Poma de Ayala, Felipe
1936 [1613] *El primer nueva coronica y buen gobierno*. Institut d-ethnologie, Paris.
1956 [1613] *El primer nueva coronica y buen gobierno* Talleres del Servicio de Prensa, Propaganda y Publicaciones Militares, Lima.
1980 [1613] *El primer nueva corónica y buen gobierno*. 3 vols. Edited by John V. Murra and Rolena Adorno. Siglo XXI Editores, Mexico City.

Hamilton, Joe
1978 Plebe and Potentate: History and Society of Prehispanic North Central Coast Peru. Unpublished MS, Austin.

Hart, Elizabeth Ann
1983 Prehispanic Political Organization on the Peruvian North Coast. Unpublished Ph.D. dissertation, Department of Anthropology, University of Michigan, Ann Arbor.

Herrera Calderón, Américo
1986 Complejos hidrológicos y agricultura costa prehispánica. *Alternativa* 2(3):9–15.

Hyslop, John
1976 An Archaeological Investigation of the Lupaca Kingdom and Its Origins. Unpublished Ph.D. dissertation, Department of Anthropology, Columbia University, New York.

Lothrop, Samuel K.
1938 *Inca Treasure as Depicted by Spanish Historians*. Museo Sudeste, Los Angeles.

Martínez Cereceda, José Luis

1995 *Las Autoridades en los Andes: Los atributos del Señor*. Pontificia Universidad Católica del Perú, Lima.

Morúa, Fray Martín de

1946 [1590] *Historia del origen y genealogía real de los Reyes Incas del Perú*. Instituto Santo Toribio de Mogrovejo, Consejo Superior de Investigaciones Científicas, Madrid.

Moseley, Michael E.

1975 Prehistoric Principles of Labor Organization in the Moche Valley, Peru. *American Antiquity* 40:191–196.

Murra, John V.

1972 El "control vertical" de un máximo de pisos ecológicos en la economia de las sociedades andinas. In *Visita de la provincia de León de Huánuco (1562)*, by Iñigo Ortiz de Zúñiga, Vol. 2:429–476. Universidad Nacional Hermilio Valdizán, Huánuco, Perú.

1978 Los olleros del Inka: Hacia una historia y Arqueología del Qollasuyu. In *Historia: Problema y promesa*, edited by Francisco Miro Quesada C., Franklin Pease, and David Sobrevilla A., pp. 415–423. Pontificia Universidad Católica del Perú, Lima.

Netherly, Patricia

1977a Local Level Lords on the North Coast of Peru. Unpublished Ph.D. dissertation, Department of Anthropology, Cornell University.

1977b Organization through Opposition: Dual Division and Quadripartition on the North Coast of Peru. Paper presented at the Annual Meeting of the American Anthropological Association, November 30–December 3, Houston.

Orrego H., Augusto

1958 Palabras del Mochica. *Revista del Museo Nacional* 27:80–95.

Pease, Franklin

1991 *Los últimos Incas del Cuzco*. Alianza Editorial, Madrid.

Pizarro, Pedro

1978 [1571] *Relacion del descubrimiento y conquista de los reinos del Peru*. Pontificia Universidad Católica del Perú, Lima.

Polo de Ondegardo, Lic. Juan

1917 [1571] Del linage de los ingas y cómo conquistaron. In *Colección de libros y documentos referentes a la historia del Perú*, edited by Horacio H. Urteaga, Vol. 4:45–94. Imprenta y Librería de Sanmartí, Lima.

Ramírez, Susan Elizabeth

1981 La organización económica de la costa norte: Un análisis preliminar del período prehispánico tardío. In *Etnohistoria y antropología andina*, edited by Amalia Castelli and Marcia Koth de Paredes, pp. 281–297. Centro de Proyección Cristiana, Lima.

1982 Retainers of the Lords or Merchants: A Case of Mistaken Identity? In *El hombre y su ambiente en los Andes Centrales*, edited by Luis Millones and Hiroyasu Tomoeda, pp. 123–136. Senri Ethnological Studies 10. Museo Nacional de Etnología, Osaka.

1986 Notes on Ancient Exchange: A Plea for Collaboration. In *Andean Archaeology: Papers in Memory of Clifford Evans*, edited by Ramiro Matos M., Solveig A. Turpin, and Herbert H. Eling Jr., pp. 225–238. University of California Press, Los Angeles.

1990 The Inca Conquest of the North Coast: A Historian's View. In *The Northern Dynasties: Kingships and Statecraft in Chimor*, edited by Michael E. Mosely and Alana Cordy-Collins, pp. 507–537. Dumbarton Oaks Research Library and Collection, Washington, D.C.

1994 Ethnohistorical Dimensions of Mining and Metallurgy in Sixteenth-Century Northern Peru. In *Quest of Mineral Wealth: Aboriginal and Colonial Mining and Metallurgy in Spanish America*, edited by Alan K. Craig and Robert C. West, pp. 93–108. Geoscience and Man, Vol. 33. Louisiana State University, Baton Rouge.

1995 De pescadores y agricultores: Una historia local de la gente del valle de Chicama antes de 1565. *Bulletin de l'Institut Français d'Etudes Andines (Lima)* 24:245–276.

1996 *The World Upside Down: Cross-Cultural Contact and Conflict in Sixteenth Century Peru*. Stanford University Press, Stanford.

1997 "Un mercader...es un pescador": Reflexiones sobre las relaciones económicas y las múltiples roles de los indios americanos en el Perú del siglo XVI. In *Arqueología, antropología, e Historia en los Andes. Homenaje a María Rostworowski*, edited by Rafael Varon Gabai and Javier Flores Espinoza, pp. 729–735. Instituto de Estudios Peruanos, Lima.

1998 Rich Man, Poor Man, Beggar Man, or Chief: Material Wealth as a Basis of Power in Sixteenth-Century Peru. In *Dead Giveaways: Indigenous Testaments of Colonial Mesoamerica and the Andes*, edited by Susan Kellogg and Matthew Restall, pp. 215–248. University of Utah Press, Salt Lake City.

2005 *To Feed and Be Fed: The Cosmological Bases of Authority and Identity*. Stanford University Press, Stanford.

In press La minería y la metalurgía nativa en el norte peruano (sigloes XVI–XVIII). *Anuario de Estudios Americanos (Seville)*.

Ravines, Rogger (editor)

1978 *Tecnología andina*. Instituto de Estudios Peruanos, Lima.

Rostworowski de Diez Canseco, María

1961 *Curacas y sucesiones, costa norte*. Imprenta Minerva Miraflores, Lima.

1970 Mercaderes del valle de Chincha en la época prehispánica: Un documento y unos comentarios. *Revista española de antropología americana (Madrid)* 5:135–178.

1975 Pescadores, artesanos y mercaderes costeños en el Perú prehispánico. *Revista del Museo Nacional [del Perú]* 41:311–349.

1977a Mercaderes del valle de Chincha en la época prehispánica: Un documento y unos comentarios. In *Etnía y sociedad: Costa peruana prehispánica*, pp. 97–140. Instituto de Estudios Peruanos, Lima.

1977b Pescadores, artesanos y mercaderes costeños en el Perú prehispánico. In *Etnía y sociedad: Costa peruana prehispánica*, pp. 211–263. Instituto de Estudios Peruanos, Lima.

1978 Estructuras políticas y económicas de la costa central del Perú precolombino. *Revista histórica* 31:203–218.

1979 Breves notas sobre la estructura socioeconómica en la costa peruana precolombina. In *Estudios americanistas: Libro jubilar en homenaje a Hermann Trimborn*, edited by Roswith Hartmann and Udo Oberem, Vol. 2:207–211. Haus Völker und Kulturen, Anthropos-Institut, San Agustin, Germany.

1981 Los pescadores del litoral peruana en el siglo XVI "Ynga Guaxma." *Nova Americana (Torino, Italy)* 4:11–42.

Rowe, Ann Pollard

1984 *Costumes and Featherwork of the Lords of Chimor: Textiles from Peru's North Coast*. Textile Museum, Washington, D.C.

Santa Cruz Pachacuti, Juan

1927 [1613] Relación de Antiguedades. In *Historia de los Incas*, edited by Horacio H. Urteaga, pp. 179–268. Imprenta y Librería Sanmartí, Lima.

Spalding, Karen

1975 Hacienda–Village Relations in Andean Society to 1830. *Latin American Perspectives* 2(4):107–121.

Trimborn, Hermann

1979 *El reino de Lambayeque en el antiguo Perú*. Haus Völker und Kulturen-Anthropos Institut, St. Augustin, Germany.

Vargas Ugarte, Ruben

1940 Método de la compañía de Jesús en la educación del indígena. *Mercurio Peruano (Lima)* 163:554–566.

1942 [1939] Los mochicas y el cacicazgo de Lambayeque. *Actas y trabajos científicos del 27 Congeso Internacional de Americanistas* 2:475–482.

Watanabe, Shinya

2002 El reino de Cuismancu: Origenes y transformación. *Boletín de arqueología* 6: 107–136.

Zevallos Quiñones, Jorge

1947 *Un diccionario castellano-Yunga*. Imprenta del Ministerio de Educación Pública, Lima.

1977 *Notas para la prehistórica tardia de Ica: Onomastica y toponomia*. Imprenta Universitaria, Trujillo.

Contributors

Gary M. Feinman,
The Field Museum,
Chicago

David J. Goldstein,
Department of Anthropology,
Southern Illinois University,
Carbondale

Takeshi Inomata,
Department of Anthropology,
University of Arizona

William H. Isbell,
Department of Anthropology,
Binghampton University

Jonathan Mark Kenoyer,
Department of Anthropology,
University of Wisconsin,
Madison

Yung-ti Li,
Institute of History and Philology,
Academia Sinica,
Taipei

Heather M.-L. Miller,
Department of Anthropology,
University of Toronto

Barbara J. Mills,
Department of Anthropology,
University of Arizona

Linda M. Nicholas,
The Field Museum,
Chicago

Susan E. Ramírez,
History Department,
Texas Christian University

Izumi Shimada,
Department of Anthropology,
Southern Illinois University,
Carbondale

Barbara L. Stark,
School of Human Evolution and Social Change,
Arizona State University

Peter S. Wells,
Department of Anthropology,
University of Minnesota

Index

Numbers in *italics* refer to figures.